THE KINEMATICS OF VORTICITY

THE KINEMATICS OF VORTICITY

CLIFFORD TRUESDELL

DOVER PUBLICATIONS, INC.
MINEOLA, NEW YORK

Bibliographical Note

This Dover edition, first published in 2018, is an unabridged republication of the work originally printed by Indiana University Press, Bloomington, Indiana, in 1954.

International Standard Book Number
ISBN-13: 978-0-486-82364-5
ISBN-10: 0-486-82364-4

Manufactured in the United States by LSC Communications
82364401 2018
www.doverpublications.com

Dedicated to my grandmother,

ALICE F. WALKER

Acknowledgments

I am grateful to the Naval Research Laboratory for encouragement and furtherance. The Applied Mathematics Branch, under the direction of Dr. H. M. Trent, has directly supported this work; the Library (particularly through Mrs. Hazel Mortenson) and the Graphic Arts Branch have given most generous and valuable assistance. I am obliged to Drs. Van Tuyl and Prim for helpful discussions in 1947–1948, and to my colleagues the late Dr. Neményi and Professor Gilbarg for numerous suggestions. My colleague Professor Hlavatý has most kindly and patiently gone over the work in detail; to him I owe the correction of many an error and the clearing of many a muddy passage. Most of all, it is the continual and ever-present effort of my assistant, formerly Miss Charlotte Brudno and now my wife, which has made completion of this work possible.

TRUESDELL

July 15, 1953

Contents

PREFACE

The general quality of this work. Inherent in this classical subject is the presentation of a large number of formulae and the manipulations necessary to obtain them. So as to smooth the reader's path I have so ordered the matter that, the preceding gradual climb being granted, but a few easy steps are required to attain any one particular eminence, and each individual section contains but a few equations. The equations themselves are of value and of interest, but, since our study is directed toward the most general continuous motions, too stubborn to yield to the methods of mathematical or numerical integration now known, it is the grasp and picture of the results which form our primary goal. Indeed, I hope that this essay will prove a not unworthy offering to the great tradition of the geometry of nature.

But from these pages the reader will see at once that up to the present time only the simplest concepts and methods have been applied. For him equipped with more refined geometrical tools, the land has been cleared and the future promises a copious harvest.

To the analyst it may be of interest that this work, unlike most recent research in fluid mechanics, does not fall within the accustomed domain of partial differential equations: our entire study is put out upon under-determined systems.

Scope of this work. There has been a constant temptation to swell the work into a general exposition of the kinematics of continuous media. This temptation I have resisted, however, so as to point up the properties of vorticity, and only a few prerequisite or supplementary matters are included. In particular, the kinematics of rate of deformation is merely summarized, and there is no reference whatever to the theory of strain and rate of strain.[1] Omitted also is any treatment of the older type of kinematical problem, where motions preserving the similarity of certain classes of geometrical figures are sought, and I have made no attempt to include results which merely attach to kinematical objects the words in use in differential geometry, but do not forward the science of kinematics in itself. While I believe the approach and some of the conclusions here presented to be new, in the working many already known results naturally offer themselves for a quick and feat proof, and thus I have attempted to incorporate all that was hitherto known touching the subject of vorticity.

---

[1] These topics are briefly developed in [1952, 1, §§12–22].

Attributions. I have tried to ascertain full and accurate attributions for all the previously known results which are presented here. I should be thankful to be told of any errors or oversights. It is perhaps necessary to mention that reference to certain recent papers which contain no new matter has been omitted intentionally, and that in many cases the proofs are entirely other than those in the original memoirs. Frequently I have not hesitated to name a general proposition after the discoverer of its main special case, since most often the generalizations included in the wordings indicate mere thrift or a wider view, rather than any bettering of the ground thought.

Originality. Judging by certain recent reviews, the reward of scruples toward forebears and contemporaries is neglect of what the author himself has added, perhaps graced by the reviewer's glib citation of the references which it has cost the author such pains to trace. For the benefit of any curious reviewer of this work, I suggest that it is safe to assume that, even though I have refrained from naming any particular proposition after myself, a result for which *no* reference is given appears here for the first time. Moreover, *some* of the results for which references *are* given appeared for the first time in my various papers on the subject, which can be traced without labor by looking up my name in the "Index of Authors Quoted."

Notation. The theory of vectors and tensors is plagued with a variety of notations and with their respective enthusiasts. For the present work it was necessary to use *some* notation, and thus to give grounds for criticism to the adherents of the others. I assure the reader that notation is largely a matter of indifference to me, and I beg him, laying symbolic prudery aside, to strip off from our subject the casual tires of fashion and to join me in contemplation of the naked beauty of its essentials. Originally I had written the whole in the Gibbs symbolism, but my colleagues have convinced me that use of some of the operations on dyadics would discourage those few readers who might otherwise overleaf these pages. On the other hand, believing that at least for simple vectorial formulae the straightforwardness of Gibbs's notations is very helpful in picturing the results, especially for physical applications, I have adopted a compromise scheme, which is explained in §1.

Circumstances of composition. This study grew out of an address [1948, **2**] delivered to the Colloquium in Aeronautical Sciences of the Graduate School of Engineering, Harvard University, March 2, 1948. The material was further developed in a lecture given at the University of Toronto, February 3, 1949, and at the Sorbonne, June 8, 1949 [1949, **2**, 1re Conf.]; portions were presented to the American Mathematical Society at Athens (Georgia), December 30, 1947; Madison (Wisconsin),

September 8, 1948; Columbus (Ohio), December 30, 1948; Philadelphia (Pennsylvania), April 29, 1949; New York City, December 27, 1949 and February 25, 1950; Washington, D. C., April 28, 1950; Gainesville (Florida), December 27, 1950. Some of the results have already been published as parts of the papers [1948, **1** & **3**], [1949, **3**], [1950, **1–3**], [1951, **1–7**], [1952, **2**], [1953, **1**]. Much of the detailed historical matter included in earlier versions of this work, circulated in manuscript since 1948, has been removed and will appear in my history of fluid mechanics in the eighteenth century [1954, **1**].

Quoique j'envisage ici une si grande généralité, tant par rapport à la nature du fluide, qu'aux forces qui agissent sur chacune de ses particules, je ne crains point les reproches, qu'on a souvent faits avec raison à ceux qui ont entrepris de porter à une plus grande généralité les recherches des autres. Je conviens qu'une trop grande généralité obscurcit souvent plutôt, qu'elle n'éclaire, & qu'elle mene quelquefois à des calculs si embrouïllés, qu'il est extrèmement difficile d'en déduire des conséquences pour les cas les plus simples. Quand les généralisations sont assujetties à cet inconvenient, il est bien certain qu'il vaudroit infiniment mieux s'en abstenir entièrement, & borner ses recherches à des cas particuliers.

Mais, dans le sujet que je me propose d'expliquer, il arrive précisement le contraire: la généralité que j'embrasse, au lieu d'éblouïr nos lumieres, nous découvrira plutôt les véritables loix de la Nature dans tout leur éclat, & on y trouvera des raisons encore plus fortes, d'en admirer la beauté & la simplicité. Ce sera une instruction importante d'apprendre que des principes, qu'on aura cru attachés à quelque cas particulier, ont une plus grande étenduë. Ensuite ces recherches ne demanderont presque point un calcul plus embarassant; & il sera aisé d'en faire l'application à tous les cas particuliers, qu'on puisse se proposer.

———EULER [1757, **1**, §§III-IV].

Durch diese Sätze wird die Reihe der Bewegungsformen, welche in der nicht behandelten Klasse der hydrodynamischen Gleichungen verborgen sind, wenigstens für die Vorstellung zugänglich, wenn auch die vollständige Ausführung der Integration nur in wenigen einfachsten Fällen möglich ist . . . .

———HELMHOLTZ [1858, **1**, Introd.].

The great clarity which geometrical investigation lends to the study of the dynamics of solids leads us to expect significant success in hydrodynamics through a study of the kinematics of variable systems.

———ZHUKOVSKI [1876, **3**].

Die Theorie dieser allgemeinen Bewegungserscheinungen in kontinuirlichen Medien hat eine noch unbegrenzte Entwicklungsfähigkeit. Doch ist es notwendig, ganz ohne vorgefaßte Vorstellungen an dieselben heranzutreten . . . .

———JAUMANN [1905, **1**, Introd.].

THE KINEMATICS OF VORTICITY

Introduction

All real fluid motions are rotational. Even in nearly irrotational flows the relatively small amount of vorticity present may be of central importance in determining major flow characteristics, and even some of those whose interest in fluid dynamics is only of the practical sort are now beginning to learn that the hitherto largely neglected questions of vorticity must at last be faced.

The more deeply one penetrates the general character of fluid motions, the more apparent it becomes that the dynamical properties of fluids in the main are but names, interpretations, and methods of measuring purely kinematical quantities, and that in general the flow of a fluid, whether perfect[1] or viscous, may be defined by purely kinematical conditions.[2] It is no accident that the greatest contributions to practical fluid dynamics were preceded by kinematical analyses which in themselves belong to pure mathematics rather than to mechanics or physics; while the work of Stokes, Helmholtz, and Kelvin is familiar, it is less well known that Euler headed his several successive presentations of the general equations of perfect fluids by increasingly detailed and accurate investigations of the possible motions of any deformable continuum, and that the same Zhukovski who discovered the artifice by which perfect fluid theory can be turned to practical use in plane wing theory began his career with a long memoir on the kinematics of continuous media.[3]

In the realization that the kinematics of rotational motions contains the essence of fluid dynamics the present essay was conceived. Many a theorem generally regarded as dynamical will here be found in a purer form, presented at its proper station in a consecutive development. In particular, classical hydrodynamics may be characterized by the kinematical statement of Kelvin's circulation theorem, and in this way all the general properties of barotropic flows of inviscid fluids subject to conservative extraneous force (properties which necessarily hold equally for a special class of flows of viscous liquids) will appear as

---

[1] In describing the theory of perfect fluids the physicist Rowland [1880, **8**, p. 262] remarked that the d'Alembert-Euler vorticity equations (94.1) "contain the whole dynamics of the subject . . . ."

[2] An exception is constituted by the beautiful general theorems of gas dynamics, where kinematics, mechanics, and thermodynamics truly co-operate. This field is presented in another monograph [1952, **2**].

[3] [1876, **3**]. *Cf.* [1947, **7**].

special cases of certain purely kinematical theorems valid for arbitrary media. Let no one contend, however, that I have merely derived the old results in a new way. Rather, circulation-preserving motions afford but the simplest and most elegant applications of some parts of the general theory, a theory constructed in the hope that it will prove useful in understanding the behavior of complicated media whose dynamical response is more elaborate than that represented by the classical laws of viscosity. All dynamical statements I have relegated to parenthetical sections, appendices, or footnotes, not in a foolish attempt to diminish their physical importance, but rather to let the argument course freely, uninterrupted by merely interpretative remarks, and to leave the propositions free for application to such special dynamical situations as may be of interest either now or in the future—for I cannot too strongly urge that a kinematical result is a result valid forever, no matter how time and fashion may change the "laws" of physics.

Chapter I. Geometrical Preliminaries

1. Algebraic operations with vectors and tensors. In this monograph I employ regularly the single cross operation of Gibbs,[1] both for vectors and for tensors. My principal reason for adopting it is that my subject is exclusively Euclidean and three-dimensional, since a vorticity vector, vortex-lines, and vortex-tubes cannot exist in two-dimensional or (for general motions) in four-dimensional spaces. While for any dyadic $\mathbf{\Sigma}$ in a space of any number of dimensions we have

$$\mathbf{\Sigma} = \tfrac{1}{2}(\mathbf{\Sigma} + \mathbf{\Sigma}_c) + \tfrac{1}{2}(\mathbf{\Sigma} - \mathbf{\Sigma}_c), \tag{1.1}$$

only in a space of *three* dimensions do we have the ***Gibbsian decomposition:***[2]

$$\mathbf{\Sigma} = \tfrac{1}{2}(\mathbf{\Sigma} + \mathbf{\Sigma}_c) - \tfrac{1}{2}\mathbf{I} \times \mathbf{\Sigma}_\times, \tag{1.2}$$

whereby the skew part of the *dyadic* is expressed in terms of the *axial vector* $\mathbf{\Sigma}_\times$, and it is the consequences of this identity which make three-dimensional kinematics a peculiar discipline and give rise to the *vorticity vector.*

Let us now make our stand quite clear by explaining our symbolism, including that used in the two foregoing equations. First, the reader is assumed to be familiar with the elements of tensor analysis[3] and with the notations for vectors commonly in use in works on mathematical physics. When we denote a vector by the bold face letter **a**, we shall mean that **a** is to be regarded as a short name for the set of contravariant components a^i or covariant components a_i, whichever is appropriate. Thus the single Gibbs equation

$$\alpha = \mathbf{a} \cdot \mathbf{b} \tag{1.3}$$

stands for *both* the equivalent explicit equations

$$\alpha = a^i b_i, \qquad \alpha = a_i b^i. \tag{1.4}$$

Latin indices $i, j, \ldots$ are always assumed to range over the values 1, 2, 3. It goes without saying that the covariant and contravariant components of any given field are related by $a^i = g^{ij}a_j$, $a_i = g_{ij}a^j$, where the g_{ij} and g^{ij} are the covariant and contravariant components of the metric tensor. By a *dyadic* $\mathbf{\Sigma}$ we shall mean any one of the sets

[1] [1881, 3] [1884, 1] [1909, 1].

[2] [1884, 1, §137].

[3] For definitions of the terms in use in tensor analysis see, *e.g.*, [1927, 5].

of covariant, contravariant, or mixed components of a tensor of the second order. Thus the one Gibbs equation

$$\mathbf{a} = \mathbf{b}\cdot\mathbf{\Sigma} \tag{1.5}$$

stands for *all* the equations

$$a_i = b_j\Sigma^j{}_i = b^j\Sigma_{ji}, \qquad a^i = b_j\Sigma^{ji} = b^j\Sigma_j{}^i, \tag{1.6}$$

while

$$\mathbf{a} = \mathbf{\Sigma}\cdot\mathbf{b} \tag{1.7}$$

stands for

$$a_i = \Sigma_{ij}b^j = \Sigma_i{}^jb_j, \qquad a^i = \Sigma^{ij}b_j = \Sigma^i{}_jb^j. \tag{1.8}$$

The symbol $\mathbf{I}$ denotes the dyadic whose components are g_{ij}, $\delta_i{}^j$, g^{ij}, *i.e.*, the metric tensor itself.

Let $g \equiv \det g_{ij}$. Only those co-ordinate systems in which $g > 0$ are admitted in this work. Let ϵ_{ijk} and ϵ^{ijk} have their usual significance as permutation symbols. Then, given two vectors $\mathbf{a}$ and $\mathbf{b}$, the equations

$$c_i \equiv \sqrt{g}\,\epsilon_{ijk}a^jb^k \tag{1.9}$$

define covariant components c_i of a vector $\mathbf{c}$, whose contravariant components are

$$c^i = \frac{\epsilon^{ijk}}{\sqrt{g}}\,a_jb_k. \tag{1.10}$$

We write *both* (1.9) and (1.10) in the single form

$$\mathbf{c} = \mathbf{a}\times\mathbf{b}. \tag{1.11}$$

For the magnitude of the vector $\mathbf{b}$ we shall use the customary symbols:

$$b \equiv |\,\mathbf{b}\,| \equiv \sqrt{b_ib^i} = \sqrt{g_{ij}b^ib^j} = \sqrt{g^{ij}b_ib_j}. \tag{1.12}$$

The *dot* or *scalar* $\Sigma.$ of a dyadic $\mathbf{\Sigma}$ is the trace of the matrix of its mixed components:

$$\Sigma. \equiv \Sigma^i{}_i. \tag{1.13}$$

Its *cross* or *vector* $\mathbf{\Sigma}_\times$ is the axial vector field whose covariant components are defined by

$$\Sigma_{\times i} \equiv \sqrt{g}\,\epsilon_{ijk}\Sigma^{jk}. \tag{1.14}$$

The existence of this type of vector field, peculiar to a space of three dimensions, we shall later find to be the foundation of the doctrine of vorticity.

It is easy to verify the interesting identity

$$(\mathbf{a}\cdot\mathbf{\Sigma})\times\mathbf{b} - (\mathbf{b}\cdot\mathbf{\Sigma})\times\mathbf{a} = -\mathbf{\Sigma}\cdot(\mathbf{a}\times\mathbf{b}) + \Sigma.\,\mathbf{a}\times\mathbf{b}. \tag{1.15}$$

The cross products $\mathbf{a} \times \mathbf{\Sigma}$ and $\mathbf{\Sigma} \times \mathbf{a}$ are defined as the dyadics $\mathbf{\Psi}$, $\mathbf{\Theta}$ whose covariant components are given by

$$\Psi_{ij} \equiv \sqrt{g}\, \epsilon_{ikl} a^k \Sigma^l{}_j, \qquad \Theta_{ij} \equiv \sqrt{g}\, \epsilon_{jkl} \Sigma_i{}^k a^l. \tag{1.16}$$

We may write our first identity (1.1) in the form

$$\begin{aligned} \Sigma_{ij} &= \tfrac{1}{2}(\Sigma_{ij} + \Sigma_{ji}) + \tfrac{1}{2}(\Sigma_{ij} - \Sigma_{ji}), \\ &= \tfrac{1}{2}(\Sigma_{ij} + \Sigma_{ji}) + \tfrac{1}{2}\delta_{ij}^{mn}\Sigma_{mn}. \end{aligned} \tag{1.17}$$

In a space of *three* dimensions, however, we have

$$\begin{aligned} \Sigma_{ij} &= \tfrac{1}{2}(\Sigma_{ij} + \Sigma_{ji}) + \tfrac{1}{2}\, \epsilon_{ijl}\epsilon^{lmn}\Sigma_{mn}, \\ &= \tfrac{1}{2}(\Sigma_{ij} + \Sigma_{ji}) - \tfrac{1}{2}\sqrt{g}\, \epsilon_{jkl}\delta_i{}^k\left(\frac{\epsilon^{lmn}}{\sqrt{g}}\Sigma_{mn}\right), \end{aligned} \tag{1.18}$$

which is the covariant form of the Gibbs decomposition (1.2). Its rather awkward appearance when written in tensor notation obscures the importance, evident from the direct form (1.2), of the axial vector $\mathbf{\Sigma}_\times$.

While it is easily possible, besides being logically preferable, to define the Gibbs operations directly, rather than (as we have done) merely regarding each equation in the Gibbsian symbolism as an abbreviation for a number of tensorial equations, we shall not attempt to use any one system of notation exclusively. Thus we shall not use Gibbs's symbols $\overset{\times}{\times}$, :, $\mathbf{\Sigma}_c$, $\mathbf{\Sigma}_2$, and $\mathbf{\Sigma}_3$, and in many expressions involving tensors of order greater than one we shall employ the common indicial notations of tensor analysis.

We shall find that in a few cases it makes for clarity to employ the *physical components* of vectors and tensors, provided the co-ordinate system be orthogonal. These components, which are explained in detail elsewhere,[4] we shall denote by indices set in the middle of the line:

$$a^{\langle i \rangle}, \qquad \Sigma^{\langle ij \rangle}, \qquad \text{etc.}$$

2. Two algebraic theorems. Another special property of three-dimensional space, discovered in principle by Cauchy,[1] concerns transformations leaving one co-ordinate fixed: $\bar{x}^1 = x^1$, $\bar{x}^\alpha = \bar{x}^\alpha(x^2, x^3)$, $\alpha = 2, 3$. *Let* $\mathbf{\Sigma}$ *be any three-dimensional dyadic*, i.e., *let* Σ^{ij}, $i, j = 1, 2, 3$ *be contravariant components of a tensor field of second order with respect to transformations* $\bar{x}^i = \bar{x}^i(x^1, x^2, x^3)$. *Let* $g' \equiv \det g_{\alpha\beta}$ *where* $\alpha, \beta = 2, 3$,

[4] Though not mentioned in purely geometrical works, the physical components of vectors and tensors are always employed in works on hydrodynamics. For an explanation of the connection between physical, contravariant, and covariant components, see [1931, **3**, Appendix, §2] [1953, **2**].

[1] [1841, **1**, Th. VIII].

and consider only those co-ordinate systems in which $g' > 0$. Then the quantities

$$ {}_1\Sigma^\alpha{}_\beta \equiv \sqrt{g'}\,\epsilon_{\beta\gamma}\Sigma^{\alpha\gamma}, \qquad \alpha, \beta = 2, 3, \tag{2.1}$$

are mixed components of a two-dimensional dyadic ${}_1\mathbf{\Sigma}$, which we may call the skew projection *of $\mathbf{\Sigma}$ onto the direction* 1. In proof of the theorem just stated we need only observe that the quantities $\Sigma^{\gamma\alpha}$, $\gamma, \alpha = 2, 3$, are contravariant components of a tensor field of second order with respect to transformations $\bar{x}^\alpha = \bar{x}^\alpha(x^2, x^3)$. The dyadic ${}_1\mathbf{\Sigma}$ is an axial dyadic, just as $\mathbf{\Sigma}_\times$ is an axial vector. Connecting the two there is a singular identity, also discovered in principle by Cauchy:

$$\Sigma_{\times i} = \sqrt{\frac{g}{g'}}\,{}_i\Sigma_\alpha{}^\alpha = \sqrt{\frac{g}{g'}}\,{}_i\Sigma_{\cdot}: \tag{2.2}$$

the magnitude of the i component of the vector $\mathbf{\Sigma}_\times$ is proportional to the scalar ${}_i\Sigma_{\cdot}$ of the skew projection ${}_i\mathbf{\Sigma}$ of $\mathbf{\Sigma}$ onto the i direction. In the case of a triply orthogonal co-ordinate system the factor $\sqrt{g/g'}$ in (2.2) may be cancelled, provided we refer $\Sigma_\times$ to its physical components:

$$\Sigma_\times{}^i = {}_i\Sigma_{\cdot}. \tag{2.3}$$

By the *proper values* of a dyadic Σ we shall mean the roots of the characteristic equation

$$\det\,(\Sigma^i{}_j - \lambda\delta^i{}_j) = 0. \tag{2.4}$$

We shall require two results from algebra, the familiar theorem of Cauchy[2] that *the proper values of a real symmetric dyadic are real, and the principal directions corresponding to two distinct proper values are orthogonal,* and its converse, given in principle by Kelvin & Tait:[3] *if all the proper values of a real dyadic be real, and if the principal directions corresponding to any pair of distinct proper values be orthogonal, then the dyadic is symmetric.*

3. Differential operations on vector and tensor fields. The symbols grad χ, div $\mathbf{b}$, curl $\mathbf{b}$, $\nabla^2\chi$, will be employed in their usual senses. More generally, by grad $\mathbf{b}$ we denote the tensor whose covariant components Ψ_{ij} are given by

$$\Psi_{ij} \equiv b_{j,i}, \tag{3.1}$$

the comma denoting covariant differentiation. Thus the equation

$$\mathbf{c} = \mathbf{a}\cdot\text{grad}\,\mathbf{b} \tag{3.2}$$

stands for

$$c_i = a^j b_{i,j} \tag{3.3}$$

---

[2] [1828, 1].
[3] [1867, 1, §183].

and all equations obtainable from it by the raising and lowering of indices. Thus we have

$$\operatorname{div}\mathbf{b} = \operatorname{grad}\mathbf{b}_{\cdot}, \qquad \operatorname{curl}\mathbf{b} = \operatorname{grad}\mathbf{b}_{\times}. \tag{3.4}$$

Similarly

$$\mathbf{b} = \operatorname{div}\mathbf{\Sigma} \tag{3.5}$$

stands for

$$b_i = \Sigma^j{}_{i,j}, \qquad \text{etc.}, \tag{3.6}$$

so that

$$\mathbf{b} = \nabla^2\mathbf{a} \tag{3.7}$$

represents

$$b_i = g^{kj}a_{i,kj} = a_i{}^{,j}{}_{,j}. \tag{3.8}$$

Since our three dimensional space is assumed to be Euclidean, we have the familiar identities

$$\operatorname{div}\operatorname{curl}\mathbf{b} = 0,\ \operatorname{curl}\operatorname{grad}\chi = 0,\ \operatorname{curl}\operatorname{curl}\mathbf{b} = \operatorname{grad}\operatorname{div}\mathbf{b} - \nabla^2\mathbf{b}, \tag{3.9}$$

which the reader shall use, along with several others found in textbooks on vector analysis, without explicit reference.

An extremely important example of the use of the cross product is furnished by the identity

$$\mathbf{b}\cdot\operatorname{grad}\mathbf{c} - (\operatorname{grad}\mathbf{c})\cdot\mathbf{b} = \operatorname{curl}\mathbf{c}\times\mathbf{b}. \tag{3.10}$$

4. Special notations. In the foregoing sections we have explained our interpretations for certain symbols already in more or less common use. We now introduce a few symbols peculiar to the present study.

We shall use a *bold face* $\mathbf{x}$ to denote the three co-ordinates x^1, x^2, x^3. Thus $\mathbf{x}$ does *not* denote a vector field. It is not to be confused in general with the field of *radius vectors* $\mathbf{r}$, since only in a rectangular Cartesian system do the two symbols represent the same quantities x, y, z. Thus *e.g.* in a spherical co-ordinate system $\mathbf{x}$ stands for the three co-ordinates r, θ, ϕ, while the components of $\mathbf{r}$ are $r_1 = r^1 = r\langle 1\rangle = r$, $r_2 = r^2 = r\langle 2\rangle = 0$, $r_3 = r^3 = r\langle 3\rangle = 0$. By $d\mathbf{x}$ we shall denote the vector field whose contravariant components are dx^i.

When in an identity the symbol $\emptyset$ occurs, it shall represent a scalar, vector, or tensor of any order, providing only that the resulting formula have a sense. The reader may easily formulate definitions of $\mathbf{b}\cdot\emptyset$, $\emptyset\cdot\mathbf{b}$, $\mathbf{b}\times\emptyset$, $\emptyset\times\mathbf{b}$ extending (1.6) and (1.16). By grad $\emptyset$ we mean the tensor whose covariant components $\Psi_{ij\cdots k}$ are given by

$$\Psi_{ij\cdots k} \equiv \emptyset_{j\cdots k,i}; \tag{4.1}$$

by div $\emptyset$ we mean the tensor whose contravariant components $\Theta^{ij\cdots k}$ are given by

$$\Theta^{ij\cdots k} \equiv \emptyset^{lij\cdots k}{}_{,l}; \tag{4.2}$$

while by curl $\emptyset$ we mean the tensor whose mixed components $\Xi^{i}{}_{l\cdots m}$ are given by

$$\Xi^{i}{}_{l\cdots m} \equiv \frac{\epsilon^{ijk}}{\sqrt{g}} \emptyset_{jl\cdots m,k}. \tag{4.3}$$

With these definitions we have always as a consequence of the Euclidean character of space the identities

$$\text{div curl } \emptyset = 0, \qquad \text{curl grad } \emptyset = 0. \tag{4.4}$$

We shall require some special notations for projections. Let $\mathbf{n}$ be a unit vector. Then the *normal projection* $\emptyset_{\mathbf{n}}$ and the *tangential projection* $\emptyset_{\mathbf{t}}$ of $\emptyset$ onto the direction of $\mathbf{n}$ are defined by

$$\emptyset_{\mathbf{n}} \equiv \mathbf{n} \cdot \emptyset, \qquad \emptyset_{\mathbf{t}} \equiv \mathbf{n} \times \emptyset. \tag{4.5}$$

We shall usually employ these notations in the case when $\mathbf{n}$ is the unit normal to a surface. For the directional derivative $\mathbf{n}\cdot$grad we shall often write d/dn; for $\mathbf{c}\cdot$grad, d/dc, etc.

In many theorems we shall wish to state conditions concerning the vanishing of certain integrals at infinity. Consider a region of space wholly or partially extending to infinity. Let there be described a sphere of radius r about the origin and let $\mathfrak{S}_r$ be that portion of the surface of this sphere which lies in the interior of the given region (Fig. 4.1).

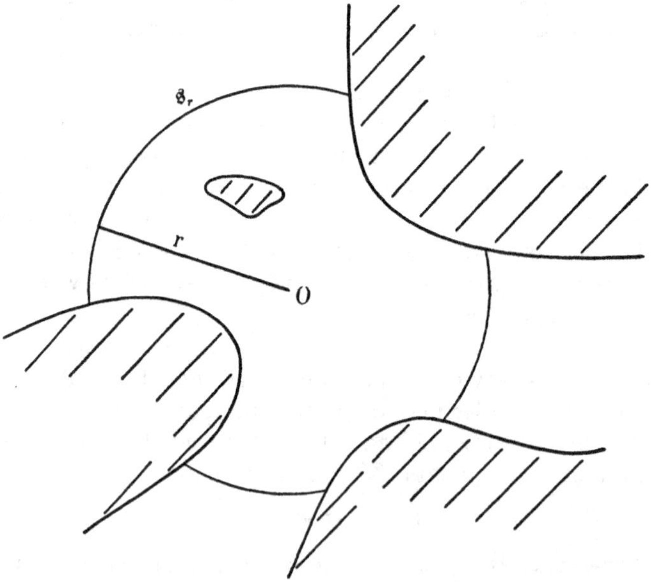

FIG. 4.1

If

$$\lim_{r\to\infty} \int_{\mathfrak{S}_r} d\mathbf{s}\cdot\varnothing r^m = 0, \tag{4.6}$$

we shall write

$$\varnothing_{\mathbf{n}} = \bar{o}(r^{-m-2}). \tag{4.7}$$

For (4.7) to hold it is evidently sufficient that the numerical magnitude of each component of $\varnothing_{\mathbf{n}}$ shall be $o(r^{-m-2})$ as $r \to \infty$, this fact serving as motivation for the symbol (4.7). Similarly, if

$$\lim_{r\to\infty} \int_{\mathfrak{S}_r} d\mathbf{s} \times \varnothing r^m = 0, \tag{4.8}$$

we shall write

$$\varnothing_{\mathbf{t}} = \bar{o}(r^{-m-2}), \tag{4.9}$$

sufficient for this relation to hold being that the numerical magnitude of each component of $\varnothing_{\mathbf{t}}$ shall be $o(r^{-m-2})$. Finally, if

$$\lim_{r\to\infty} \int_{\mathfrak{S}_r} d\mathbf{s}\, \varnothing r^m = 0, \tag{4.10}$$

we shall write

$$\varnothing = \bar{o}(r^{-m-2}), \tag{4.11}$$

sufficient again being the stronger condition that each component of $\varnothing$ shall be $o(r^{-m-2})$.

The symbol $\bar{o}$ may be read "smaller mean order than."

The nth *power* $\mathbf{b}^{(n)}$ of a vector $\mathbf{b}$ is defined as that tensor whose contravariant components $\Psi^{i_1\cdots i_n}$ are given by

$$\Psi^{i_1\cdots i_n} \equiv b^{i_1}b^{i_2}\ldots b^{i_n}. \tag{4.10}$$

This definition is extended and put into inductive form as follows:

$$\mathbf{b}^{(-1)} \equiv 0,\ \mathbf{b}^{(0)} \equiv 1,\ \mathbf{b}^{(n+1)} \equiv \mathbf{b}^{(n)}\mathbf{b}, \qquad n = 0, 1, 2, \ldots. \tag{4.11}$$

The symbol $\{\mathbf{b}^{(n)}\varnothing\}$ shall denote the symmetrized expression[1]

$$\{\mathbf{b}^{(n)}\varnothing\} \equiv \mathbf{b}^{(n)}\varnothing + \mathbf{b}^{(n-1)}\varnothing\mathbf{b} + \ldots + \varnothing\mathbf{b}^{(n)}. \tag{4.12}$$

Thus *e.g.*

$$\begin{aligned} \{\mathbf{b}^{(0)}\varnothing\} &\equiv \varnothing, \qquad \{\mathbf{b}^{(1)}\varnothing\} \equiv \mathbf{b}\varnothing + \varnothing\mathbf{b}, \\ \{\mathbf{b}^{(2)}\varnothing\} &\equiv \mathbf{bb}\varnothing + \mathbf{b}\varnothing\mathbf{b} + \varnothing\mathbf{bb}. \end{aligned} \tag{4.13}$$

5. Assumptions of smoothness. The terms *curve*, *surface*, and *region* will be used freely in this work. We tacitly assume sufficient smoothness

[1] [1949, 3] [1951, 1 & 2].

that the transformations of Green (§7) and Stokes (§8) be valid for the usual class of fields and that existence and uniqueness of the solution to the Dirichlet problem hold for the usual class of boundary values. Thus a surface possesses a continuously turning tangent plane except upon a set of surface measure zero, and a curve possesses a continuously turning tangent except upon a set of linear measure zero; these exceptional points will henceforth in most cases be tacitly disregarded, since they may be omitted in the formation of line and surface integrals, respectively.

Precise definitions and sufficient conditions of smoothness are available in the literature.[1] It is not from mere carelessness that we do not state a particular set of them here. Rather, the quality of this work is for the most part formal. Questions of regularity are foreign to its purpose, but furnish a constant subject of research in pure analysis; whenever conditions sufficient for the truth of the fundamental theorems of analysis are weakened as a result of one of these researches, the range of validity of the results in this work is thereby automatically extended. In the few places where degree of smoothness is a matter of interest, special attention will be given to it.

6. Circulation, flux, total, and moments. Let $\emptyset$ be an arbitrary integrable tensor field. The line integral

$$\int_{\mathfrak{c}} d\mathbf{x}\cdot\emptyset = \int_{\mathfrak{c}} dx^i\, \emptyset_{ij\cdots k} \tag{6.1}$$

along a curve $\mathfrak{c}$ is called the *flow of* $\emptyset$ *along* $\mathfrak{c}$. A closed curve is called a *circuit*, and the integral (6.1) is then called the *circulation of* $\emptyset$ *around*[1] $\mathfrak{c}$ and is written

$$\oint_{\mathfrak{c}} d\mathbf{x}\cdot\emptyset. \tag{6.2}$$

Two curves which can be continuously deformed one into another while remaining in a given point set are *reconcileable*[2] in that set. A

[1] It is sufficient, for example, that the curves, (two-sided) surfaces, and regions be *regular* in the sense of Kellogg [1929, 2, Ch. IV, §§7–9]. These conditions may be supplemented by the definition of an *infinite region* as an unbounded point set whose intersection with every sphere is a region. A typical lightening of the requirements permits surfaces to have a finite number of conical points.

[1] These names derive from Kelvin [1869, 1, §60(a)].

[2] This term was introduced by Kelvin [1869, 1, §58] and is generally employed in hydrodynamics; the later and less descriptive word "homotopic" is used by the topologists. Kelvin stated that he employed Riemann's theory of multiple connectivity as presented by Helmholtz [1858, 1], but neither in Helmholtz's paper nor in Riemann's [1857, 2] is the concept of reconcileable circuits formulated or explicitly employed.

circuit which can be continuously shrunk down to a point and in the process remain in a given set is said to be *reducible* in that set.

The surface integral

$$\int_{\mathfrak{s}} d\mathbf{s} \cdot \emptyset \tag{6.3}$$

over a surface $\mathfrak{s}$ is called the *flux of* $\emptyset$ *across* $\mathfrak{s}$; when $\mathfrak{s}$ is a closed surface the integral is called[3] the *flux of* $\emptyset$ *out of* $\mathfrak{s}$ and is written

$$\oint_{\mathfrak{s}} d\mathbf{s} \cdot \emptyset, \tag{6.4}$$

it being understood that $d\mathbf{s}$ points outward.

A closed surface which can be shrunk down continuously to a point and in the process remain within a given region is said to be *reducible* in that region.

The volume integral

$$\int_{\mathfrak{v}} \emptyset \, dv \tag{6.5}$$

is called[4] the *total* $\emptyset$ *in* $\mathfrak{v}$. More generally, employing the notation (4.12) we define the nth *moment* $\emptyset_{(n)}$ of $\emptyset$ with respect to the origin by

$$\emptyset_{(n)} \equiv \int_{\mathfrak{v}} \{r^{(n)} \emptyset\} \, dv. \tag{6.6}$$

The total $\emptyset$ is thus the zeroth moment $\emptyset_{(0)}$.

7. Green's transformation. We shall employ ***Green's transformation***[1] in the forms

$$\int_{\mathfrak{v}} \operatorname{grad} \emptyset \, dv = \oint_{\mathfrak{s}} d\mathbf{s} \, \emptyset, \tag{7.1}$$

$$\int_{\mathfrak{v}} \operatorname{div} \emptyset \, dv = \oint_{\mathfrak{s}} d\mathbf{s} \cdot \emptyset, \tag{7.2}$$

$$\int_{\mathfrak{v}} \operatorname{curl} \emptyset \, dv = \oint_{\mathfrak{s}} d\mathbf{s} \times \emptyset, \tag{7.3}$$

---

[3] This name derives from Maxwell [1873, 2, §§12–13], who called (6.3) "the surface integral of the flux."

[4] This quantity was introduced under the name "Feldsumme" by A. Föppl [1897, 2, §4].

---

[1] An equivalent result was given by Green [1828, 2, pp. 23–26]. Special cases were stated previously by Lagrange [1762, 2, §45] and by Gauss [1813, 1, §§3–5]. The general case was put in a rather obscure way as a formula for differentiating a triple integral by Ostrogradsky [1831, 2].

it being assumed that each component of $\emptyset$ is continuous throughout the closure of the finite region $\mathfrak{v}$, whose boundary is the surface $\mathfrak{s}$, and continuously differentiable throughout the interiors of each of a finite number of regions, of which $\mathfrak{v}$ is the sum, and that the volume integrals are convergent.[2]

The verbal expression of (4.2) in the terminology of §6 is: *the total divergence of a quantity in a region equals the flux of the quantity out of the boundary of the region.*

Another form of Green's transformation is[3]

$$\int_{\mathfrak{v}} [\mathbf{b}\{\mathbf{c}\mathbf{r}^{(n-1)}\} + \mathbf{c}\{\mathbf{b}\mathbf{r}^{(n-1)}\} - \mathbf{b}\cdot\mathbf{c}\{\mathbf{r}^{(n-1)}\mathbf{I}\}$$
$$+ (\text{curl}\,\mathbf{c} \times \mathbf{b} + \text{curl}\,\mathbf{b} \times \mathbf{c} + \mathbf{b}\,\text{div}\,\mathbf{c} + \mathbf{c}\,\text{div}\,\mathbf{b})\mathbf{r}^{(n)}]\,dv$$
$$= \oint_{\mathfrak{s}} [d\mathbf{s}\cdot(\mathbf{bc} + \mathbf{cb})\mathbf{r}^{(n)} - d\mathbf{s}\,\mathbf{b}\cdot\mathbf{c}\mathbf{r}^{(n)}]. \tag{7.4}$$

Here $\mathbf{b}$ and $\mathbf{c}$ are arbitrary continuously differentiable fields and n is an arbitrary non-negative integer.

8. Kelvin's transformation ("Stokes's theorem"). Let the field $\mathbf{c}$ be continuous and piecewise continuously differentiable in a closed region whose boundary contains the surface $\mathfrak{s}$, bounded by the curve $\mathfrak{c}$. Then[1]

$$\oint_{\mathfrak{c}} d\mathbf{x}\cdot\mathbf{c} = \int_{\mathfrak{s}} d\mathbf{s}\cdot\text{curl}\,\mathbf{c}, \tag{8.1}$$

subject to the usual right-handed screw convention connecting the sense of description of $\mathfrak{c}$ with the sense of the normal field $d\mathbf{s}$. We shall call (8.1) ***Kelvin's transformation.***[2] Its verbal expression in the termi-

---

[2] [1929, **2**, Ch. IV, §11].

[3] [1949, **3**] [1951, **2**]. The special case $n = 0$ is given by Burgatti [1931, **2**].

---

[1] For our purpose it is essential to notice that *the field* $\mathbf{c}$ *need not be defined at all on one side of the surface* $\mathfrak{s}$, so that later we shall be able to apply Kelvin's transformation to circuits lying on a bounding surface of a motion. See [1929, **1**, Kap. 2, §3], where references to works proving the result subject to weaker conditions are given.

[2] This celebrated integral transformation was employed in the special case of a plane area by Ampère [1826, **1**]. The general theorem was discovered independently by Kelvin (1850) and Hankel [1861, **1**, §7] (see also [1863, **1**, §4]). Hankel's attention was attracted to the problem by Riemann's extensive consideration of Ampère's result. Kelvin's proofs were given in [1867, **1**, §190(j)] [1869, **1**, §60(q)], and in the latter he claimed priority for the former. In the 1879 edition of [1867, **1**] Kelvin mentioned that the theorem was set by Stokes [1854, **1**, Question 8] as a Smith's prize examination question, as had been remarked by Maxwell [1873, **2**, §24], after whose custom the name *Stokes's theorem* has been generally adopted. That the theorem is actually due to Kelvin is evidenced by his letter to Stokes of date July 2, 1850, in which the

nology of §3 is: *the circulation of a quantity around a closed circuit equals the flux of its curl across any surface bounded by the circuit.*

A formal equivalent of (8.1) is

$$\oint_c d\mathbf{x} \times \mathbf{c} = \int_s [d\mathbf{s} \cdot \operatorname{grad} \mathbf{c} - d\mathbf{s} \operatorname{div} \mathbf{c}]. \tag{8.2}$$

9. Vector-lines, vector-sheets, and vector-tubes. A curve everywhere tangent to a given continuous vector field $\mathbf{c}$ is a *vector-line* of $\mathbf{c}$. Necessary and sufficient that a given curve be a vector-line of $\mathbf{c}$ is that it admit a representation $\mathbf{x} = \mathbf{x}(l)$ satisfying the differential equation

$$\frac{d\mathbf{x}}{dl} \times \mathbf{c} = 0. \tag{9.1}$$

A field which is continuous in a closed region possesses at least one vector-line through each interior point of the region; if moreover the field $\mathbf{c}$ satisfy a Lipschitz condition,[1] there is exactly one vector-line through each point where $c \neq 0$.

Consider a region in which $c \neq 0$, and let $\mathbf{t}$, $\mathbf{n}$, $\mathbf{b}$ be the unit tangent, principal normal, and binormal to the vector-line. The field $\mathbf{t}$ is especially useful in questions concerning the geometry of the vector-lines of $\mathbf{c}$. Put

$$\Theta \equiv \operatorname{div} \mathbf{t}, \qquad \mathbf{\Psi} \equiv \operatorname{curl} \mathbf{t}, \qquad \Omega \equiv \mathbf{t} \cdot \operatorname{curl} \mathbf{t}. \tag{9.2}$$

We have then

$$\Omega = \mathbf{t} \cdot \operatorname{curl} \frac{\mathbf{c}}{c} = \frac{\mathbf{t}}{c} \cdot \operatorname{curl} \mathbf{c} = \frac{\mathbf{c} \cdot \operatorname{curl} \mathbf{c}}{c^2}. \tag{9.3}$$

---

theorem is fully stated in a postscript [Larmor's annotation to the 1905 reprint of [1854, 1]].

That Kelvin did not repeat his assertion of discovery after the theorem had been attributed publicly to Stokes may be explained by the warm friendship between these two savants, but that Stokes made no disclaimer is more difficult to understand. The general literature has followed the British school in completely overlooking the quite legitimate priority of Hankel, who could hardly be expected to follow Cambridge examination schedules.

The attribution of names to theorems is but too often unjust: even aside from discovery, it was Kelvin who first realized the significance of (8.1) and who first showed how it could be used in the proofs of important vorticity theorems. While it is futile to attempt to turn common usage, in the present work it is unfitting to refer to a central proposition concerning vorticity by the name of a person who neither discovered, claimed, nor employed it, but who merely set it as an examination question. If further names are to be attached, they should be those of Ampère and Hankel. Stokes was indeed one of the discoverers of an even more important kinematical theorem (§27), which in its turn is often attributed to Helmholtz.

[1] *E.g.* [1930, 2, ¶¶72, 78]. Uniqueness is proved here subject to a condition weaker than the Lipschitz condition.

Since

$$\text{grad } \mathbf{c} = \text{grad } (c\mathbf{t}) = (\text{grad } c)\mathbf{t} + c \text{ grad } \mathbf{t}, \tag{9.4}$$

we have

$$\begin{aligned} \text{div } \mathbf{c} = (\text{grad } \mathbf{c})_{.} &= \mathbf{t}\cdot\text{grad } c + c \text{ div } \mathbf{t}, \\ &= \frac{dc}{ds} + c\Theta, \end{aligned} \tag{9.5}$$

$$\text{curl } \mathbf{c} = (\text{grad } \mathbf{c})_{\times} = -\mathbf{t} \times \text{grad } c + c\mathbf{\Psi},$$

where d/ds denotes the derivative with respect to arc-length along the vector-line.

An intrinsic expression for curl **c** has been obtained by Bjørgum[2] in the following way. For any unit vector **e** and any vector **k** we have identically

$$\mathbf{k} = \mathbf{e}\mathbf{e}\cdot\mathbf{k} - \mathbf{e} \times (\mathbf{e} \times \mathbf{k}), \tag{9.6}$$

and hence in particular

$$\text{curl } \mathbf{c} = \mathbf{t}\mathbf{t}\cdot\text{curl } \mathbf{c} - \mathbf{t} \times (\mathbf{t} \times \text{curl } \mathbf{c}). \tag{9.7}$$

But

$$\begin{aligned} \text{curl } \mathbf{c} &= \text{curl } (c\mathbf{t}) = \text{grad } c \times \mathbf{t} + c \text{ curl } \mathbf{t}, \\ \mathbf{t} \times \text{curl } \mathbf{c} &= \mathbf{t} \times (\text{grad } c \times \mathbf{t}) + c\mathbf{t} \times \text{curl } \mathbf{t}, \\ &= \text{grad } c - \mathbf{t}\mathbf{t}\cdot\text{grad } c + c\mathbf{t} \times \mathbf{\Psi}, \end{aligned} \tag{9.8}$$

where the last step follows by (9.7). Since

$$\text{grad} = \mathbf{t}\mathbf{t}\cdot\text{grad} + \mathbf{n}\mathbf{n}\cdot\text{grad} + \mathbf{b}\mathbf{b}\cdot\text{grad}, \tag{9.9}$$

we may put (9.8) into the form

$$\mathbf{t} \times \text{curl } \mathbf{c} = \mathbf{n}\mathbf{n}\cdot\text{grad } c + \mathbf{b}\mathbf{b}\cdot\text{grad } c - c\mathbf{t} \times \mathbf{\Psi}, \tag{9.10}$$

whence by one of the Serret-Frenet formulae follows

$$\mathbf{t} \times \text{curl } \mathbf{c} = \left(\frac{dc}{dn} - \kappa c\right)\mathbf{n} + \frac{dc}{db}\mathbf{b}, \tag{9.11}$$

where κ is the curvature of the vector-line. Hence

$$-\mathbf{t} \times (\mathbf{t} \times \text{curl } \mathbf{c}) = \frac{dc}{db}\mathbf{n} + \left(c\kappa - \frac{dc}{dn}\right)\mathbf{b}. \tag{9.12}$$

Putting this result and $(9.3)_2$ into (9.7) yields the desired expression:

$$\text{curl } \mathbf{c} = c\Omega\mathbf{t} + \frac{dc}{db}\mathbf{n} + \left(c\kappa - \frac{dc}{dn}\right)\mathbf{b}. \tag{9.13}$$

---

[2] [1951, **12**, §2.3]. *Cf.* [1952, **6**, §2].

By putting $c = 1$ we obtain the corollary:

$$\mathbf{\Psi}^{\cdot} = \Omega \mathbf{t} + \kappa \mathbf{b}. \tag{9.14}$$

The quantity Ω which appears in the foregoing expressions is an important invariant of the system of vector-lines. We shall call it the *abnormality*[3] of the field, this term being suggested by the following geometric interpretation, which Lecornu[4] has attributed to Bertrand. Consider any regular surface element containing the point P in its interior and having unit normal $\mathbf{t}$ at P. At points other than P, the surface need have no special orientation. Consider on the surface a region $\mathfrak{s}$ of area s, containing P and bounded by the circuit $\mathfrak{c}$, reducible on the surface. Form the circulation of $\mathbf{t}$ around $\mathfrak{c}$, by Kelvin's transformation obtaining the identity

$$\frac{\oint_{\mathfrak{c}} d\mathbf{x}\cdot\mathbf{t}}{s} = \frac{\int_{\mathfrak{s}} \mathbf{N}\cdot\operatorname{curl}\mathbf{t}\, ds}{s}, \tag{9.15}$$

where $\mathbf{N}$ is the unit normal to $\mathfrak{s}$. Now we reduce $\mathfrak{c}$ to the point P. Then $s \to 0$, $\mathbf{N} \to \mathbf{t}$, and by $(9.2)_3$ we get

$$\Omega\,|_P = \lim_{s\to 0} \frac{\oint_{\mathfrak{c}} d\mathbf{x}\cdot\mathbf{t}}{s}. \tag{9.16}$$

If the vector-lines possess a normal congruence of surfaces, by letting $\mathfrak{s}$ be one of these we get from (9.16) $\Omega = 0$. Thus the value of Ω in general may be regarded as a measure of the departure of the field $\mathbf{c}$ from the property of having a normal congruence.

A surface everywhere tangent to the field $\mathbf{c}$ is a *vector-sheet* of $\mathbf{c}$. If it be supposed that at least one vector-line of $\mathbf{c}$ passes through each point upon a certain curve $\mathfrak{c}$, these vector-lines sweep out a vector-sheet; when $\mathfrak{c}$ is a circuit, the vector-sheet is called a *vector-tube* of $\mathbf{c}$. In order that a surface $f(x) = \text{const.}$ be a vector-sheet of $\mathbf{c}$ it is necessary and sufficient that

$$\mathbf{c}\cdot\operatorname{grad} f = 0 \tag{9.17}$$

except at a set of points of surface measure zero. A field possessed of a unique vector-line through each interior point of a region is endowed with a unique vector-sheet through each sufficiently short curve interior to the region.

---

[3] The terms *torsion of the curve system* and *torsion of neighboring vector-lines* have been proposed [1951, **12**, §2.4].

[4] [1919, **1**].

Consider now a given vector-tube $\mathfrak{s}$ of $\mathbf{c}$ whose boundary consists of two simple circuits $\mathfrak{c}_1$ and $\mathfrak{c}_2$, each once embracing the tube. Let $\mathfrak{s}_1$ and $\mathfrak{s}_2$ be two surfaces whose complete boundaries are $\mathfrak{c}_1$ and $\mathfrak{c}_2$, respectively (Fig. 9.1). Assuming that $\mathbf{c}$ be integrable upon $\mathfrak{s}_1$ and $\mathfrak{s}_2$, we may call

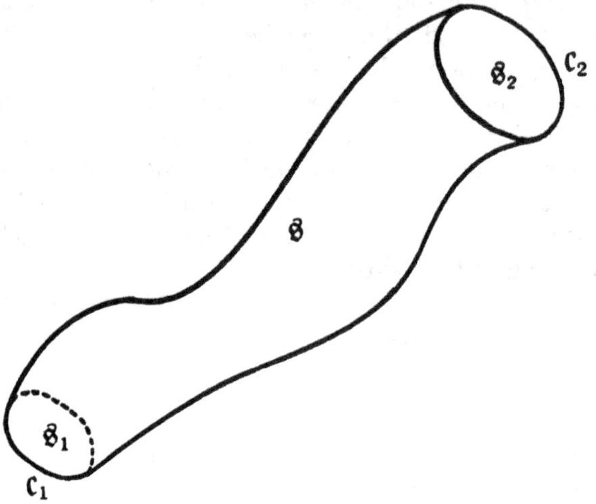

Fig. 9.1

its flux through these surfaces the *strengths* of the vector-tube at the respective *cross-sections*. Let $\mathfrak{s}_3$ denote the closed surface $\mathfrak{s} + \mathfrak{s}_1 + \mathfrak{s}_2$. Then, since $d\mathbf{s}\cdot\mathbf{c} = 0$ upon the vector-tube, we have

$$\oint_{\mathfrak{s}_3} d\mathbf{s}\cdot\mathbf{c} = \int_{\mathfrak{s}_1} d\mathbf{s}\cdot\mathbf{c} + \int_{\mathfrak{s}_2} d\mathbf{s}\cdot\mathbf{c}. \tag{9.18}$$

In (9.18) the normals to $\mathfrak{s}_1$ and $\mathfrak{s}_2$ are both taken outward. If that upon $\mathfrak{s}_2$ be taken inward, we have

$$\oint_{\mathfrak{s}_3} d\mathbf{s}\cdot\mathbf{c} = \int_{\mathfrak{s}_1} d\mathbf{s}\cdot\mathbf{c} - \int_{\mathfrak{s}_2} d\mathbf{s}\cdot\mathbf{c}: \tag{9.19}$$

the flux of $\mathbf{c}$ *out of a surface formed by a vector-tube and two cross-sections equals the difference between the strengths of the tube at these two cross-sections.*

Let $\mathfrak{v}$ be the region bounded by the surface $\mathfrak{s}_3$. If now $\mathbf{c}$ be continuous throughout the closure of $\mathfrak{v}$ and continuously differentiable throughout the interiors of a finite number of regions of which $\mathfrak{v}$ is the sum, then by Green's transformation (7.2) we have

$$\int_{\mathfrak{v}} \operatorname{div} \mathbf{c}\, dv = \int_{\mathfrak{s}_1} d\mathbf{s}\cdot\mathbf{c} - \int_{\mathfrak{s}_2} d\mathbf{s}\cdot\mathbf{c}, \tag{9.20}$$

it being assumed that the volume integral is convergent. That is, *the total divergence of a continuously differentiable field* **c** *in a region bounded by a vector-tube and two cross-sections equals the difference between the strengths of the tube at these two cross-sections.*

10. Solenoidal fields. I. Integral properties. An integrable vector field **c** whose flux out of any reducible closed surface $\mathfrak{s}$ in its region of definition is zero is called *solenoidal.*[1] For a solenoidal field it follows at once from (9.4) that the strength of any vector-tube must be the same at all cross-sections. This property constitutes the ***characterization of Helmholtz*** (1858):[2] *a field continuous*[3] *in a closed region is solenoidal if and only if the strength of every vector-tube be the same at any two cross-sections.* To complete the proof, suppose indeed that for every vector-tube possessed of cross-sections the cross-sectional strength be constant. Then, given any sufficiently small neighborhood, it is possible by a result stated in §9 to find within it a vector-tube with two cross-sections, and by (9.4) the flux out of this small surface vanishes. Since such a surface can be found in every sufficiently small neighborhood, it follows that the flux out of any reducible closed surface is zero.

From (9.5) follows the ***characterization of Kelvin*** (1851):[4] *a continuously differentiable field is solenoidal if and only if its divergence vanishes:*

$$\operatorname{div} \mathbf{c} = 0. \tag{10.1}$$

[1] The name was introduced by Kelvin [1850, 1] [1851, 3, §68]. See *e.g.* [1933, 1, §48].

[2] [1858, 1, §2]. In many works on hydrodynamics, *e.g.* [1932, 1, §§36, 145] may be found a misleading and incomplete statement deriving from Helmholtz [1858, 1, end of §2], who from the fact that the vorticity field is solenoidal concludes "Es folgt hieraus auch, daß ein Wirbelfaden nirgends innerhalb der Flüßigkeit aufhören dürfe, sondern entweder ringförmig innerhalb der Flüßigkeit in sich zurücklaufen, oder bis an die Grenzen der Flüßigkeit reichen müße. Denn wenn ein Wirbelfaden innerhalb der Flüßigkeit irgend wo endete, würde sich eine geschloßene Fläche construien laßen, für welche das Integral $\int q \cos \theta \, d\omega$ nicht den Werth Null hätte." This statement becomes incorrect if "vortex-line" rather than "vortex-filament" be substituted for "Wirbelfaden"; *cf.* [1929, 2, Ch. II, §6]. By the well known result mentioned in §9, a vector-line of a single-valued field cannot begin or end at a point interior to a closed region in which the *field itself* satisfies a Lipschitz condition and is not zero. In fact, whether or not the *divergence* is zero is quite irrelevant to the behavior of the vector-lines. We shall show below that the vector-lines of any continuously differentiable field are also the vector-lines of an infinite number of solenoidal fields, so that the vector-lines of solenoidal fields can possess no special properties. Furthermore, as was remarked by Hadamard [1903, 2, ¶67], the case of a *closed* vector-line is rare, for the vector-lines, being trajectories of ordinary differential equations, even in the case of a region of finite total measure are generally of infinite length and may well pass and repass arbitrarily near to a given point infinitely many times without ever actually touching it.

[3] The example of the field $\mathbf{c} = \mathbf{i}$ when $x < 0$, $\mathbf{c} = \mathbf{j}$ when $x \geqq 0$ shows that this result does not hold for piecewise smooth fields.

[4] [1851, 3, §74].

For a continuously differentiable solenoidal field **c** it is easy to show that

$$\operatorname{div}(\mathbf{c}\mathbf{r}^{(n+1)}) = \{\mathbf{r}^{(n)}\mathbf{c}\}, \tag{10.2}$$

where the notations are those defined in §4. Hence by (7.2) we obtain the following expression[5] for the moments (6.6):

$$\mathfrak{C}_{(n)} \equiv \int_{\mathfrak{v}} \{\mathbf{r}^{(n)}\mathbf{c}\}\, dv = \oint_{\mathfrak{s}} d\mathbf{s}\cdot\mathbf{c}\mathbf{r}^{(n+1)}. \tag{10.3}$$

Thus the moments of a continuously differentiable solenoidal field over a volume are completely determined by the normal projection $\mathbf{c}_n$ *of* **c** *upon the boundary* $\mathfrak{s}$. In particular, *if the normal projection of a continuously differential solenoidal field* **c** *upon* $\mathfrak{s}$ *be zero, then all the moments of* **c** *over the volume* $\mathfrak{v}$ *bounded by* $\mathfrak{s}$ *are zero.*

This result is a special case of a more general one, contained by implication in an example given by Kelvin.[6] For a solenoidal field **c** we have

$$\operatorname{div}(\mathbf{c}\emptyset) = \mathbf{c}\cdot\operatorname{grad}\emptyset. \tag{10.4}$$

By application of Green's transformation (7.2) we thus obtain

$$\int_{\mathfrak{v}} \mathbf{c}\cdot\operatorname{grad}\emptyset\, dv = \oint_{\mathfrak{s}} d\mathbf{s}\cdot\mathbf{c}\emptyset. \tag{10.5}$$

The previous result (10.3) is obtained by putting $\emptyset = \mathbf{r}^{(n+1)}$. From the more general formula (10.5) we conclude that *if* **c** *be a continuously differentiable solenoidal field and* $\emptyset$ *be twice continuously differentiable, then the total* $\mathbf{c}\cdot\operatorname{grad}\emptyset$ *in a volume* $\mathfrak{v}$ *is completely determined by the values of* $\emptyset$ *and of the normal projection* $\mathbf{c}_n$ *of* **c** *upon the boundary* $\mathfrak{s}$. In particular, *if the normal projection of a continuously differentiable solenoidal field* **c** *upon* $\mathfrak{s}$ *be zero, then the total* $\mathbf{c}\cdot\operatorname{grad}\emptyset$ *in* $\mathfrak{v}$ *is zero:*[7]

$$\int_{\mathfrak{v}} \mathbf{c}\cdot\operatorname{grad}\emptyset\, dv = 0. \tag{10.6}$$

Still more important properties of a solenoidal field may be deduced by putting $\mathbf{b} = \mathbf{c}$ in (7.4) and employing (10.1):

---

[5] [1949, **3**, §4]. The special case $n = 0$ was employed in [1897, **2**, §4] and stated in [1941, **1**, eq. (7)].

[6] [1849, **1**, §7]. *Cf.* [1951, **1**, §9].

[7] By means of Kelvin's transformation (§8), this result may be seen to be equivalent to the lemma 1 of Weyl [1940, **4**]. A special case is given by Berker [1949, **7**, Th. III].

$$\oint_{\mathfrak{s}} [d\mathbf{s}\cdot\mathbf{c}\mathbf{c}\mathbf{r}^{(n)} - \tfrac{1}{2}c^2\, d\mathbf{s}\, \mathbf{r}^{(n)}] = \int_{\mathfrak{v}} [\mathbf{c}\{\mathbf{c}\mathbf{r}^{(n+1)}\} - \tfrac{1}{2}c^2\{\mathbf{r}^{(n-1)}\mathbf{I}\} + \operatorname{curl}\mathbf{c} \times \mathbf{c}\mathbf{r}^{(n)}]\, dv. \quad (10.7)$$

The case $n = 0$ yields Poincaré's identity[8]

$$\oint_{\mathfrak{s}} [d\mathbf{s}\cdot\mathbf{c}\mathbf{c} - \tfrac{1}{2}c^2\, d\mathbf{s}] = \int_{\mathfrak{v}} \operatorname{curl}\mathbf{c} \times \mathbf{c}\, dv. \quad (10.8)$$

By formulating conditions sufficient for the vanishing of the surface integral we conclude that *if a field* $\mathbf{c}$ *which is continuously differentiable in a region satisfy the following conditions:*

1. $\mathbf{c}$ *is solenoidal;*
2. *Upon each finite boundary of* $\mathfrak{v}$,

$$\mathbf{c}_n = 0 \qquad \textit{and} \qquad c = \text{const.};$$

3. *In any portion of* $\mathfrak{v}$ *which extends to* ∞,

$$\mathbf{c}\mathbf{c}_n = \bar{o}(r^{-2}), \qquad c^2 = \bar{o}(r^{-2});$$

then

$$\int_{\mathfrak{v}} \operatorname{curl}\mathbf{c} \times \mathbf{c}\, dv = 0. \quad (10.9)$$

Sufficient conditions stronger than 2 and 3 are, respectively,

2′. *Upon each finite boundary of* $\mathfrak{v}$,

$$c = 0;$$

3′. *In any portion of* $\mathfrak{v}$ *which extends to* ∞,

$$c = o(r^{-1}).$$

The case $n = 1$ in (10.7) yields

$$\oint_{\mathfrak{s}} [d\mathbf{s}\cdot\mathbf{c}\mathbf{c}\mathbf{r} - \tfrac{1}{2}c^2\, d\mathbf{s}\, \mathbf{r}] = \int_{\mathfrak{v}} [\mathbf{c}\mathbf{c} - \tfrac{1}{2}c^2\mathbf{I} + \operatorname{curl}\mathbf{c} \times \mathbf{c}\mathbf{r}]\, dv. \quad (10.10)$$

By taking the dot of (10.10) we obtain an identity of Lamb and J. J. Thomson[9]

$$\tfrac{1}{2}\int_{\mathfrak{v}} c^2\, dv = \oint_{\mathfrak{s}} d\mathbf{s}\cdot(\tfrac{1}{2}c^2\mathbf{r} - \mathbf{c}\mathbf{c}\cdot\mathbf{r}) + \int_{\mathfrak{v}} \mathbf{r}\cdot\operatorname{curl}\mathbf{c} \times \mathbf{c}\, dv, \quad (10.11)$$

---

[8] [1893, 1, §5]. *Cf.* §64[1].
[9] See §59[9] for references.

from which we conclude that *if a field* $\mathbf{c}$ *which is continuously differentiable in a region* $\mathfrak{v}$ *satisfy the following conditions:*

1. $\mathbf{c}$ *is solenoidal;*
2. *Upon each finite boundary of* $\mathfrak{v}$, $c = 0$;
3. *In any portion of* $\mathfrak{v}$ *which extends to* ∞,

$$c^2 = \bar{o}(r^{-3}), \qquad \mathbf{r}\cdot\mathbf{c}c_{\mathbf{n}} = \bar{o}(r^{-3});$$

4. *Throughout the interior of* $\mathfrak{v}$, curl $\mathbf{c} \times \mathbf{c} = 0$;

then $\mathbf{c} = 0$ *throughout* $\mathfrak{v}$.

By taking the cross of (7.10) we similarly obtain[10]

$$\oint_{\mathfrak{s}} [d\mathbf{s}\cdot\mathbf{c}\mathbf{r} \times \mathbf{c} + \tfrac{1}{2}c^2\, d\mathbf{s} \times \mathbf{r}] = \int_{\mathfrak{v}} \mathbf{r} \times (\text{curl } \mathbf{c} \times \mathbf{c})\, dv. \qquad (10.12)$$

From this formula we similarly conclude that *if a field* $\mathbf{c}$ *which is continuously differentiable in a region* $\mathfrak{v}$ *satisfy the following conditions:*

1. $\mathbf{c}$ *is solenoidal;*
2. *Upon each finite boundary of* $\mathfrak{v}$,

$$c_{\mathbf{n}} = 0, \qquad c = \text{const.};$$

3. *In any portion of* $\mathfrak{v}$ *which extends to* ∞,

$$\mathbf{r} \times \mathbf{c}c_{\mathbf{n}} = \bar{o}(r^{-2});$$

then

$$\int_{\mathfrak{v}} \mathbf{r} \times (\text{curl } \mathbf{c} \times \mathbf{c})\, dv = 0. \qquad (10.13)$$

Sufficient conditions stronger than 2 and 3 in both of the two foregoing results are:

2′. *Upon each finite boundary of* $\mathfrak{v}$,

$$c = 0;$$

3′. *In any portion of* $\mathfrak{v}$ *which extends to* ∞,

$$c = o(r^{-\frac{3}{2}}),$$

these conditions being also sufficient that (10.9) shall hold.

For the validity of the foregoing results it is not necessary that the region be simply-connected.

11. Solenoidal fields. II. Differential properties. *Let* $\mathbf{c}$ *be any continuously differentiable vector field; then there exist an infinite number of scalar functions* m *such that* $m\mathbf{c}$ *is solenoidal.* To establish this result we need only observe that any function m satisfying

[10] See §64[2] for references.

$$\operatorname{div}(m\mathbf{c}) = 0 \tag{11.1}$$

will serve the purpose. Equivalently, log m must satisfy the first order linear partial differential equation

$$\mathbf{c}\cdot\operatorname{grad}\log m = -\operatorname{div}\mathbf{c}, \tag{11.2}$$

whose coefficients are continuous functions and which thus possesses an infinite number of solutions. As a corollary it follows that *the vector-lines of any continuously differentiable field* $\mathbf{c}$ *are also the vector-lines of an infinite number of solenoidal fields*,[1] for the fields $\mathbf{c}$ and $m\mathbf{c}$ have the same vector-lines.

It is shown in works on vector analysis[2] that *any continuously differentiable solenoidal vector field* $\mathbf{c}$ *may be represented locally in Euler's form*[3]

$$\mathbf{c} = \operatorname{grad} f \times \operatorname{grad} g, \tag{11.3}$$

where f and g are scalar functions. Thus the vector-lines of a smooth solenoidal field in a sufficiently small region are always the curves of intersection of two families of surfaces, f = const. and g = const. Since from (11.3) follows

$$\mathbf{c}\cdot\operatorname{grad} f = 0, \qquad \mathbf{c}\cdot\operatorname{grad} g = 0, \tag{11.4}$$

the surfaces f = const. and g = const. are necessarily vector-sheets of $\mathbf{c}$. One or the other function, say g, may be taken as any function such that the surfaces g = const. are vector-sheets of $\mathbf{c}$. Then the second, f, is uniquely determined to within a constant in a sufficiently small region by the formula

$$f = -\int \frac{\mathbf{e} \times \mathbf{c}\cdot d\mathbf{x}}{\mathbf{e}\cdot\operatorname{grad} g}, \tag{11.5}$$

where $\mathbf{e}$ is *any* continuous vector field such that $\mathbf{e}\cdot\operatorname{grad} g \neq 0$ and where the path of integration lies wholly upon one of the surfaces g = const. From the result of the preceding paragraph we then conclude that *the vector-lines of any continuously differentiable field* $\mathbf{c}$ *are locally the curves of intersection of two families of surfaces*, and further that any such field may be represented locally in the form

$$\mathbf{c} = h \operatorname{grad} f \times \operatorname{grad} g. \tag{11.6}$$

In the representation (11.3), as we have seen, while g may be chosen as any function such that the surfaces g = const. are vector-sheets of $\mathbf{c}$, the second function f is not at our disposal. More generally, let

[1] [1897, 1, §5].
[2] *E.g.* [1933, 1, §49] [1947, 2, §104, Th. 3].
[3] [1770, 1, §§26, 49] [1806, 1, §142].

f = const. and g = const. be any two independent families of vector-sheets, so that (11.6) is valid. If further $\mathbf{c}$ is to be solenoidal,

$$0 = \operatorname{div} \mathbf{c} = \frac{\partial(h, f, g)}{\partial(x, y, z)}. \tag{11.7}$$

Since f and g are by hypothesis independent, it follows from (11.7) that we must have $h = h(f, g)$. Hence and by (11.6) we conclude ***Euler's theorem on solenoidal fields:***[4] *if* f = const. *and* g = const. *be any two independent families of vector-sheets of a continuously differentiable solenoidal field* $\mathbf{c}$, *then* $\mathbf{c}$ *may be represented locally in the form*

$$\mathbf{c} = h(f, g) \operatorname{grad} f \times \operatorname{grad} g. \tag{11.8}$$

Since

$$\operatorname{curl} (f \operatorname{grad} g + \operatorname{grad} h) = \operatorname{grad} f \times \operatorname{grad} g, \tag{11.9}$$

by (11.3) it follows that *any continuously differentiable solenoidal field* $\mathbf{c}$ *can be represented as the* curl *of a* vector potential $\mathbf{p}$:

$$\mathbf{c} = \operatorname{curl} \mathbf{p}, \tag{11.10}$$

the field $\mathbf{p}$ *being indeterminate up to an additive gradient.* Conversely, since div curl $\mathbf{p} = 0$, *the* curl *of any twice continuously differentiable field* $\mathbf{p}$ *is solenoidal.*[5]

From $(9.5)_3$ it follows that div $\mathbf{c} = 0$ if and only if at points where $\mathbf{c} \neq 0$

$$\frac{d \log c}{ds} = -\Theta. \tag{11.11}$$

Hence by (10.1) follows the ***characterization of Bjørgum:***[6] *a non-vanishing continuously differentiable field* $\mathbf{c}$ *is solenoidal if and only if its magnitude* c *is related to its vector-lines by* (11.11), *or, equivalently,*

$$c = c_0 \exp \left[-\int_0^s \Theta \, ds \right]. \tag{11.12}$$

It follows as a corollary that a non-vanishing solenoidal field is completely determined by its vector-lines and by its magnitude at one point on each; moreover, in a region where Θ is finite c cannot vanish unless it does so all along a vector-line.

---

[4] [1757, 3, §§XLVII–XLIX].

[5] Rowland [1880, 8] observed that from any field $\mathbf{b}$ an infinite number of solenoidal fields $\operatorname{curl}^n \mathbf{b}$ can be constructed; since $\operatorname{curl}^n \mathbf{b} = -\nabla^2 \operatorname{curl}^{n-2} \mathbf{b}$ if $n \geq 3$, volume integrals yielding $\operatorname{curl}^{n-2} \mathbf{b}$ in terms of $\operatorname{curl}^n \mathbf{b}$ are easily derived. Rowland gave a kinematical interpretation for these fields.

[6] [1951, 12, §3.1].

12. Lamellar, complex-lamellar, and Beltrami fields. A field $\mathbf{c}$ is *lamellar* in a region if

$$\oint_{\mathfrak{c}} d\mathbf{x} \cdot \mathbf{c} = 0 \qquad (12.1)$$

for any reducible circuit $\mathfrak{c}$ in the region. It follows that *a continuous field* $\mathbf{c}$ *is lamellar in a region* $\mathfrak{v}$ *if and only if there exists a function* χ *such that*

$$\mathbf{c} = -\text{grad}\,\chi; \qquad (12.2)$$

the function χ, *which is called the* potential *of* $\mathbf{c}$, *is single-valued if* $\mathfrak{v}$ *be simply-connected, in general infinitely many-valued if* $\mathfrak{v}$ *be multiply-connected.* Thus a lamellar field is everywhere normal to the *equipotential surfaces* χ = const. From Kelvin's transformation (8.1) it follows that *a continuously differentiable field* $\mathbf{c}$ *is lamellar if and only if*

$$\text{curl}\,\mathbf{c} = 0. \qquad (12.3)$$

The property of being *everywhere normal to a one parameter family of surfaces* is not limited to lamellar fields. Necessary and sufficient for the existence of such a normal congruence is the local representation

$$\mathbf{c} = \psi\,\text{grad}\,\chi. \qquad (12.4)$$

Fields of this type are called *complex-lamellar.* The terms *lamellar* and *complex-lamellar* were introduced by Kelvin,[1] who proved the foregoing results, as well as the following:[2] *a continuously differentiable field* $\mathbf{c}$ *is complex-lamellar if and only if it be either lamellar or normal to its* curl:

$$\mathbf{c} \cdot \text{curl}\,\mathbf{c} = 0. \qquad (12.5)$$

---

[1] [1850, **1**] [1851, **3**, §§68–69, 75]. The geometry of the surfaces χ = const. in relation to the field $\mathbf{c}$ is studied in [1924, **2**] [1925, **3**] [1927, **4**].

In the present work, differentiability is of the essence of our subject, so more general definitions of our terms are not useful. For completeness, however, we should notice the definitions of H. Weyl [1940, **4**] in the spirit of "generalized solutions." Let $F(G)$ be the class of scalars (vectors) which vanish upon the boundary $\mathfrak{s}$ of $\mathfrak{v}$ and which are continuous and endowed with continuous first derivatives at interior points of $\mathfrak{v}$. Then Weyl calls a field $\mathbf{c}$ solenoidal or lamellar if and only if

$$\int_{\mathfrak{v}} \mathbf{c} \cdot \text{grad}\, f\, dv = 0 \qquad \text{for all } f \in F,$$

$$\int_{\mathfrak{v}} \mathbf{c} \cdot \text{curl}\, \mathbf{g}\, dv = 0 \qquad \text{for all } \mathbf{g} \in G,$$

respectively. His Theorem 1 asserts that a field $\mathbf{c}$ which is both lamellar and solenoidal is equal a.e. to a field $\mathbf{c}'$ possessing derivatives of all orders and satisfying div $\mathbf{c}' = 0$, curl $\mathbf{c}' = 0$.

[2] [1851, **3**, §75]. These fields are discussed also by Veltmann [1870, **2**, pp. 453–456]. *Cf. e.g.* [1947, **2**, §105].

An equivalent criterion had been derived much earlier by Euler.[3] From (9.13) and (9.14) we conclude at once that *a non-vanishing continuously differentiable field is complex-lamellar if and only if its abnormality* Ω *is zero*, or, equivalently, $\Psi = \kappa \mathbf{b}$. From (12.4) it is plain that the magnitude of a complex-lamellar field in general is in no way restricted by the pattern of its vector-lines. For a lamellar field, however, we obtain from (9.13) the much more severe necessary and sufficient conditions of Bjørgum:[4] $\Omega = 0$ wherever $c \neq 0$, the magnitude c does not change in the direction of the binormal, and moreover

$$c = c_0 \exp \int_0^n \kappa \, dn, \tag{12.6}$$

where the integration is carried out along the principal normal lines, which lie in the normal congruence. Consequently, since c_0 is constant on each binormal line, a lamellar field in a sufficiently small region is determined completely by its vector-lines and by its magnitude at a single arbitrary point on each equipotential surface.

Two common examples of complex-lamellar fields are *plane* fields, which in a suitable rectangular Cartesian system x, y, z assume the form

$$c_x = f(x, y), \qquad c_y = g(x, y), \qquad c_z = 0, \tag{12.7}$$

and *rotationally-symmetric* fields, which in a suitable cylindrical system r, θ, r assume the form

$$c_r = f(r, z), \qquad c_z = g(r, z), \qquad c_\theta = 0. \tag{12.8}$$

The opposite extreme from (12.5) is the case when the field is parallel to its own curl:

$$\mathbf{c} \times \operatorname{curl} \mathbf{c} = 0, \qquad \operatorname{curl} \mathbf{c} \neq 0. \tag{12.9}$$

Such a field we shall call a *Beltrami field*.[5] There is an extensive literature on these fields, and recently they have been made the subject of a special treatise by Bjørgum & Godal.[6] We shall require here to know only a few of their many interesting properties.

The expression of $\mathbf{c}$ in terms of scalar functions is elaborate and is not required for this work.[7] From (9.13) it follows that $\mathbf{c}$ is a Beltrami field if and only if

$$\operatorname{curl} \mathbf{c} = \Omega \mathbf{c}, \qquad \Omega \neq 0, \tag{12.10}$$

---

[3] [1770, 2, §1].

[4] [1951, 12, §3.4].

[5] See §52[1–5] for the history of these fields.

[6] [1951, 12] [1953, 3].

[7] Such an expression was shown me by Dr. Prim in 1947; his result and others are obtained by Bjørgum [1951, 12, Sect. 5]. A vector differential equation characterizing

so that *the abnormality of a Beltrami field is the factor of proportionality between the* curl *and the field.* Now the abnormality Ω, being a property of the vector-lines only, must for a Beltrami field be expressible in terms of curl $\mathbf{c}$ alone (since $\mathbf{c}$ and curl $\mathbf{c}$ have the same vector-lines). In fact

$$\Omega = \frac{\Omega\mathbf{c}\cdot\text{curl}\left(\dfrac{\text{curl }\mathbf{c}}{\Omega}\right)}{\Omega c^2} = \frac{\text{curl }\mathbf{c}\cdot\text{curl}\left(\dfrac{\text{curl }\mathbf{c}}{\Omega}\right)}{\Omega c^2}, \qquad (12.11)$$
$$= \frac{\text{curl }\mathbf{c}\cdot\text{curl curl }\mathbf{c}}{(\text{curl }\mathbf{c})^2},$$

a formula noticed by Lecornu.[8]

Now let σ be a scalar function such that $\sigma\mathbf{c}$ is solenoidal (§11). Then

$$0 = \text{div }\sigma\mathbf{c} = \text{div}\left(\frac{\sigma}{\Omega}\text{ curl }\mathbf{c}\right) = \text{grad }\frac{\sigma}{\Omega}\cdot\text{curl }\mathbf{c} = \Omega\mathbf{c}\cdot\text{grad }\frac{\sigma}{\Omega}, \qquad (12.12)$$

so that the surfaces σ/Ω = const. are vector-sheets of $\mathbf{c}$. But since $\Omega = \text{curl }\mathbf{c}/\mathbf{c}$, the surfaces in question are the surfaces curl $\mathbf{c}/\sigma\mathbf{c}$ = const. Conversely, suppose we are given a Beltrami field for which these surfaces are vector-sheets. Then $\Omega\mathbf{c}\cdot\text{grad }(\sigma/\Omega) = 0$, and from $(12.12)_4$ we may work backward and derive $(12.12)_1$. Summary of these results yields a generalization of a theorem of Neményi & Prim,[9] itself a generalization of a theorem of Beltrami (1889),[10] which in turn had

---

Beltrami fields is obtained by Ballabh [1948, **12**, §4] by taking the curl of (12.10) and eliminating Ω from the result by (12.11). A simpler differential equation is obtained by putting (12.11) into (12.10) as it stands. A pair of simultaneous differential equations for the Monge potentials (§13) of a Beltrami field $\mathbf{c}$ was obtained by Morera [1889, **2**].

Another property of Beltrami fields was shown to be necessary by Appell [1921, **2**, §763], sufficient by Carstoiu [1946, **4**], *viz.*, if $\lambda \equiv |\text{curl }\mathbf{c}|/c$, then $\mathbf{c}$ is a Beltrami field if and only if $\lambda = \Omega$. The proof is immediate from (9.13).

[8] [1919, **1**].

[9] [1949, **4**, Theorem 1].

[10] [1889, **1**]. The proof of Morera [1889, **2**] is as follows. Let f, g, h be the Monge potentials (§13) of $\mathbf{c}$; then if $\mathbf{c}$ is a Beltrami field,

$$\begin{aligned}\text{div }\mathbf{c} &= \text{div }(\Omega^{-1}\text{ curl }\mathbf{c}) = \text{grad }\Omega^{-1}\cdot\text{curl }\mathbf{c},\\ &= \text{grad }\Omega^{-1}\cdot\text{grad }f\times\text{grad }g\\ &= \frac{\partial(\Omega^{-1}, f, g)}{\partial(x, y, z)}.\end{aligned}$$

Since curl $\mathbf{c} \neq 0$, f and g are independent. Thus if $\mathbf{c}$ is solenoidal it follows that $\Omega = \Omega(f, g)$. Since the vector-lines of $\mathbf{c}$ are the curves of intersection of f = const., g = const., it follows that Ω is constant along them.

been given previously in a major special case by Gromeka (1881):[11] *in a twice continuously differentiable Beltrami field, the surfaces* curl $\mathbf{c}/\sigma c$ = const. *are vector-sheets if and only if the field* $\sigma\mathbf{c}$ *be solenoidal.* It follows *a fortiori* that σ/Ω *is constant on each vector-line of a Beltrami field* $\mathbf{c}$, *where* σ *is any function such that* div $\sigma\mathbf{c} = 0$. Putting $\sigma = 1$ yields the corollary: *the surfaces of constant abnormality* (Ω = const.) *are vector-sheets of a Beltrami field* $\mathbf{c}$ *if and only if* $\mathbf{c}$ *be solenoidal*, and, in particular, *the abnormality of a solenoidal Beltrami field is constant on each vector-line.*[12] Thus the solenoidal character of a Beltrami field may be determined from its vector-lines alone, independently of its magnitude. The case $\sigma = 1$ in the identity (12.12) may be reduced to the form

$$\frac{1}{c}\operatorname{div}\mathbf{c} = -\frac{d}{ds}\log|\Omega|. \tag{12.13}$$

Writing (12.11) in the form

$$(\operatorname{curl}\mathbf{c})^2 = \mathbf{c}\cdot\operatorname{curl}\operatorname{curl}\mathbf{c} \tag{12.14}$$

shows that a Beltrami field and the curl of its curl always meet at an acute angle. This angle may be zero; then curl $\mathbf{c}$ is also a Beltrami field. By taking the curl of (12.10) we easily show that *the* curl *of a Beltrami field is again a Beltrami field* $\mathbf{c}$ *if and only if the abnormality of* $\mathbf{c}$ *is uniform; in this case* $\mathbf{c}$ *is solenoidal, all consecutive* curls *of* $\mathbf{c}$ *are Beltrami fields, and all have the same abnormality.*[13]

A relation between a Beltrami field and its vector-lines may be read off from (9.13). As has been remarked by Bjørgum,[14] this relation results from that given above for lamellar fields if we replace $\Omega = 0$ by $\Omega \neq 0$. Equivalently, we may derive from (9.5) and (12.12) the consequence

$$c\Omega = c_0\Omega_0 \exp\left(-\int_0^s \Theta\, ds\right) \tag{12.15}$$

in place of $\Omega = \Omega_0 = 0$ for the lamellar case.

---

[11] [1881, **4**, Gl. 2, §9].

[12] The work of Gromeka concerns almost solely the case $\Omega = k$ = const. The condition of the above corollary is then trivially satisfied, so that $\mathbf{c}$ is solenoidal. Then taking the curl of (12.9) yields

$$\nabla^2\mathbf{c} + k^2\mathbf{c} = 0,$$

an equation which has been extensively studied and is generally called *the equation of Pockels*. The field $\mathbf{c}$ is rendered determinate by suitable boundary conditions. The problem of finding such fields might well be called "Gromeka's problem." *Cf.* [1908, **1**, §§39–52] [1940, **2**, §§5–7] [1953, **3**]. Some interesting new hydrodynamical applications are given by Popov [1948, **11**].

[13] [1949, **4**, Th. 3]. *Cf.* the foregoing footnote.

[14] [1951, **12**, §3.3].

13. Monge's potentials and Stokes's potentials. Since the field curl **c** is always solenoidal, we may represent it locally in terms of Euler's potentials (11.3):

$$\text{curl } \mathbf{c} = \text{grad } f \times \text{grad } g. \tag{13.1}$$

Then

$$\text{curl } (c - f \text{ grad } g) = 0. \tag{13.2}$$

By a result from §12 it now follows that there exists a scalar h such that

$$\mathbf{c} = \text{grad } h + f \text{ grad } g. \tag{13.3}$$

We shall call the three scalars f, g, h the *Monge potentials*[1] of **c**. The representation (13.3), stating that any continuously differentiable field may be obtained by superposition of a suitable lamellar field upon a suitable complex-lamellar field, enjoys only a local validity.[2]

From (13.3) and (13.1) we obtain

$$\mathbf{c} \cdot \text{curl } \mathbf{c} = \frac{\partial(h, f, g)}{\partial(x, y, z)} \tag{13.4}$$

a generalization of (12.5).

In works on vector analysis[3] it is shown that any vector field **c** which is piecewise differentiable in a finite region $\mathfrak{v}$ may be represented in Stokes's form[4]

$$\mathbf{c} = -\text{grad } u + \text{curl } \mathbf{p}. \tag{13.5}$$

Thus an arbitrary field **c** may be represented as the sum of a lamellar field and a solenoidal field.[5]

Functions u and **p** satisfying (13.5) are said to be, respectively, a

---

[1] The result was implied by Monge [1787, 1, §§XVI–XVIII, XX] and Pfaff [1818, 1, §4], although not explicitly stated by either. The elegant derivation given above is that of Hankel [1861, 1, §11]. See also [1871, 1, §13] [1879, 2, Ch. III, §III] [1901, 2, §1]. This subject is treated from a hydrodynamical point of view in [1859, 1] [1881, 1] [1881, 2, pp. 6–11]. For a modern exposition, see *e.g.* [1927, 1, §2.83] [1947, 2, §105].

[2] [1903, 2, ¶68].

[3] *E.g.* [1928, 2, ¶¶109–113] [1933, 1, §83].

[4] [1851, 1, Part I, Sect. 1, §§3–8]. The modifications in the Poisson integrals for u and **p** necessary when **c** is discontinuous across certain surfaces were given by Beltrami [1871, 1, §15].

[5] Four stronger decomposition theorems are obtained by Weyl [1940, 4, Th. II]. From (13.5) and (11.3) it follows that an arbitrary field **c** may be represented also in the form

$$\mathbf{c} = -\text{grad } s + \text{grad } f \times \text{grad } g.$$

scalar potential and a *vector potential* of $\mathbf{c}$. For a field $\mathbf{c}$ which is continuously differentiable a pair of potentials is given by

$$u = \frac{1}{4\pi} \int \frac{\operatorname{div} \mathbf{c}}{r} \, dv + h, \qquad \mathbf{p} = \frac{1}{4\pi} \int \frac{\operatorname{curl} \mathbf{c}}{r} \, dv, \tag{13.6}$$

where h is a suitably selected harmonic function. In the case when $\mathbf{c}$ is defined in a region which extends to ∞ in all directions, if $c = O(r^{-2})$ the formulae (13.6) continue to hold with $h = 0$.

The potentials u and $\mathbf{p}$ are by no means uniquely determined, and the pair (13.6) is only one of an infinite number of possible pairs of potentials. In this work we shall always suppose that any vector potential employed is solenoidal, as is indeed the case with $(13.6)_2$.

In contrast to the representation (13.3) in terms of Monge's potentials, (13.5) may be shown to enjoy a global validity.[6]

[6] [1903, 2, ¶68].

Chapter II. Kinematical Preliminaries

14. Material co-ordinates and spatial co-ordinates. By a *continuum* we shall mean a region (§5) in a Euclidean three-dimensional space, subject only to the proviso that the region be possessed of a positive volume. For the most part we shall be concerned with arbitrary curves, surfaces, and regions which are subsets of a continuum. By a *motion* of a continuum we shall mean a one parameter family of mappings of the continuum onto other continua. The real parameter t we identify with the *time*, and we suppose its domain of variation to be $-\infty < t < +\infty$, where $t = 0$ is an arbitrary initial instant.

At $t = 0$ let X^I be the co-ordinates of a typical point, or, as we shall often call it, a *particle* $\mathbf{X}$ of a continuum. Let the motion be the family of mappings

$$x^i = x^i(X, t), \qquad i = 1, 2, 3, \qquad \text{or} \qquad \mathbf{x} = \mathbf{x}(\mathbf{X}, t), \tag{14.1}$$

where x^i are the co-ordinates of the point $\mathbf{x}$ into which $\mathbf{X}$ is carried at time t. We shall always visualize the motion as carrying the set of particles into various configurations in a stationary space, suitable for the imposition of the laws of Newtonian mechanics, although no mechanical questions are considered in this work except as occasional illustrations. The functional relation (14.1) then traces the history of each particle $\mathbf{X}$ throughout all time.

We shall require that except possibly upon certain singular surfaces, lines, or points, the motion shall be *continuous*. By this term we intend that the transformation (14.1) be *single-valued and thrice continuously differentiable* with respect to space and time variables, and that moreover *its inverse*

$$\mathbf{X} = \mathbf{X}(\mathbf{x}, t) \tag{14.2}$$

shall exist and enjoy these same properties. In particular, two particles once distinct shall ever remain distinct. Thus we do not treat of the kinematics of wave motions, for which the reader is referred to the elegant analysis of Hadamard.[1] Unless there be some especial interest attached to a case when continuity in this broad sense is violated, we shall not restate the assumption, and thus *in all our subsequent results, the reader must understand that we refer only to continuous motions.*

---

[1] [1903, **2**, ¶¶69–123]. See also [1929, **1**, Ch. VI], and the more recent work on slip surfaces by Moreau [1949, **13**] [1952, **3**, §8]. For analysis of singular points see the thesis of Zhukovski [1876, **3**, Gl. II].

The co-ordinates $\mathbf{X}$, which in a rectangular Cartesian system coincide with the initial radius vectors of the particles and may be written in the form

$$\mathbf{X} = \mathbf{R} = \mathbf{i}X + \mathbf{j}Y + \mathbf{k}Z, \tag{14.3}$$

are the *material co-ordinates*[2] of the particles, and the variables $\mathbf{X}$, t are the *material variables*. Throughout all time a given co-ordinate $\mathbf{X}$ remains attached to a single particle, and serves as a name for it. The co-ordinates $\mathbf{x}$, which in a rectangular Cartesian system coincide with the radius vectors of the points of space and may be written in the form

$$\mathbf{x} = \mathbf{r} = \mathbf{i}x + \mathbf{j}y + \mathbf{k}z, \tag{14.4}$$

---

[2] In this work we eschew the general misnomer by which X, Y, Z are called "Lagrangian" co-ordinates, while x, y, z are called "Eulerian" co-ordinates. The origin of this incorrect usage is as follows.

By the middle nineteenth century the history of fluid dynamics in the eighteenth century had apparently sunk into obscurity. Euler's papers were not often read, of his results which were not forgotten several were attributed to more recent authors who had appropriated them without acknowledgement or discovered them afresh, and indeed his supreme achievements in mathematics, mechanics, and mathematical physics were undervalued then, though not so much as now. The erroneous terminology still current was introduced in the posthumous memoir of Dirichlet [1860, 1, Introd.], edited by Dedekind, where [1757, 2] was quoted as the source of the "Eulerian" method, while it was stated that Lagrange in the *Méchanique Analitique* [1788, 1, Part II, Sect. II, ¶¶4–7] had introduced the "Lagrangian" method, but had immediately converted the resulting equations to "Eulerian" form. Although in the next year Hankel [1861, 1, §1] stated that his teacher Riemann had told him that Euler had introduced the "Lagrangian" method in [1770, 1], one year's priority has been sufficient to perpetuate the error.

Riemann's attribution is correct, but the references quoted are not the earliest, either for Euler or for Lagrange. Subsequent writers on hydrodynamics have followed Hankel in adopting the printer's error on the title page by which [1770, 1] is dated 1759, while the correct date is 1769; Lagrange's first exposition of the "Lagrangian" description is not in the *Méchanique Analitique* but actually in [1762, 3, Chs. XL, XLIV, XLVIII, LII]. The whole matter is easily clarified, however. In a letter [1862, 2], written to Lagrange under the date 27 October 1759, Euler after expressing his admiration for Lagrange's first memoir on the propagation of sound stated that one had reason to doubt that propagation in two or three dimensions would follow the same law as in the one dimensional case, since he had already found the fundamental equations to be of different form. The equations he gives are the linearized equations of plane flow of a perfect fluid expressed in terms of the variables X, Y. (That the date of Euler's discovery of the material description is 1759 or earlier is shown also by [1766, 1, §§4–13, 31–40], a memoir dated 1759. In [1767, 1], written in 1750–1751, Euler for plane motions had used a description partly spatial and partly material.)

In his next letter, date 1 January 1760, which Lagrange himself caused to be published [1762, 1], Euler gave a detailed summary of the whole theory of perfect fluids expressed in material variables. Further on in the same volume appears Lagrange's treatment [1762, 3], which is in fact largely a paraphrase of Euler's, employing even

are the *spatial co-ordinates* of the points,[3] and the variables **x**, *t* are the *spatial variables*. Throughout all time a given co-ordinate **x** remains attached to a fixed place, serving as an identification for it.

The *material description* of the motion chronicles the history of each particle **X**, while the *spatial description* presents the state of motion at each place **x**. The spatial description was formulated by d'Alembert (1749)[4] in two special cases, shortly thereafter being generalized by Euler (1752-1755),[5] who later (1759)[6] discovered the material description also. Thus d'Alembert and Euler share[7] in the brilliant discovery of the field description of media in motion—the very foundation stone of modern physics, though seldom so much as mentioned in histories of that subject—but to Euler alone we owe the idea of *formulating a field theory as the theory of the integrals of a set of partial differential*

---

the same notation, but without acknowledgement. Despite his great personal and mathematical debt to Euler, in the first edition of the *Méchanique Analitique* [1788, 1, 2e partie, sect. 10] Lagrange did not mention Euler's name in connection with fluid dynamics, instead attributing the whole science to d'Alembert; the corresponding passage in the second edition [1815, 1], published after Lagrange's death, gives a different impression, for between the reference to d'Alembert and the high praise bestowed on the analytical theory have been inserted two sentences attributing the definitive formulation to Euler.

[3] Our use of x, y, z, X, Y, Z reverts to the original notation of Euler [1762, 1]. The literature in general follows the later usage of Lagrange [1788, 1, Part II, Sect. 11, ¶4] in which a, b, c are written for X, Y, Z.

[4] [1752, 1, §43]. The first attempt to discuss any local features of the motion of a continuous medium is an isolated passage in the *Hydrodynamica* [1738, 1, Sect. 13, §13], where Daniel Bernoulli by an argument which he admitted to be incomplete calculated the force acting on an infinitesimal fluid element. The basic difficulty lies in his failure to introduce the velocity field at all. The first author to do so was Euler, whose first and incomplete formulation [1745, 1, Cap. II, Satz 1, Anm. 3], which must have been known to d'Alembert, uses intrinsic co-ordinates and divides the fluid mass into stream fillets.

[5] Euler's earliest surviving exposition of the spatial description is [1761, 1, §§1–41], composed between 1752 and 1755. The first to appear in print, however, was [1757, 2, §§I–XXI], a paper (dated 1755) which has been widely read and justly celebrated. See my history [1954, 1, Parts V, VII-X, XII].

[6] Euler's full exposition of the material description is contained in [1770, 1, §§100–118].

[7] In 1734 Euler [1736, 1, §98] had implied that his *Mechanica* was but the first part of a planned treatise in six parts, the sixth of which was to concern the motion of fluids. At this time he cannot have had a plan of his later researches, but only an intent to undertake them. The gradual development of Euler's ideas on fluids, as well as the connection of his work with that of Daniel and John (I) Bernoulli and d'Alembert, is traced in my history [1954, 1]. His planned treatise was never completed, but the portion concerning fluids finally appeared as the great memoirs [1769, 1] [1770, 1] [1771, 1] [1772, 1], German translations of which were published in revised form as a single volume [1806, 1] long after his death.

equations.[8] In the nearly 200 years following Euler's profound and decisive innovation there have been but few important deserters from his viewpoint, despite the plethora of physical phenomena since discovered.[9] The aim of the present work is to chart the mathematical bed rock upon which the Eulerian structure is founded, in the special case of phenomena in real Euclidean three-dimensional space.

Any function of the spatial variables $\mathbf{x}$, t is also a function of the material variables $\mathbf{X}$, t, and conversely, for by (14.1) we have $g(\mathbf{x}, t) = g(f(\mathbf{X}, t), t)$, while by (14.2) we have $G(\mathbf{X}, t) = G(F(\mathbf{x}, t), t)$. We shall distinguish between differential operations with respect to the two sets of variables by employing lower case letters for the former and small capitals for the latter. Thus

---

[8] The modern reader will trace several of the basic ideas of fluid dynamics, singly and often in a form now not immediately recognizable, to Euler's predecessors, but he will find their exposition faltering and irregular, incomplete and even obscure, and interspersed with passages now obsolete, irrelevant, or of subsidiary interest. The masterful papers [1757, **1–3**] [1769, **1**] [1770, **1**] of Euler, however, are thorough, orderly expositions of the foundations of fluid mechanics in unrivalled clarity, elegance, and precision; to this day, as doubtless for the future, they provide a lucid introduction for the beginner and an inexhaustible source of inspiration for the proficient. A notion of the magnificent power of [1757, **2**] may be had from the fact that the description of Dugas [1950, **5**, Ch. VIII, §6], "mémoire si parfait qu'il n'a pas vieilli d'une ligne," is almost literally true. Euler's analysis, being mainly formal, indeed requires supplement and amplification in respect to rigor, but in the development of the basic concepts and principles Euler displays an insight, a spirit of careful inquiry, and a finality which is to be sought in vain in modern treatments.

The theory of fluids has progressed in the two centuries since Euler, but none of his contributions has required correction or essential revision. As in so many other mathematical sciences so also in the theory of fluid motion it is Euler's formulation, not his predecessors', which has been universally adopted and upon which further researches have been founded. As Lagrange tardily acknowledged [1815, **1**, 2e partie, sect. 10], Euler's great discovery is that the partial differential equations define the subject. In the expression of Fourier [1833, **2**], "On est parvenu à exprimer par des équations générales à différences partielles les conditions du mouvement des fluides. Cette découverte, qui est un des plus beaux resultats de la Géometrie moderne, est due à d'Alembert et à Euler . . . . Euler . . . donne ces équations sous une forme simple et distincte qui embrasse tous les cas possibles, et il les démontre avec cette clarté admirable qui est le caractère principal de tous ses écrits." In the older researches the description of the motion is rather assumed than constructed, but Euler before stating any physical assumptions formulated the concept of a continuous medium and analysed the pure geometry of its motion, or *kinematics*, with such thoroughness that little was added to this discipline until the work of Cauchy seventy years later.

[9] The theories of fluid dynamics, acoustics, thermal conduction, elasticity, thermodynamics, electricity and magnetism, and relativity are successive triumphs of the direct Eulerian method, and even to describe strictly molecular phenomena the Hamilton-Jacobi theory, the kinetic theory of gases, the Gibbsian mechanics, and quantum mechanics have recourse to indirect and more elaborate field descriptions.

$$\frac{DF}{Dt}, \qquad \text{GRAD } F, \qquad \text{etc.}$$

are material differentiations, while

$$\frac{\partial F}{\partial t}, \qquad \text{grad } F, \qquad \text{etc.}$$

are spatial.

In nearly every instance the reader of this work may picture our results in terms of a single rectangular Cartesian frame, to which both the spatial and the material co-ordinates are referred. The results remain valid, however, under much more general circumstances, when the x^i and the X^I are taken as co-ordinates in two quite independent arbitrary Euclidean co-ordinate systems with metric tensors g_{ij} and G_{IJ}, respectively. Thus, for example, it is easy to see that the quantities

$$X^I{}_{,i} \equiv \frac{\partial X^I}{\partial x^i} \tag{14.5}$$

transform as contravariant components of a vector field with respect to change of the material co-ordinates X^I when the spatial co-ordinates x^i are kept fixed, but as covariant components of a vector field with respect to change of the spatial co-ordinates x^i when the material co-ordinates X^I are kept fixed. Similarly, the quantities

$$x^i{}_{,I} \equiv \frac{\partial x^i}{\partial X^I} \tag{14.6}$$

transform in the fashion indicated by the positions of the indices i and I. More generally, the notation "$_{,I}$" shall denote covariant differentiation with respect to X^I with metric tensor G_{IJ}, while "$_{,i}$" shall denote covariant differentiation with respect to x^i with metric tensor g_{ij}. Such quantities as $X^I{}_{,i}$ and $x^i{}_{,I}$ are components of two point vector fields. Since the components of such a field form a matrix, no confusion can result in the present work by representing such fields by dyadic symbols. Thus we shall often write grad $\mathbf{X}$ to represent the matrix whose components are $X_i{}^I \equiv X^I{}_{,i}$ and GRAD $\mathbf{x}$ for the matrix whose components are $x_I{}^i \equiv x^i{}_{,I}$.

The quantity

$$\begin{aligned} J \equiv \sqrt{\frac{g}{G}} \det \text{GRAD } \mathbf{x} &= \frac{1}{3!}\sqrt{\frac{g}{G}}\, \epsilon_{ijk}\epsilon^{IJK} x^i{}_{,I} x^j{}_{,J} x^k{}_{,K} \\ &= \sqrt{\frac{g}{G}}\, \frac{\partial(x^1, x^2, x^3)}{\partial(X^1, X^2, X^3)}, \end{aligned} \tag{14.7}$$

which in the case when both co-ordinate systems are rectangular Cartesian reduces simply to the Jacobian of the x^i with respect to the X^I, is an absolute scalar with respect to each type of transformation. So also is

$$j \equiv \sqrt{\frac{G}{g}} \frac{\partial(X^1, X^2, X^3)}{\partial(x^1, x^2, x^3)}, \tag{14.8}$$

and we have

$$Jj = 1. \tag{14.9}$$

The requirement of continuity of motion, stated above, is expressed in part by

$$0 < J < +\infty, \qquad 0 < j < +\infty. \tag{14.10}$$

From (14.5) and (14.6) we have

$$X^I{}_{,i}x^i{}_{,J} = \delta^I{}_J, \qquad x^i{}_{,I}X^I{}_{,j} = \delta^i{}_j. \tag{14.11}$$

Regarding the first of these as a linear system, we see by (14.10) that a unique solution exists and is given by

$$\begin{aligned} X^I{}_{,i} &= \frac{3!}{2!} \frac{\epsilon^{IJK}\epsilon_{ijk}x^j{}_{,J}x^k{}_{,K}}{\epsilon^{LMN}\epsilon_{lmn}x^l{}_{,L}x^m{}_{,M}x^n{}_{,N}}, \\ &= \frac{\dfrac{\partial(x^j, x^k)}{\partial(X^J, X^K)}}{\dfrac{\partial(x^1, x^2, x^3)}{\partial(X^1, X^2, X^3)}} \qquad \begin{pmatrix} i, j, k = (123) \\ I, J, K = (123) \end{pmatrix}. \end{aligned} \tag{14.12}$$

This formula expresses $X^I{}_{,i}$ as a function of the material co-ordinates **X** only. There is an analogous formula expressing $x^i{}_{,I}$ in terms of the spatial co-ordinates **x**. With the aid of these formulae any spatial differentiation may be expressed in terms of material differentiations, and conversely. For example,

$$\begin{aligned} F_{,i} &= F_{,I}X^I{}_{,i} \\ &= \frac{3!}{2!} \frac{\epsilon^{IJK}\epsilon_{ijk}F_{,I}x^j{}_{,J}x^k{}_{,K}}{\epsilon^{LMN}\epsilon_{lmn}x^l{}_{,L}x^m{}_{,M}x^n{}_{,N}}, \\ &= \frac{\displaystyle\sum_{I=1}^{3} \frac{\partial F}{\partial X^I} \frac{\partial(x^j, x^k)}{\partial(X^J, X^K)}}{\dfrac{\partial(x^1, x^2, x^3)}{\partial(X^1, X^2, X^3)}} \qquad \begin{pmatrix} i, j, k = (123) \\ I, J, K = (123) \end{pmatrix}. \end{aligned} \tag{14.13}$$

These conversion formulae were discovered by Euler (1760).[10]

From (14.12) follow the easy consequences

$$\frac{\epsilon^{ijk}}{\sqrt{g}} X^J{}_{,i} = \frac{\epsilon^{IJK}}{J\sqrt{G}} x^i{}_{,I} x^k{}_{,K},$$
$$\sqrt{g}\, \epsilon_{ijk} x^j{}_{,J} = \frac{\sqrt{G}\, \epsilon_{IJK}}{j} X^I{}_{,i} X^K{}_{,k}. \tag{14.14}$$

15. The transformation of material arc, surface, and volume elements. Let a curve in space be given at $t = 0$ by the equation $\mathbf{X} = \mathbf{X}(l)$, where l is a parameter. As the motion proceeds, the curve is carried into $\mathbf{x} = \mathbf{x}(\mathbf{X}(l), t) = \mathbf{x}(l, t)$, l serving as a material co-ordinate for points on the initial curve. A moving curve consisting thus of ever the same particles is called a *material curve.* The relation between spatial and material elements of arc along the curve was first given by Euler (1760):[1]

$$d\mathbf{x} = d\mathbf{X} \cdot \text{GRAD}\, \mathbf{x}, \qquad d\mathbf{X} = d\mathbf{x} \cdot \text{grad}\, \mathbf{X},$$

or

$$dx^i = x^i{}_{,I}\, dX^I, \qquad dX^I = X^I{}_{,i}\, dx^i. \tag{15.1}$$

Now let a surface in space be given by $\mathbf{x} = \mathbf{x}(l, m, t)$, where l and m are material parameters which, like l above, remain attached to the same particle. A moving surface consisting thus of always the same particles is called a *material surface.* Initially an element of surface $d\mathbf{S}$ is given by

$$d\mathbf{S} = \frac{\partial \mathbf{X}}{\partial l} \times \frac{\partial \mathbf{X}}{\partial m}\, dl\, dm, \quad \text{or} \quad dS_I = \sqrt{G}\, \epsilon_{IJK} \frac{\partial X^J}{\partial l} \frac{\partial X^K}{\partial m}\, dl\, dm, \tag{15.2}$$

where $\mathbf{X} = \mathbf{x}(l, m, 0)$. The element of surface $d\mathbf{s}$ at an arbitrary time t, computed from

$$d\mathbf{s} = \frac{\partial \mathbf{x}}{\partial l} \times \frac{\partial \mathbf{x}}{\partial m}\, dl\, dm, \tag{15.3}$$

stands in simple relation to $d\mathbf{s}$. Since

$$ds_i = \sqrt{g}\, \epsilon_{ijk} x^j{}_{,J} x^k{}_{,K} \frac{\partial X^J}{\partial l} \frac{\partial X^K}{\partial m}\, dl\, dm, \tag{15.4}$$

---

[10] [1762, 1] [1766, 1, §36] [1770, 1, §§105–111] [1806, 1, §§223–224]. They are given also by Lagrange [1762, 3, Ch. XLIV] [1788, 1, Part II, Sect. 11, §5]. For the case of two variables, Euler had derived equivalent results in 1751 [1767, 1, §15].

[1] [1762, 1] [1770, 1, §105] [1806, 1, §217]. See also [1762, 3, Ch. XLIV] [1788, 1, Part II, Sect. 11, ¶5].

by known identities concerning permutations and determinants and by the definitions (15.2) and (14.7) we have

$$\begin{aligned} x^i{}_{,I}\, ds_i &= \sqrt{g}\,\epsilon_{ijk} x^i{}_{,I} x^j{}_{,J} x^k{}_{,K} \frac{\partial X^J}{\partial l}\frac{\partial X^K}{\partial m}\,dl\,dm, \\ &= \tfrac{1}{2}\sqrt{g}\,\epsilon_{ijk} x^i{}_{,I} x^j{}_{,J} x^k{}_{,K} \delta^{JK}_{LM} \frac{\partial X^L}{\partial l}\frac{\partial X^M}{\partial m}\,dl\,dm, \\ &= \tfrac{1}{2}\sqrt{g}\,\epsilon_{ijk} x^i{}_{,I} x^j{}_{,J} x^k{}_{,K} \epsilon^{NJK}\epsilon_{NLM} \frac{\partial X^L}{\partial l}\frac{\partial X^M}{\partial m}\,dl\,dm, \qquad (15.5) \\ &= \tfrac{1}{2}\sqrt{\frac{g}{G}}\, x^i{}_{,I}\delta^{NJK}_{ijk} x^j{}_{,J} x^k{}_{,K}\, dS_N, \\ &= \sqrt{\frac{g}{G}}\,\delta^N_I \frac{\partial(x^1, x^2, x^3)}{\partial(X^1, X^2, X^3)}\, dS_N, \\ &= J\, dS_I. \end{aligned}$$

That is,

$$\text{GRAD}\ \mathbf{x}\cdot d\mathbf{s} = J\, d\mathbf{S}, \qquad \text{grad}\ \mathbf{X}\cdot d\mathbf{S} = j\, d\mathbf{s}. \qquad (15.6)$$

Formulae equivalent to these were discovered by Nanson (1878).[2]

From the ordinary formula for the element of volume we have

$$dv = J\, dV, \qquad dV = j\, dv, \qquad (15.7)$$

where dv is a spatial, dV a corresponding material element of volume. Cauchy[3] remarked that this relation, which is due to Euler (1755),[4] shows that the portion of the hypothesis of continuity of motion which is expressed by (14.10) is simply a condition that an element of volume cannot be created or destroyed. From (15.1) and (15.5) it follows that in a continuous motion similarly an element of arc or element of surface can never be created nor destroyed.

A motion is called *isochoric* if the volume of space occupied by any material volume remains invariant throughout the motion,[5] *i.e.* dv = const. = dV. From (15.7) it follows that a necessary and sufficient condition for isochoric motion is

$$J = j = 1. \qquad (15.8)$$

---

[2] [1878, 1]. See also [1932, 1, §146].

[3] [1827, 1, 1st part, Sect. 1, §8].

[4] [1761, 1, §§1–38] [1762, 1] [1766, 1, §35] [1770, 1, §§118, 123–129].

[5] A medium dynamically susceptible only of isochoric motions is called *incompressible*; media not incompressible are called *compressible*. Of course it is quite possible for a compressible medium to be set into isochoric motion. Classical hydrodynamics deals mainly but not exclusively with incompressible fluids: *cf.* [1752, 1, §59] [1761, 1, §§1–38].

The transformations (15.1), (15.6), and (15.7) present the precise history of an initial element of arc, element of surface, and element of volume as the curve, surface, and volume, respectively, are carried by the motion. For those who prefer to fancy small elements, these formulae show approximately how a small portion of a material curve, surface, or volume is distorted in the course of a motion.

16. Velocity and acceleration. The ***velocity*** $\dot{\mathbf{x}}$ of a particle is given by the definition

$$\dot{\mathbf{x}} \equiv \frac{D\mathbf{x}}{Dt} \qquad \left(\dot{x}^i \equiv \frac{Dx^i(X^1, X^2, X^3, t)}{Dt}\right). \tag{16.1}$$

From this definition and from (14.1) it follows that

$$\dot{\mathbf{x}} = \dot{\mathbf{x}}(\mathbf{X}, t), \tag{16.2}$$

but if the spatial description be preferred we may in principle eliminate $\mathbf{X}$ by (14.2), obtaining

$$\dot{\mathbf{x}} = \dot{\mathbf{x}}(\mathbf{x}, t). \tag{16.3}$$

Indeed, the spatial description is usually considered complete when the functional form of (16.3) is known, the velocity field alone being sufficient for most purposes.[1]

The vector-lines, vector-sheets, and vector-tubes (§9) of the velocity field are called *stream-lines*, *stream-surfaces*, and *stream-tubes*, respectively. The stream-lines are generally distinct from the *paths* of the particles, which are obtained by holding $\mathbf{X}$ fixed and letting t vary from $-\infty$ to $+\infty$ in (14.1). At any given instant the stream-line and path-line through a given point have a common tangent, so that for these two systems of lines to coincide it is necessary and sufficient that they be stationary. The motion is then said to be a *motion with steady stream-lines*. To this end it is sufficient, though not necessary, that the time t be altogether absent from the spatial description (12.3):

$$\dot{\mathbf{x}} = \dot{\mathbf{x}}(\mathbf{x}), \tag{16.4}$$

so that the velocity field suffers no change in time when viewed by an observer at rest with respect to the spatial co-ordinate system; in this case the motion is *steady* in the particular frame employed. More generally, any quantity whose spatial time derivative vanishes will be said to be *steady*.

Motions whose velocity fields are respectively complex-lamellar,

---

[1] The velocity field in the form (16.3) was taken as a primitive concept in the work of d'Alembert [1752, 1, §43] and in the first papers of Euler [1745, 1, Cap. II, Satz 1, Anm. 3] [1761, 1, §§11, 23] [1757, 2, §IX].

plane, or rotationally-symmetric, are called *complex-lamellar motions,*[2] *plane motions,*[3] or *rotationally-symmetric motions.*[4]

The ***acceleration*** $\ddot{\mathbf{x}}$ is defined by

$$\ddot{x}^i = \frac{D\dot{x}^i}{Dt} + \Gamma^i_{jk}\dot{x}^j\dot{x}^k, \tag{16.5}$$

the Γ^i_{jk} being Christoffel symbols. In a rectangular Cartesian system (16.5) reduces to

$$\ddot{\mathbf{x}} = \frac{D\dot{\mathbf{x}}}{Dt} = \frac{D^2\mathbf{x}}{Dt^2}\,. \tag{16.6}$$

From the definition of covariant differentiation it is easy to show that (16.5) is equivalent to ***the d'Alembert-Euler acceleration formula*** (1749-1752)[5]

---

[2] Complex-lamellar motion of fluids was first mentioned by Earnshaw [1837, **1**, §§2–5] (*cf.* §98), who called it "motion in wave surfaces." In a long series of papers ([1842, **3**] *et seqq.*), whose content is almost totally wrong, Challis claimed that all fluid motions are complex-lamellar. The only correct new result he obtained may be expressed as follows: for a complex-lamellar field $\mathbf{c}$ we have

$$\operatorname{div} \mathbf{c} = \frac{dc}{ds} - \bar{J}c,$$

where $\bar{J}$ is the mean curvature of the normal surface (*cf.* [1850, **2**]). Challis's result follows by inspection from $(9.5)_3$, since for a complex-lamellar field the unit tangent $\mathbf{t}$ is the unit normal to the normal surface, whence $\Theta = -\bar{J}$ (*e.g.* [1947, **2**, §131]).

Caldonazzo [1924, **2**, §6] proved that in an isochoric motion whose velocity field is complex-lamellar, if the speed be constant along each stream-line the normal surfaces are minimal surfaces. The special case when the velocity is steady and lamellar and the speed is constant along each stream-line has been completely characterized in a celebrated and difficult analysis of Hamel [1937, **1**]; the method of proof and the extension of the result to n dimensions is discussed by Howard [1953, **6**]. Caldonazzo's theorem is derived anew by Castoldi [1947, **5**], who attempts to find conditions characterizing such motions of inviscid incompressible fluids. The shortest proof of Caldonazzo's result is that of Prim [1948, **6**, §3] [1952, **4**, Ch. V, Sect. B], which consists in putting $\mathbf{c} = \dot{\mathbf{x}}$ in the above formula of Challis and noting that the condition $\bar{J} = 0$ defines a minimal surface. This method of proof shows that, conversely, if the normal surfaces be minimal then the speed must be constant on each stream-line. These authors state their results unnecessarily weakly in terms of inviscid fluids. See §46[6].

[3] Plane motions were introduced by Euler [1745, **1**, Cap. II, Satz 1, Anm. 3] and d'Alembert [1752, **1**, §21]. Previous authors, beginning with Newton [1687, **1**, Lib. II, Prop. XXXVI] and including both these in their earlier works, had confined their attention to motions satisfying the "hypothesis of parallel sections," in which the velocity vector is assumed to have almost the same value at all points of each plane z = const., and to be almost normal to these planes. *Cf.* also §14[4].

[4] Rotationally-symmetric motions were introduced by d'Alembert [1752, **1**, §26 *et seqq.*], Svanberg [1841, **2**, §2], and Stokes [1842, **2**, pp. 12–13].

[5] d'Alembert obtained formulae equivalent to (16.7) for a special class of plane

$$\ddot{x}^i = \frac{\partial \dot{x}^i}{\partial t} + \dot{x}^i{}_{,j}\dot{x}^j, \qquad \text{or} \qquad \ddot{\mathbf{x}} = \frac{\partial \dot{\mathbf{x}}}{\partial t} + \dot{\mathbf{x}}\cdot \operatorname{grad} \dot{\mathbf{x}}. \tag{16.7}$$

The first term in (16.7) is called the *local* acceleration, for it represents the change in the velocity field apparent to a fixed observer; in a steady motion it vanishes. The second term is called the *convective* acceleration, for it represents the acceleration necessary in a steady motion which coincides with $\dot{\mathbf{x}}$ at time t in order to direct the particles along their appointed paths at the required speed. More generally, for any quantity $\emptyset$, we shall call $\dot{\mathbf{x}}\cdot \operatorname{grad} \emptyset$ the *convection* of $\emptyset$.

According to our definition of a continuous motion (§14), the field $\dot{\mathbf{x}}$ is twice continuously differentiable, $\ddot{\mathbf{x}}$ once continuously differentiable.

Suppose that it be desired to relate the velocities and accelerations apparent to observers stationary with respect to two different spatial frames which are themselves in relative motion. Let rectangular Cartesian axes be employed in each case, let primes denote quantities associated with the second system, and let $d\mathbf{i}'/dt$, $d\mathbf{j}'/dt$, and $d\mathbf{k}'/dt$ be the time rates of change of the unit vectors $\mathbf{i}'$, $\mathbf{j}'$, $\mathbf{k}'$ in the second frame as apparent to an observer stationary in the first frame. Then the *angular velocity* $\boldsymbol{\omega}$ of the second frame with respect to the first is defined by

$$\boldsymbol{\omega} \equiv \mathbf{i}' \frac{d\mathbf{j}'}{dt}\cdot \mathbf{k}' + \mathbf{j}' \frac{d\mathbf{k}'}{dt}\cdot \mathbf{i}' + \mathbf{k}' \frac{d\mathbf{i}'}{dt}\cdot \mathbf{j}', \tag{16.8}$$

and if $\mathbf{c}$ be the radius vector with respect to the first frame of the point which is the origin of the second frame, so that $\mathbf{r} = \mathbf{c} + \mathbf{r}'$, it is easy to show that

$$\dot{\mathbf{x}} = \frac{d\mathbf{c}}{dt} + \boldsymbol{\omega} \times \mathbf{r}' + \dot{\mathbf{x}}', \tag{16.9}$$

$$\ddot{\mathbf{x}} = \frac{d^2\mathbf{c}}{dt^2} + \frac{d\boldsymbol{\omega}}{dt} \times \mathbf{r}' + \boldsymbol{\omega} \times (\boldsymbol{\omega} \times \mathbf{r}') + 2\boldsymbol{\omega} \times \dot{\mathbf{x}}' + \ddot{\mathbf{x}}'. \tag{16.10}$$

The latter formula, expressing the acceleration of a material point observed in a moving co-ordinate system, is due to Coriolis (1835).[6] The term $d\boldsymbol{\omega}/dt \times \mathbf{r}'$ is called the *Euler acceleration*; $\boldsymbol{\omega} \times (\boldsymbol{\omega} \times \mathbf{r}')$, the *centripetal acceleration*; $2\boldsymbol{\omega} \times \dot{\mathbf{x}}$, the *Coriolis acceleration*.

We shall have such frequent use for the representation of $\dot{\mathbf{x}}$ and $\ddot{\mathbf{x}}$ in

---

[1752, 1, §§73, 86] and rotationally-symmetric [1752, 1, §§43–44, 86] motions. The general case was given by Euler [1761, 1, §§40–41, 56] [1757, 2, §XIX] [1770, 1, §18] [1806, 1, §136] [1862, 1, §157].

[6] [1835, 2].

terms of Monge's or Stokes's potentials (§13) that we shall henceforth reserve the following ***special notations*** for them:

$$\dot{\mathbf{x}} = -\operatorname{grad}\phi + \operatorname{curl}\boldsymbol{\pi}, \tag{16.11}$$

$$= \operatorname{grad} h + f \operatorname{grad} g. \tag{16.12}$$

$$\ddot{\mathbf{x}} = -\operatorname{grad}\phi^* + \operatorname{curl}\boldsymbol{\pi}^*, \tag{16.13}$$

$$= \operatorname{grad} h^* + f^* \operatorname{grad} g^*. \tag{16.14}$$

As noted in §13, the functions occurring in these decompositions are not uniquely determined.

A motion is said to be *slow* or *small* if its convective acceleration be negligible:

$$\ddot{\mathbf{x}} \approx \frac{\partial \dot{\mathbf{x}}}{\partial t} \approx \frac{\partial^2 \mathbf{x}}{\partial t^2} \,. \tag{16.15}$$

These motions originated in the theory of sound, and indeed the whole sciences of acoustics and of dynamic elasticity are virtually limited to the case of slow motions. There is also a linearized theory of viscous fluids based upon (16.15). For a slow motion we have

$$\operatorname{div}\ddot{\mathbf{x}} \approx \frac{\partial}{\partial t}\operatorname{div}\dot{\mathbf{x}}, \qquad \operatorname{curl}\ddot{\mathbf{x}} \approx \frac{\partial}{\partial t}(\operatorname{curl}\dot{\mathbf{x}}). \tag{16.16}$$

In a steady slow motion the acceleration is considered altogether negligible, $\ddot{\mathbf{x}} \approx 0$.

17. Parenthesis: the dynamical equations for fluids. At this point we pause in our kinematical exposition to notice that *the dynamical properties of a special type of continuous medium generally are defined by specifying the form of the* acceleration $\ddot{\mathbf{x}}$ *in terms of dynamical variables in a preferred spatial frame of reference.* In classical mechanics the preferred frame is called a *Galilean, Newtonian,* or *inertial* frame, and the class of all such frames consist of those obtainable from the given one by a transformation under which $\ddot{\mathbf{x}}$ is invariant. Thus any two Galilean frames are in relative translatory motion at a constant velocity. A property invariant with respect to the transformation from one Galilean frame to another is called a *Galilean invariant.* The concept of steadiness of motion, for example, is *not* a Galilean invariant.

The dynamical equation for a *viscous fluid* of uniform *viscosities* λ, μ is ***Poisson's dynamical equation*** (1829):[1]

$$\rho\ddot{\mathbf{x}} = \rho\mathbf{f} - \operatorname{grad} p + (\lambda + 2\mu)\operatorname{grad}\operatorname{div}\dot{\mathbf{x}} - \mu \operatorname{curl}\operatorname{curl}\dot{\mathbf{x}}, \tag{17.1}$$

[1] [1831, 1, ¶64].

where ρ is the *density*, $\mathbf{f}$ is the *extraneous force*, and p is the *pressure*. Writing this equation in the form

$$\rho(\mathbf{f} - \ddot{\mathbf{x}}) = -\text{grad}\,[-p + (\lambda + 2\mu)\,\text{div}\,\dot{\mathbf{x}}] + \text{curl}\,(\mu\,\text{curl}\,\dot{\mathbf{x}}) \qquad (17.2)$$

shows that it is a *specification of a pair of Stokes potentials for the sum of the extraneous and inertial forces,* the potentials themselves being linear functions of p, div $\dot{\mathbf{x}}$, and curl $\dot{\mathbf{x}}$.

In two important special cases (17.2) yields potentials for the acceleration itself rather than for the effective force $\rho(\mathbf{f} - \ddot{\mathbf{x}})$. In both these cases the extraneous force is *conservative*:

$$\mathbf{f} = -\text{grad}\, \upsilon, \qquad \upsilon = \upsilon(\mathbf{x}, t). \qquad (17.3)$$

First, for a homogeneous incompressible fluid (§15[5]) ρ = const. and it can be shown (§22) that div $\dot{\mathbf{x}} = 0$; hence when the extraneous force is conservative (17.2) becomes ***Navier's dynamical equation*** (1822):[2]

$$\ddot{\mathbf{x}} = -\text{grad}\left(\frac{p}{\rho} + \upsilon\right) - \text{curl}\left(\frac{\mu}{\rho}\,\text{curl}\,\dot{\mathbf{x}}\right), \qquad (17.4)$$

which specifies Stokes potentials for $\ddot{\mathbf{x}}$:

$$\phi^* = \frac{p}{\rho} + \upsilon, \qquad \pi^* = -\frac{\mu}{\rho}\,\text{curl}\,\dot{\mathbf{x}}. \qquad (17.5)$$

Hence if the field $\mathbf{f}$ be solenoidal (as is the Newtonian gravitational field, for example) we obtain

$$\text{div}\,\ddot{\mathbf{x}} = \nabla^2\phi^* = \frac{1}{\rho}\nabla^2 p, \qquad (17.6)$$

whether or not the fluid be viscous.

Second, for an *inviscid* fluid ($\lambda = \mu = 0$) subject to conservative extraneous force (17.2) reduces to ***Euler's dynamical equation*** (1752-1755):[3]

$$\ddot{\mathbf{x}} = -\text{grad}\,\upsilon - \frac{1}{\rho}\,\text{grad}\,p, \qquad (17.7)$$

which specifies Monge potentials for $\ddot{\mathbf{x}}$:

$$h^* = -\upsilon, \qquad f^* = -\frac{1}{\rho}, \qquad g^* = p. \qquad (17.8)$$

A flow is said to be *barotropic* if a relation $f(p, \rho, t) = 0$ holds. Barotropy is equivalent to the coincidence of the surfaces p = const. with the surfaces ρ = const. at any given time t. If neither density nor

---

[2] [1827, **3**, pp. 400–414].

[3] [1761, **1**, §§43–45, 61] [1757, **2**, §§XX–XXI] [1770, **1**, §33] [1862, **1**, §156].

pressure be uniform, then, subject to moderate assumptions regarding the smoothness of the function f, barotropy is equivalent to a functional relation $p = p(\rho, t)$. For a homogeneous fluid possessed of an *equation of state* we have in any motion whatever $p = p(\rho, \delta)$, where δ is a suitable *state variable, e.g.*, the *specific internal energy* ϵ, the *specific entropy* η, or the *temperature* θ.

More generally, suppose that for any reason whatever it is possible to find a scalar ξ such that for the flow in question we have $p = p(\rho, \xi, t)$. Then we may write Euler's equation (17.7) in the form

$$\ddot{\mathbf{x}} = -\operatorname{grad} \upsilon - \frac{1}{\rho}\left[\frac{\partial p}{\partial \rho} \operatorname{grad} \rho + \frac{\partial p}{\partial \xi} \operatorname{grad} \xi\right],$$

$$= -\operatorname{grad}\left[\upsilon + \int \frac{1}{\rho}\frac{\partial p}{\partial \rho}\, d\rho\right] - \left[\frac{1}{\rho}\frac{\partial p}{\partial \xi} - \int \frac{1}{\rho}\frac{\partial^2 p}{\partial \rho\, \partial \xi}\, d\rho\right] \operatorname{grad} \xi, \quad (17.9)$$

yielding a different and often more interesting trio of Monge potentials:

$$h^* = -\upsilon - \int \frac{1}{\rho}\frac{\partial p}{\partial \rho}\, d\rho, \quad f^* = -\frac{1}{\rho}\frac{\partial p}{\partial \xi} + \int \frac{1}{\rho}\frac{\partial^2 p}{\partial \rho\, \partial \xi}\, d\rho, \quad g^* = \xi. \quad (17.10)$$

18. Material derivative. If the motion be referred to a rectangular Cartesian frame, then the derivative $D\varnothing/Dt$ of a quantity $\varnothing(\mathbf{X}, t)$, introduced in §14, expresses the rate of change of $\varnothing$ as apparent to an observer situate upon a moving particle. If $\varnothing$ be a vector or dyadic referred to a curvilinear co-ordinate system, however, $D\varnothing/Dt$ is not generally a vector or dyadic, respectively. We therefore introduce the ***material derivative***[1] $\delta\varnothing/\delta t$ of the quantity $\varnothing(\mathbf{X}, t)$ by the sequence of definitions

$$\frac{\delta F}{\delta t} \equiv \frac{DF}{Dt}, \qquad \frac{\delta b^i}{\delta t} \equiv \frac{Db^i}{Dt} + \Gamma^i_{jk} b^j \dot{x}^k,$$
$$\frac{\delta \Sigma^{ij}}{\delta t} \equiv \frac{D\Sigma^{ij}}{Dt} + \Gamma^i_{lm}\Sigma^{jl}\dot{x}^m + \Gamma^i_{lm}\Sigma^{li}\dot{x}^m, \quad (18.1)$$

etc., the Γ^i_{jk} being the Christoffel symbols appropriate to the particular system of spatial co-ordinates employed. In a rectangular Cartesian system of spatial co-ordinates the material derivative reduces to the ordinary partial derivative: $\delta\varnothing/\delta t = D\varnothing/Dt$. It is easy to see that all cases of (18.1) are included in the single spatial expression

$$\frac{\delta \varnothing^{\cdots}_{\cdots}}{\delta t} = \frac{\partial \varnothing^{\cdots}_{\cdots}}{\partial t} + \dot{x}^i \varnothing^{\cdots}_{\cdots, i}, \quad (18.2)$$

[1] In the hydrodynamical literature the general results are usually derived in Cartesian co-ordinates, where there is no distinction between D/Dt and $\delta/\delta t$. In some cases, however, the notation D/Dt of Stokes [1845, 1, §5] [1851, 2, §49] really indicates the material derivative.

which we may write in the form

$$\frac{\delta \emptyset}{\delta t} = \frac{\partial \emptyset}{\partial t} + \dot{x}\cdot \text{grad } \emptyset, \tag{18.3}$$

a result used implicitly both by Euler (1769)[2] and by Lagrange (1781).[3] In the acceleration $\ddot{\mathbf{x}}$ we have already encountered one special case of the material derivative, for we may write either (16.5) or Euler's formula (16.7) as

$$\ddot{\mathbf{x}} = \frac{\delta \dot{\mathbf{x}}}{\delta t}. \tag{18.4}$$

More generally, when it does not lead to confusion we shall use a dot to indicate the material derivative:

$$\dot{\emptyset} \equiv \frac{\delta \emptyset}{\delta t}. \tag{18.5}$$

For any continuously differentiable field $\emptyset$ we have by the mere independence of the variables the commutation rules

$$\frac{\partial}{\partial t} \text{grad } \emptyset = \text{grad } \frac{\partial \emptyset}{\partial t}, \qquad \frac{D \text{ GRAD } \emptyset}{Dt} = \text{GRAD } \frac{D\emptyset}{Dt}. \tag{18.6}$$

Recall that quantities $\Phi_{K\cdots L}^{I\cdots J}$ forming a tensor field with respect to transformations of the material co-ordinates X^I are scalars with respect to transformations of the spatial co-ordinates x^i. Then by $(18.1)_1$ we have $\dot{\Phi}_{K\cdots L}^{I\cdots J} = D\Phi_{K\cdots L}^{I\cdots J}/Dt$, whence by $(18.6)_2$ we obtain

$$\text{GRAD } \dot{\Phi}_{K\cdots L}^{I\cdots J} = \frac{\delta}{\delta t} \text{GRAD } \Phi_{K\cdots L}^{I\cdots J}. \tag{18.7}$$

While no such result holds in general for quantities $\Phi_{K\cdots Lk\cdots l}^{I\cdots Ji\cdots j}$ forming two-point fields, nevertheless for the special case $x^i{}_{,I}$ we have

$$\frac{\delta x^i{}_{,I}}{\delta t} = \dot{x}^i{}_{,I}, \qquad \text{or} \qquad \frac{\delta}{\delta t} \text{GRAD } \mathbf{x} = \text{GRAD } \dot{\mathbf{x}}. \tag{18.8}$$

To prove this tensor result it is sufficient to consider the special case when the spatial frame is rectangular Cartesian, *viz.* $Dx^i{}_{,I}/Dt = \dot{x}^i{}_{,I}$, but this formula follows at once from $(18.6)_2$ and the definition (16.1).

From (18.3) and (18.6) it is easy to see that for a continuously differentiable field $\emptyset$ of any character we have

$$\frac{\delta \text{ grad } \emptyset}{\delta t} = \text{grad } \dot{\emptyset} - \text{grad } \dot{\mathbf{x}}\cdot \text{grad } \emptyset. \tag{18.9}$$

---

[2] [1770, 1, §6].

[3] [1783, 1, §§10–11] [1788, 1, Part II, Sect. 11, ¶¶11–12].

The simplicity of (18.3) is somewhat deceiving, since the definition of the operator "grad" (covariant derivative) in a curvilinear co-ordinate system depends upon the order of the tensor to which it is applied. Another way of saying this same thing is to remark that if **e** be a unit vector tangent to one of the co-ordinate curves, we shall have in general $\dot{\mathbf{e}} \neq 0$, since to an observer situate upon a moving particle **e** appears to change. An example illustrating the difference between $\delta\emptyset/\delta t$ and $D\emptyset/Dt$ is furnished by the acceleration referred to cylindrical co-ordinates r, θ, z. From the definition (16.5) we have

$$\begin{aligned} \ddot{x}^\theta &= \frac{D\dot{x}^\theta}{Dt} + \frac{2\dot{x}^r\dot{x}^\theta}{r}, \\ &= \frac{D\dot{x}^\theta}{Dt} + \frac{2\dot{x}^\theta}{r}\frac{Dr}{Dt} = \frac{1}{r^2}\frac{D}{Dt}(r^2\dot{x}^\theta), \end{aligned} \tag{18.10}$$

or, in terms of physical components,

$$\ddot{x}_\theta = \frac{1}{r}\frac{D}{Dt}(r\dot{x}_\theta). \tag{18.11}$$

Consequently $\ddot{x}^\theta = 0$ if and only if $r^2\dot{x}^\theta =$ const. for each particle. Interpretation of this result yields the ***law of areas***:[4] *for the azimuthal acceleration of a particle to vanish it is necessary and sufficient that the projection of the radius vector of that particle upon a base plane sweep out equal areas in equal times.*

19. Boundaries. We may now find a condition that a given moving surface $F(\mathbf{x}, t) = 0$ be a material surface. Let a typical surface point be endowed with velocity $\dot{\mathbf{x}}^*$, not necessarily the velocity $\dot{\mathbf{x}}$ of the particle instantly occupying that same point. If, continually observing a particular moving point upon the surface, we differentiate the equation $F(\mathbf{x}, t) = 0$ with respect to time, we obtain

$$\frac{\partial F}{\partial t} + \dot{\mathbf{x}}^* \cdot \text{grad}\, F = 0. \tag{19.1}$$

Hence, employing the notation $(4.5)_1$ for the projection of $\dot{\mathbf{x}}^*$ onto the normal to the surface $F(\mathbf{x}, t) = 0$, we have

$$\dot{\mathbf{x}}_n^* = -\frac{\dfrac{\partial F}{\partial t}}{|\,\text{grad}\, F\,|}. \tag{19.2}$$

A necessary condition that $F(\dot{\mathbf{x}}, t) = 0$ be a material surface is

$$\dot{\mathbf{x}}_n^* = \dot{\mathbf{x}}_n. \tag{19.3}$$

[4] This proposition is familiar in the kinematics of a single point. The present derivation, employing field concepts, is due in principle to Svanberg [1841, 2, §3].

But

$$\dot{\mathbf{x}}_n = \frac{\dot{\mathbf{x}} \cdot \text{grad}\, F}{|\,\text{grad}\, F\,|}, \tag{19.4}$$

so that

$$\dot{\mathbf{x}}_n^* - \dot{\mathbf{x}}_n = -\frac{\dfrac{\partial F}{\partial t} + \dot{\mathbf{x}} \cdot \text{grad}\, F}{|\,\text{grad}\, F\,|}. \tag{19.5}$$

Hence by (19.3) and (18.3) it follows that *in order for* $F(\mathbf{x}, t) = 0$ *to be a material surface it is necessary that its material derivative vanish:*

$$\dot{F} = 0. \tag{19.6}$$

This condition is also *sufficient.* For if (19.6) be satisfied, it may be considered a partial differential equation for F, to solve which by Lagrange's method we set up the subsidiary system

$$\frac{d\mathbf{x}}{dt} \times \dot{\mathbf{x}} = 0, \tag{19.7}$$

in which $\mathbf{x}$ is to be regarded as a function of t. Let three independent scalar integrals of (19.7) be written in the form $\mathbf{X}' = \mathbf{X}'(\mathbf{x}, t)$, where $\mathbf{X}'$ is an arbitrary vector constant of integration. But (19.7) is a differential equation satisfied by the paths of the particles, so that the three scalar integrals may be so chosen that $\mathbf{X}' = \mathbf{X}$. The requirement that the integrals be independent then becomes exactly the continuity hypothesis (14.10). We conclude that any surface satisfying (19.6) is of the form $F(\mathbf{X}) = 0$, so that if a certain particle be at any one time situate upon such a surface, it must ever remain situate upon it. The condition (19.6) was given by Lagrange (1781), to whom is due the sufficiency proof above,[1] the necessity proof being Kelvin's.[2]

A surface is a possible *boundary* if no particle ever crosses it. Thus any material surface is a possible boundary. But conversely, in order that a surface be a possible boundary the condition (19.3) is necessary, whence follows (19.6). Thus a *surface* $F(\mathbf{x}, t) = 0$ *may be a boundary surface if and only if it be a material surface.*[3]

In the particular case when any motion of the boundary surface is tangential to itself[4] ($\dot{\mathbf{x}}_n^* = 0$), (19.6) reduces to

$$\dot{\mathbf{x}}_n = 0, \tag{19.8}$$

---

[1] [1783, 1, §§10–11] [1788, 1, Part II, Sect. 11, ¶12].

[2] [1848, 1].

[3] When continuity requirements are lightened this theorem ceases to hold. An old controversy on this matter is settled in [1951, 8].

[4] To satisfy this condition it is sufficient, though not necessary, that the boundary be rigid and stationary.

or equivalently

$$d\mathbf{s}\cdot\dot{\mathbf{x}} = 0. \tag{19.10}$$

Thus *necessary and sufficient conditions that a surface whose only motion is tangential to itself be a possible boundary are that it be either a stream-surface or a surface on which the material is at rest* ($\dot{\mathbf{x}} = 0$). In particular, a steady stream-surface is always a material surface and a possible boundary.

20. Kinematics of the element of arc. I. The basic identity. Let us now imagine that a field of elements of arc $d\mathbf{x}$ be prescribed as arbitrary continuously differentiable functions of $\mathbf{X}$ at $t = 0$. At time t these have been carried into elements $d\mathbf{x}$ given by (15.1). By (18.8) follows

$$\frac{\delta}{\delta t}\,dx^i = \frac{\delta}{\delta t}\,(x^i{}_{,I}\,dX^I) = x^i{}_{,I}\,dX^I = x^i{}_{,j}\,dx^j = d\dot{x}^i, \tag{20.1}$$

or

$$\frac{\delta}{\delta t}\,(d\mathbf{x}) = d\mathbf{x}\cdot\text{grad}\,\dot{\mathbf{x}} = d\dot{\mathbf{x}}. \tag{20.2}$$

From this basic formula we shall now deduce several consequences.

21. Kinematics of the element of arc. II. Deformation. The *distance* $d(\mathbf{X}_1, \mathbf{X}_2, t)$ between two particles $\mathbf{X}_1$ and $\mathbf{X}_2$ is defined by

$$d(\mathbf{X}_1, \mathbf{X}_2, t) = \int_{\mathbf{X}_1}^{\mathbf{X}_2} dx = \int_{\mathbf{X}_1}^{\mathbf{X}_2} \sqrt{d\mathbf{x}\cdot d\mathbf{x}} = \int_{\mathbf{X}_1}^{\mathbf{X}_2} \sqrt{g_{ij}\,dx^i\,dx^j}, \tag{21.1}$$

where the path of integration is a straight line. From the transformation properties of definite integrals follows

$$\dot{d} = \int_{\mathbf{X}_1}^{\mathbf{X}_2} \frac{\delta(dx)}{\delta t} = \int_{\mathbf{X}_1}^{\mathbf{X}_2} \frac{1}{2\,dx}\,\frac{\delta(dx^2)}{\delta t}. \tag{21.2}$$

Thus if we know the material derivative of dx^2 we can calculate the time rate of change of every distance in the continuum. We shall now evaluate $\delta(dx^2)/\delta t$.

By (20.2) we have

$$\begin{aligned}\frac{\delta}{\delta t}\,(dx^2) &= \frac{\delta}{\delta t}\,(d\mathbf{x}\cdot d\mathbf{x}) = 2\,\frac{\delta(d\mathbf{x})}{\delta t}\cdot d\mathbf{x} = 2\,d\mathbf{x}\cdot\text{grad}\,\dot{\mathbf{x}}\cdot d\mathbf{x},\\ &= (\dot{x}_{i,j} + \dot{x}_{j,i})\,dx^i\,dx^j.\end{aligned} \tag{21.3}$$

If we introduce the ***rate of deformation*** $\boldsymbol{\Delta}$ as the dyadic whose covariant components Δ_{ij} are given by

$$\Delta_{ij} \equiv \tfrac{1}{2}(\dot{x}_{i,j} + \dot{x}_{j,i}), \tag{21.4}$$

(21.3) then becomes the ***equation of Cauchy, Beltrami, and Durrande*** (1823, 1871):[1]

$$\frac{\delta}{\delta t}(dx^2) = 2\Delta_{ij}\,dx^i\,dx^j = 2\,d\mathbf{x}\cdot\mathbf{\Delta}\cdot d\mathbf{x}. \tag{21.5}$$

The symmetric part of the velocity gradient is thus a measure of the rate at which the squared element of arc length is changing, and a *necessary and sufficient condition that the motion be locally and instantaneously one of which a rigid body is susceptible* is ***Euler's criterion***:[2]

$$\mathbf{\Delta} = 0. \tag{21.6}$$

Euler[3] showed also that in the general case when $\mathbf{\Delta} \neq 0$ the diagonal components Δ_{xx}, Δ_{yy}, Δ_{zz} in a rectangular Cartesian system are the limiting values of rates of extension per unit length of line segments along the subscript directions, while the off-diagonal components Δ_{xy}, Δ_{yz}, Δ_{zx} are the rates at which two line segments in the subscript directions are approaching each other. Since $\mathbf{\Delta}$ is symmetric, by the algebraic theorem of Cauchy (§2) it possesses at each point (at least) three mutually orthogonal principal directions, which are called *principal axes of extension* at that point, such that when referred to these axes it becomes diagonal. We shall write Δ_1, Δ_2, Δ_3 for the proper values of $\mathbf{\Delta}$, which are called the *principal rates of extension*. Elements of arc parallel to these directions at a point are experiencing extremal rates of change, relative to other elements at the same point. For an element $d\mathbf{x}$ pointing along the first of these axes (21.5) yields

$$\frac{\delta \log dx}{\delta t} = \Delta_1. \tag{21.7}$$

The three *principal invariants* I, II, and III of $\mathbf{\Delta}$ are given by

$$\begin{aligned} I &\equiv \Delta_{\cdot} = \Delta_1 + \Delta_2 + \Delta_3 = \operatorname{div}\dot{\mathbf{x}}, \\ II &\equiv \tfrac{1}{2}\delta^{ij}_{kl}\Delta^k{}_i\Delta^l{}_j = \Delta_2\Delta_3 + \Delta_3\Delta_1 + \Delta_1\Delta_2, \\ III &\equiv \det \Delta^i{}_j = \Delta_1\Delta_2\Delta_3. \end{aligned} \tag{21.8}$$

These invariants were introduced by Cauchy.[4] The quantity

$$\Delta^{ij}\Delta_{ij} = (\Delta_1)^2 + (\Delta_2)^2 + (\Delta_3)^2 \tag{21.9}$$

[1] The analysis of Cauchy [1823, 1] [1827, 2] [1828, 1] is phrased in terms of infinitesimal strain. Beltrami [1871, 1, §4] and Durrande [1871, 2] give the result in substantially the form above.

[2] [1761, 1, §§75–77] [1770, 1, §13] [1806, 1, §133]. The generalization of this result in differential geometry is called *Killing's equation* [1892, 1, p. 167].

[3] [1770, 1, §§9–12] [1806, 1, §§130–132]. We do not reproduce the analysis here, but the reader may easily construct it from the formulae of §31. An expression for $\mathbf{\Delta}$ according to the material description is given by Duhem [1903, 4].

[4] [1827, 2, eq. (20)].

is essentially positive and cannot be zero unless every component of $\mathbf{\Delta}$ be zero; thus it represents the total amount of deformation, and its square root is called the *intensity of deformation.* From the evident identity

$$\Delta^{ij}\Delta_{ij} = -2II + I^2 \tag{21.10}$$

we see that an increase in the magnitude of the first invariant I (whether positive or negative) at constant intensity decreases II.

The most general motion *rigid* throughout a volume is obtained by regarding Euler's condition (21.6) as a first order differential equation for $\dot{\mathbf{x}}$. An easy integration yields

$$\dot{\mathbf{x}} = \boldsymbol{\omega} \times \mathbf{r} + \dot{\mathbf{x}}_0, \tag{21.11}$$

where $\boldsymbol{\omega}$ and $\dot{\mathbf{x}}_0$ are arbitrary functions of time only. Comparison of (21.11) with (16.9) shows that $\boldsymbol{\omega}$ is the angular velocity of a frame with respect to which the medium is instantaneously stationary or in rigid translatory motion. Consequently $\boldsymbol{\omega}$ is called the *angular velocity of the rigid motion.* From the manner in which (21.11) was derived it is plain that *the angular velocity of a rigid motion characterizes the whole motion,* being neither associated with any particular point nor dependent upon any particular choice of origin.

22. Kinematics of the element of arc. III. The theorem of Bertrand and Kelvin & Tait. In order for the element $d\mathbf{x}$ to be suffering no rotation at the instant t we must have

$$\frac{\delta}{\delta t}(d\mathbf{x}) = k\, d\mathbf{x} = k\, d\mathbf{x}\cdot\mathbf{I}, \tag{22.1}$$

where k is a real factor of proportionality. By (20.2) follows the equivalent condition

$$d\mathbf{x}\cdot \text{grad}\,\dot{\mathbf{x}} = k\, d\mathbf{x}. \tag{22.2}$$

That is, k must be a real proper value of the matrix grad $\dot{\mathbf{x}}$, and hence a real solution of the real cubic

$$\det\,(k\mathbf{I} - \text{grad}\,\dot{\mathbf{x}}) = 0. \tag{22.3}$$

The discriminant D of this cubic is

$$D = -18IKL - 4I^3L + I^2K^2 - 4K^3 - 27L^2, \tag{22.4}$$

where

$$K \equiv \tfrac{1}{2}\delta^{ij}_{kl}\dot{x}^k{}_{,i}\dot{x}^l{}_{,j}, \qquad L \equiv \det \dot{x}^i{}_{,j}. \tag{22.5}$$

Recalling that a complex proper value cannot yield a real solution $d\mathbf{x}$ of (22.2), and that to distinct real proper values there correspond linearly independent real solutions $d\mathbf{x}$, from the theory of cubic equations we obtain cases 1 and 2 of the following theorem, due in principle to Bertrand and to Kelvin & Tait:[1] *in any continuous motion there is at each point at least one real direction which is suffering no instantaneous rotation. There are three possibilities*:

1. $D > 0$ *there are three and only three such directions.*
2. *If* $D < 0$ *there is one and only one such direction.*
3. *If* $D = 0$ *there may be one, two, three or an infinite number of such directions.*

The additional possibilities mentioned in case 3 are easiest seen by example. The condition $D = 0$ is requisite and sufficient for the occurrence of multiple proper values. In the rectilinear shearing flow $\dot{x} = 0$, $\dot{y} = x$, $\dot{z} = 0$ there is but the single proper value $k = 0$, thrice repeated, and any line parallel to the x-z-plane is carried always parallel to itself. In the motion[2] $\dot{x} = 0$, $\dot{y} = x$, $\dot{z} = z$ the proper values are $k = 0, 1$, the former being twice repeated, and the reader will easily verify that the only lines not being rotated are those parallel to the x-axis or to the z-axis.

The kinematical significance of the three cases $D > 0$, $D < 0$, $D = 0$ is unfortunately far from plain. We shall perforce rest content with a single observation. In §24 we shall see that $I = 0$ in an isochoric motion. Putting this result into (22.4), we may then conclude the following simple corollary: *in an isochoric motion such that* $L \neq 0$ *there are at each point exactly three instantaneously stationary directions if* $K < -\frac{3}{2}|L|$; *if* $K > -\frac{3}{2}|L|$, *exactly one, as is also the case if* $K > 0$, $L = 0$. The case of rigid motion is included, since by (21.11) we have then $I = 0$, $K = \omega^2$, $L = 0$.

Since grad $\dot{\mathbf{x}}$ is generally not symmetric, the algebraic theorem of Kelvin & Tait (§2) implies that in the case when three distinct instantaneously stationary directions exist they are not usually orthogonal.

23. Kinematics of line integrals. Let $\emptyset$ be an arbitrary continuously differentiable single-valued quantity, and let $\mathfrak{C}$ be an arbitrary material curve. For the material derivative of the flow of $\emptyset$ along $\mathfrak{C}$ we have

$$\frac{\delta}{\delta t}\int_{\mathfrak{C}} d\mathbf{x}\cdot\emptyset = \int_{\mathfrak{C}} \frac{\delta}{\delta t}(d\mathbf{x}\cdot\emptyset) = \int_{\mathfrak{C}} \left[\frac{\delta(d\mathbf{x})}{\delta t}\cdot\emptyset + d\mathbf{x}\cdot\frac{\delta\emptyset}{\delta t}\right], \qquad (23.1)$$

since the parametric equation of the curve $\mathfrak{C}$ is of the form $\mathbf{X} = \mathbf{X}(l)$,

---

[1] [1867, 2]. Kelvin & Tait [1867, 1, §181] state the result in terms of infinitesimal strains. None of these authors mentions the possibility $D = 0$.

[2] For this example I am indebted to my colleague G. Whaples.

and thus constant in time according to the material description. By (20.2) follows then

$$\frac{\delta}{\delta t}\int_{\mathfrak{C}} d\mathbf{x}\cdot\emptyset = \int_{\mathfrak{C}} d\mathbf{x}\cdot[\dot{\emptyset} + \text{grad}\,\dot{\mathbf{x}}\cdot\emptyset] = \int_{\mathfrak{C}} [d\mathbf{x}\cdot\dot{\emptyset} + d\dot{\mathbf{x}}\cdot\emptyset], \tag{23.2}$$

a result implicit in the analysis of Kelvin (1869).[1]

24. Kinematics of the element of volume. Euler's expansion formula and the d'Alembert-Euler continuity equation. If the spatial co-ordinate system be steady, so that $\partial g_{ij}/\partial t = 0$, since $g_{ij,k} = 0$ (lemma of Ricci) it follows that

$$\dot{g}_{ij} = 0. \tag{24.1}$$

Hence by (14.7) we have

$$\dot{J} = \sqrt{\frac{g}{G}}\frac{\delta}{\delta t}\frac{\partial(x^1, x^2, x^3)}{\partial(X^1, X^2, X^3)}\,. \tag{24.2}$$

Denoting by $X^I{}_i$ the cofactor of $x^i{}_{,I}$ in the Jacobian $\partial(x^1, x^2, x^3)/\partial(X^1, X^2, X^3)$, from the formula for differentiating a determinant we readily obtain

$$\begin{aligned}\dot{J} &= \sqrt{\frac{g}{G}}\frac{\delta x^i{}_{,I}}{\delta t}X^I{}_i,\\ &= \sqrt{\frac{g}{G}}\dot{x}^i{}_{,I}X^I{}_i = \sqrt{\frac{g}{G}}\dot{x}^i{}_{,j}x^j{}_{,I}X^I{}_i,\\ &= \sqrt{\frac{g}{G}}\dot{x}^i{}_{,j}\delta^j{}_i\frac{\partial(x^1, x^2, x^3)}{\partial(X^1, X^2, X^3)},\\ &= J\dot{x}^i{}_{,i}.\end{aligned} \tag{24.3}$$

We shall employ the symbol ϑ for $\dot{x}^i{}_{,i}$. By (21.8) and (15.7) we may put (24.3) into the various equivalent forms

$$\vartheta \equiv I = \dot{x}^i{}_{,i} = \mathbf{\Delta}. = \text{div}\,\dot{\mathbf{x}} = \frac{\delta \log J}{\delta t} = \frac{\delta \log dv}{\delta t}, \tag{24.4}$$

a result of basic importance. Since the value of ϑ at a point has thus been shown to be *the rate per unit volume at which an element of volume carried by the particle instantaneously occupying the point is increasing*, ϑ is called the ***expansion***, and (24.4) is ***Euler's expansion formula*** (1755).[1]

---

[1] [1869, 1, 59(a)]. The general formula is given in [1905, 1, §383], [1916, 1, §29].

---

[1] [1757, 2, §§X–XV] [1770, 1, §14] [1806, 1, §134] [1862, 1, §156].

Two other forms, generally called the ***Eulerian continuity equation,***[2] are

$$\frac{\delta \log j}{\delta t} + \vartheta = 0, \qquad \frac{\partial j}{\partial t} + \operatorname{div}(j\dot{\mathbf{x}}) = 0. \tag{24.5}$$

When the motion is *steady* (24.5) reduces to

$$\operatorname{div}(j\dot{\mathbf{x}}) = 0: \tag{24.6}$$

in a steady motion the field $j\dot{\mathbf{x}}$ *is solenoidal.* When the motion is *isochoric* we may put (15.8) into (24.5) and obtain

$$\operatorname{div}\dot{\mathbf{x}} = 0: \tag{24.7}$$

in any isochoric motion, steady or not, the velocity field is solenoidal. Conversely, if $\operatorname{div}\dot{\mathbf{x}} = 0$, from (24.4) it follows that $J = \text{const.}$ for each particle; hence, since $J = 1$ at $t = 0$, we have $J = 1$ always, and the motion is isochoric. The definition (15.8) of an isochoric motion refers to two separate co-ordinate systems, the spatial and the material, but the foregoing result, wholly spatial in character, shows that no part of the material description need be known in order to ascertain whether or not a motion be isochoric.

The theorems of §10 are all applicable to the special cases of steady or of isochoric motion. In particular, *for any isochoric motion*

1. *The simultaneous flux of* $\dot{\mathbf{x}}$ *through any two cross-sections of any stream-tube are equal:*[3]

$$\int_{\mathfrak{s}_1} d\mathbf{s}\cdot\dot{\mathbf{x}} = \int_{\mathfrak{s}_2} d\mathbf{s}\cdot\dot{\mathbf{x}}; \tag{24.8}$$

2. *All the moments of* $\dot{\mathbf{x}}$ *taken over any closed stream-tube are zero:*[4]

$$\int_{\mathfrak{v}} \{\mathbf{r}^{(n)}\dot{\mathbf{x}}\}\, dv = 0; \tag{24.9}$$

while for any steady motion

---

[2] In the usual form of these equations the Jacobian J is replaced by the ratio of the initial *density* ρ_0 to the present density ρ, quantities which it is not convenient to introduce in the present treatment. In [1752, 1] d'Alembert obtained the following special cases of (24.5): rotationally-symmetric isochoric motion [§45], plane isochoric motion [§73], steady rotationally symmetric motion [§116]. Euler derived (24.7) about 1752 [1761, 1, §§21–38]; his analysis, which includes determination of the coefficients in the secular equation, is discussed and generalized in [1953, 4].

[3] This form of the principle of continuity was laid down as a postulate in the older hydrodynamics and is still of frequent occurrence in the engineering literature.

[4] The case $n = 0$ is given by A. Föppl [1897, 2, §4]. For a homogeneous incompressible substance the total velocity of $\mathfrak{v}$ is proportional to the momentum of $\mathfrak{v}$.

1. *The simultaneous flux of $j\dot{\mathbf{x}}$ through any two cross-sections of any stream-tube are equal:*[5]

$$\int_{\mathfrak{s}_1} d\mathbf{s}\cdot\dot{\mathbf{x}}j = \int_{\mathfrak{s}_2} d\mathbf{s}\cdot\dot{\mathbf{x}}j; \tag{24.10}$$

2. *All the moments of $j\dot{\mathbf{x}}$ taken over any closed stream-tube are zero:*

$$\int_{\mathfrak{v}} \{\mathbf{r}^{(n)} j\dot{\mathbf{x}}\}\, dv = 0. \tag{24.11}$$

Comparison of (24.7) with (21.8), and (21.10) shows that *in an isochoric motion the first invariant of deformation vanishes and the second invariant is essentially negative:*

$$I = 0, \qquad II \leqq 0. \tag{24.12}$$

From (11.12) and (24.6) it follows that *j is steady if and only if the speed and the stream-line pattern are related through*

$$j\dot{x} = j_0\dot{x}_0 \exp\left[-\int_0^s \Theta\, ds\right], \tag{24.13}$$

where $\Theta \equiv \operatorname{div} \mathbf{t}$, $\mathbf{t} \equiv \dot{\mathbf{x}}/\dot{x}$. The isochoric case is obtained by putting $j = 1$. Note that Θ need not be steady, but when it is, $\dot{x}$ as given by (24.13) is steady, and hence so is $\dot{\mathbf{x}}$. Thus, in particular, any congruence of continuously differentiable curves can be stream-lines of an isochoric motion (*cf.* 10[2]), and any such congruence if steady may be the stream-lines of a steady motion.

We notice also from $(16.16)_1$ and (24.7) that *in any isochoric slow motion the acceleration is solenoidal.*

A class of motions of some interest is that in which *the expansion* is steady:

$$\frac{\partial\vartheta}{\partial t} = 0. \tag{24.14}$$

For such a motion

$$\operatorname{div}\frac{\partial\dot{\mathbf{x}}}{\partial t} = 0: \tag{24.15}$$

the local acceleration is solenoidal. Consequently by (11.10) there exists a field $\tilde{\mathbf{p}}$ such that

$$\frac{\partial\dot{\mathbf{x}}}{\partial t} = \operatorname{curl}\frac{\partial\tilde{\mathbf{p}}}{\partial t} = \frac{\partial}{\partial t}\operatorname{curl}\tilde{\mathbf{p}}, \tag{24.16}$$

and hence *in order that the expansion be steady it is necessary and sufficient that the velocity be the sum of a solenoidal field and a steady field:*

---

[5] Equivalently, the *mass-flux* $\int d\mathbf{s}\cdot\dot{\mathbf{x}}\rho$ is constant along each stream-tube.

$$\dot{\mathbf{x}}(\mathbf{x}, t) = \operatorname{curl} \tilde{\mathbf{p}}(\mathbf{x}, t) + \mathbf{u}(\mathbf{x}). \tag{24.17}$$

In particular, (24.14) is satisfied by any steady motion and by any isochoric motion.

The expansion ϑ enjoys an invariance quite unusual among the quantities occurring in mechanics. In order to compare the values of the expansion as apparent to observers in frames which are suffering acceleration one with respect to another, by (16.9) we calculate that

$$\begin{aligned}\vartheta = \operatorname{div} \dot{\mathbf{x}} &= \operatorname{div}\left(\frac{d\mathbf{c}}{dt} + \boldsymbol{\omega} \times \mathbf{r}' + \dot{\mathbf{x}}'\right),\\ &= \operatorname{div}' (\boldsymbol{\omega} \times \mathbf{r}' + \dot{\mathbf{x}}'),\\ &= -\boldsymbol{\omega} \cdot \operatorname{curl}' \mathbf{r}' + \operatorname{div}' \dot{\mathbf{x}}',\\ &= \vartheta' :\end{aligned} \tag{24.18}$$

the expansion has the same value for all observers. We have seen already that whether or not a motion be isochoric may be determined from observations in the spatial frame only; the result (24.18) then indicates further that a motion which is isochoric in one frame is also isochoric in every other frame.

25. Kinematics of volume integrals. The transport theorem of Reynolds. Let $\emptyset$ be an arbitrary continuously differentiable single-valued field. For the material derivative of its total in a material volume $\mathfrak{V}$ we have

$$\frac{\delta}{\delta t}\int_{\mathfrak{V}} \emptyset \, dv = \int_{\mathfrak{V}} \left[\frac{\delta \emptyset}{\delta t}\, dv + \emptyset \, \frac{\delta (dv)}{\delta t}\right], \tag{25.1}$$

whence by Euler's expansion formula (24.4) follows[1]

$$\frac{\delta}{\delta t}\int_{\mathfrak{V}} \emptyset \, dv = \int_{\mathfrak{V}} [\dot{\emptyset} + \emptyset \operatorname{div} \dot{\mathbf{x}}] \, dv. \tag{25.2}$$

A more interesting form of this result is

$$\frac{\delta}{\delta t}\int_{\mathfrak{V}} \emptyset \, dv = \int_{\mathfrak{V}} \left[\frac{\partial \emptyset}{\partial t} + \operatorname{div} (\dot{\mathbf{x}} \emptyset)\right] dv. \tag{25.3}$$

Hence by (7.2) follows the ***transport theorem of Reynolds*** (1903):[2]

$$\frac{\delta}{\delta t}\int_{\mathfrak{V}} \emptyset \, dv = \frac{\partial}{\partial t}\int_{\mathfrak{v}} \emptyset \, dv + \oint_{\mathfrak{s}} d\mathbf{s} \cdot \mathbf{x} \emptyset, \tag{25.4}$$

---

[1] [1903, 5, §14] [1905, 1, §383] [1916, 1, §29].

[2] [1903, 5, §9] [1916, 1, §29]. Curiously enough Reynolds saw fit to lay down this easily demonstrated transformation as one of the basic postulates of his *Sub-Mechanics of the Universe.*

a result which may be expressed in words as follows: *the rate of change of the total $\emptyset$ over a material volume $\mathfrak{V}$ equals the rate of change of the total $\emptyset$ over a fixed volume $\mathfrak{v}$ instantaneously coinciding with $\mathfrak{V}$ plus the flux of $\dot{\mathbf{x}}\emptyset$ out of the bounding surface.* The transport theorem is really an alternative formulation of Euler's expansion formula (24.4).

In the case of an isochoric motion we may put $\mathbf{c} = \dot{\mathbf{x}}$ in (10.5), obtaining the following remarkable result:

$$\int_{\mathfrak{v}} \dot{\mathbf{x}}\cdot\text{grad}\ \emptyset\ dv = \oint_{\mathfrak{s}} d\mathbf{s}\cdot\dot{\mathbf{x}}\emptyset; \tag{25.5}$$

that is, *in an isochoric motion the total convection of any continuously differentiable single-valued quantity $\emptyset$ in a region equals the flux of $\dot{\mathbf{x}}\emptyset$ out of the bounding surface*, and, in particular, *in a closed stream-tube the total convection of any such quantity is zero.*

A more general result analogous to (25.5) may be derived by putting $j\emptyset$ for $\emptyset$ in (25.4) and (25.3), comparing the two resulting formulae, and employing (24.5). We obtain

$$\int_{\mathfrak{v}} \left[-\frac{\partial j}{\partial t}\emptyset + j\dot{\mathbf{x}}\cdot\text{grad}\ \emptyset\right] dv = \oint_{\mathfrak{s}} d\mathbf{s}\cdot\dot{\mathbf{x}}j\emptyset. \tag{25.6}$$

The first term on the left vanishes in a steady motion; in particular, *for the volume within a closed stream-tube in a steady motion we have*

$$\int_{\mathfrak{v}} j\dot{\mathbf{x}}\cdot\text{grad}\ \emptyset\ dv = 0 \tag{25.7}$$

for any continuously differentiable single-valued quantity $\emptyset$. The formulae (25.5) through (25.7) are suggested by an early analysis of Kelvin.[3]

26. Kinematics of the element of surface. By differentiating Nanson's formula $(15.6)_1$ we obtain

$$\dot{x}^i{}_{,I}\ ds_i + x^i{}_{,I}\frac{\delta ds_i}{\delta t} = \frac{\delta J}{\delta t}\, dS_I, \tag{26.1}$$

whence by Euler's expansion formula (24.4) follows

$$x^i{}_{,I}\frac{\delta ds_i}{t} = -\dot{x}^i{}_{,i}x^i{}_{,I}\ ds_i + J\dot{x}^k{}_{,k}\ dS_I. \tag{26.2}$$

Now multiplying by $X^I{}_{,l}$, after reducing the result with the aid of $(14.11)_2$, $(15.6)_2$, and (14.9) we obtain the formula of Lamb (1877):[1]

[3] [1849, 1].

[1] [1877, 1].

$$\frac{\delta ds_l}{\delta t} = -\dot{x}^i{}_{,l}\, ds_i + \dot{x}^k{}_{,k}\, ds_l, \tag{26.3}$$

or

$$\frac{\delta d\mathbf{s}}{\delta t} = -\operatorname{grad} \dot{\mathbf{x}}\cdot d\mathbf{s} + \operatorname{div} \dot{\mathbf{x}}\, d\mathbf{s}. \tag{26.4}$$

The theorem of Bertrand and Kelvin & Tait given in §22 may be deduced in another way by using (26.4) to determine those elements of surface $d\mathbf{s}$ which are instantaneously in motion parallel to themselves.

27. Kinematics of surface integrals. Let $\emptyset$ be an arbitrary continuously differentiable tensor field. For the material derivative of its flux across a material surface $\mathfrak{S}$ we have

$$\frac{\delta}{\delta t}\int_{\mathfrak{S}} d\mathbf{s}\cdot\emptyset = \int_{\mathfrak{S}} \left[\frac{\delta d\mathbf{s}}{\delta t}\cdot\emptyset + d\mathbf{s}\cdot\frac{\delta\emptyset}{\delta t}\right], \tag{27.1}$$

whence by (26.4) follows

$$\frac{\delta}{\delta t}\int_{\mathfrak{S}} d\mathbf{s}\cdot\emptyset = \int_{\mathfrak{S}} [d\mathbf{s}\cdot(\dot{\emptyset} + \emptyset \operatorname{div} \dot{\mathbf{x}}) - \operatorname{grad} \dot{\mathbf{x}}\cdot d\mathbf{s}\cdot\emptyset]. \tag{27.2}$$

For the case when $\emptyset = \mathbf{c}$ we have also[1]

$$\frac{\delta}{\delta t}\int_{\mathfrak{S}} d\mathbf{s}\cdot\dot{\mathbf{c}} = \int_{\mathfrak{S}} d\mathbf{s}\cdot[\mathbf{c} - \mathbf{c}\cdot\operatorname{grad} \dot{\mathbf{x}} + \mathbf{c} \operatorname{div} \dot{\mathbf{x}}]. \tag{27.3}$$

28. Kinematics of vector-lines and vector-tubes. The Helmholtz-Zorawski criteria. We now seek a *necessary and sufficient condition that the strength of all vector-tubes of an arbitrary field* $\mathbf{c}$ *at a given cross-section remain constant as the motion proceeds.* This is precisely the condition that the flux of $\mathbf{c}$ through an arbitrary material surface $\mathfrak{S}$ be constant. From (27.3) follows ***Zorawski's criterion*** (1900):[1]

$$\dot{\mathbf{c}} - \mathbf{c}\cdot\operatorname{grad} \dot{\mathbf{x}} + \mathbf{c} \operatorname{div} \dot{\mathbf{x}} = 0, \tag{28.1}$$

or equivalently

$$\frac{\delta}{\delta t}(J\mathbf{c}) = J\mathbf{c}\cdot\operatorname{grad} \dot{\mathbf{x}},$$

$$\frac{\partial \mathbf{c}}{\partial t} + \operatorname{curl}(\mathbf{c} \times \dot{\mathbf{x}}) + \dot{\mathbf{x}} \operatorname{div} \mathbf{c} = 0. \tag{28.2}$$

---

[1] Various proofs of this result, first stated explicitly by Zorawski [1900, 1], are given in [1905, 1, §383] [1916, 1, §29] [1947, 1].

---

[1] [1900, 1].

In general, the particles constituting a vector-tube of $\mathbf{c}$ at a given instant will no longer constitute a vector-tube at subsequent instants. We may therefore seek a *necessary and sufficient condition that the vector-tubes of* $\mathbf{c}$ *be material tubes*, or, put more shortly, a condition that the vector-tubes of $\mathbf{c}$ be *permanent.* Given a particular vector-line, let $\mathbf{x} = \mathbf{x}(l, t)$ be the material line which initially coincides with the given vector-line and hence initially satisfies (9.1). If this material line is to remain a vector-line it is necessary that (9.1) be satisfied for all time, and hence

$$\frac{\delta}{\delta t}\left(\frac{\partial \mathbf{x}}{\partial l} \times \mathbf{c}\right) = 0. \tag{28.3}$$

Now for any material line $\mathbf{x} = \mathbf{x}(l, t)$, we have

$$\frac{\delta}{\delta t}\left(\frac{\partial \mathbf{x}}{\partial l} \times \mathbf{c}\right) = \left(\frac{\partial \mathbf{x}}{\partial l}\cdot\operatorname{grad}\dot{\mathbf{x}}\right) \times \mathbf{c} + \frac{\partial \mathbf{x}}{\partial l} \times \dot{\mathbf{c}}. \tag{28.4}$$

The condition (9.1) that $\mathbf{x} = \mathbf{x}(l, t)$ be initially a vector-line of $\mathbf{c}$ yields

$$\frac{\partial \mathbf{x}}{\partial l} = f\mathbf{c}, \tag{28.5}$$

where f is a scalar function. Then for vector-lines of $\mathbf{c}$ (28.4) becomes

$$\frac{\delta}{\delta t}\left(\frac{\partial \mathbf{x}}{\partial l} \times \mathbf{c}\right) = f\mathbf{c} \times [\dot{\mathbf{c}} - \mathbf{c}\cdot\operatorname{grad}\dot{\mathbf{x}}]. \tag{28.6}$$

Equivalent to the condition (28.3) then is

$$\mathbf{c} \times [\dot{\mathbf{c}} - \mathbf{c}\cdot\operatorname{grad}\dot{\mathbf{x}}] = 0. \tag{28.7}$$

In order for a material line which is initially a vector-line of $\mathbf{c}$ to remain a vector-line of $\mathbf{c}$, it is thus necessary that (28.7) be satisfied. Conversely, if (28.7) be satisfied then $\partial\mathbf{x}/\partial l \times \mathbf{c}$ is a quantity initially zero whose time derivative ($\mathbf{X}$ being held constant) is always zero, and hence itself is always zero. Thus the condition (28.7) is also sufficient for the vector-lines of $\mathbf{c}$ to be material lines. Equivalent forms of this ***Helmholtz-Zorawski criterion*** (1858, 1900)[2] for permanent vector-lines are

$$\begin{gathered}\mathbf{c} \times \left[\frac{\delta(J\mathbf{c})}{\delta t} - J\mathbf{c}\cdot\operatorname{grad}\dot{\mathbf{x}}\right] = 0,\\ \mathbf{c} \times \left[\frac{\partial \mathbf{c}}{\partial t} + \operatorname{curl}(\mathbf{c} \times \dot{\mathbf{x}}) + \dot{\mathbf{x}}\operatorname{div}\mathbf{c}\right] = 0.\end{gathered} \tag{28.8}$$

---

[2] This criterion was implicit in the analysis of Helmholtz [1858, 1, §2] and Nanson [1874, 1] but was first explicitly stated by Zorawski [1900, 1]. The present proof is that of [1947, 1] [1950, 1]; another is given in [1950, 8]. Equivalent to (28.7) is the

As a first application, we may seek a condition that the stream-lines be material lines, *i.e.* that the motion be one with steady stream-lines. Putting $\mathbf{c} = \dot{\mathbf{x}}$ in $(28.8)_2$ we obtain

$$\dot{\mathbf{x}} \times \frac{\partial \dot{\mathbf{x}}}{\partial t} = 0, \tag{28.9}$$

as indeed is geometrically evident.

Dr. Carstoiu has called my attention to the fact that from (27.3) we may conclude a generalization of Zorawski's criterion by narrowing the class of surfaces: *in order that the flux of* $\mathbf{c}$ *through a material surface whose unit normal is* $\mathbf{n}$ *remain constant in time, it is necessary and sufficient that*

$$\mathbf{n} \cdot [\dot{\mathbf{c}} - \mathbf{c} \cdot \operatorname{grad} \dot{\mathbf{x}} + \mathbf{c} \operatorname{div} \dot{\mathbf{x}}] = 0. \tag{28.10}$$

Whether, in a given motion, any material surfaces possessing this connection with the field $\mathbf{c}$ over a period of time can exist remains an open question. If the field $\mathbf{c}$ be complex-lamellar, its vector-tubes possess normal cross-sections; if the normal congruence be material, then

$$\mathbf{c} \cdot [\dot{\mathbf{c}} - \mathbf{c} \cdot \operatorname{grad} \dot{\mathbf{x}} + \mathbf{c} \operatorname{div} \dot{\mathbf{x}}] = 0 \tag{28.11}$$

is necessary and sufficient that the strength of the vector-tubes at each normal cross-section remain constant in time.

---

requirement that $\mathbf{c}\dot{\mathbf{c}} + (\dot{\mathbf{x}} \cdot \operatorname{grad} \mathbf{c})\, \mathbf{c}$ shall be symmetric. In this form the condition has been derived by Synge [1951, **9**] for the n-dimensional case. While Synge's derivation is valid only in a Euclidean space, it is easy to see that the analysis given in the text can be carried over step by step to an affine space by simply replacing each statement about the vanishing of a cross product by an evident analogue about the vanishing of the skew of part of a dyadic. The generalization to varieties is discussed in [1937, **3**] [1951, **13**].

Chapter III. VORTICITY

29. Definition of the vorticity. With the preliminaries of the foregoing chapters behind us, we may return to the opening phrases of this work and take the decisive step of putting $\mathbf{\Sigma} = \text{grad}\,\dot{\mathbf{x}}$ in the Gibbs decomposition theorem (1.2), thereby and by (21.4) obtaining the identity

$$\text{grad}\,\dot{\mathbf{x}} = \mathbf{\Delta} - \tfrac{1}{2}\mathbf{I} \times \mathbf{w}, \tag{29.1}$$

or

$$\dot{x}_{i,j} = \Delta_{ij} - \tfrac{1}{2}\sqrt{g}\,\epsilon_{ijk}w^k, \tag{29.2}$$

where the ***vorticity***[1] $\mathbf{w}$ is defined by

$$\mathbf{w} \equiv \text{curl}\,\dot{\mathbf{x}} = (\text{grad}\,\dot{\mathbf{x}})_\times, \quad \text{or} \quad w^k = \frac{\epsilon^{klm}}{\sqrt{g}}\dot{x}_{m,l}. \tag{29.3}$$

Before our eyes opens forth now the splendid prospect of three-dimensional kinematics, the mother tongue for man's perception of the

[1] This name was introduced by Lamb [1916, **2**, preface & §30]. Previous authors had employed $\frac{1}{2}\mathbf{w}$, whose components were called the *angular velocities* by Stokes [1851, **2**, §2], *Rotationsgeschwindigkeiten* by Helmholtz [1858, **1**, §1], the *component rotations* by Kelvin [1869, **1**, §60(e)], and the *molecular rotations* by Basset [1888, **1**, §18]. In the vigorous expression of Clifford [1878, **3**, Bk. II, Ch. II, pp. 122–123, Bk. III, Ch. II, pp. 193–194], $\frac{1}{2}\mathbf{w}$ is called the *spin*. The usual French terms are *vecteur tourbillon* or *rotationnel* for $\mathbf{w}$ or $\frac{1}{2}\mathbf{w}$ and *vitesses angulaires* for the components of $\frac{1}{2}\mathbf{w}$; the German is *Wirbel* or *Wirbelvektor*. For the general case of motion in which $w \neq 0$ Helmholtz used the term *Rotationsbewegung*, which Kelvin translated as *vortex motion*; more customary nowadays in English is *rotational motion*, the term *vortex motion* usually being reserved for irrotational motions induced by vortices (§35).

Helmholtz's use of the word *rotation* in this connection was attacked by Bertrand [1867, **2**], and there ensued an acrimonious public controversy [1868, **2**–**7**] between these two savants, in which Bertrand at first denied the correctness of Helmholtz's results, while Helmholtz accused Bertrand of twisting his intended meanings. Bertrand pointed out that in the shearing motion $\dot{x} = y$, $\dot{y} = 0$, $\dot{z} = 0$ the particles move in straight lines, and yet according to Helmholtz's nomenclature the motion is rotational, $w_z = -1$. While indeed the *particles*, which are but points, do not rotate in orbits like planets, any *volume*, however small, suffers rotation with respect to its initial configuration: to a stationary observer a small rigid object convected by this motion would appear to rotate. A French review of the controversy [1868, **8**], after discussing this same example not altogether satisfactorily, wholly adopted Helmholtz's position, as has posterity. Another way of expressing the rotational character of Bertrand's shearing motion was put forward by St. Venant [1869, **2**, §6, footnote]: the straight stream-lines are the only *lines* in the x-y-plane which are not suffering rotation.

changing world about him. Its peculiar and characteristic glory is the vorticity vector **w**, for whose existence it is both requisite and sufficient that the number of dimensions be three.

The first treatment of vorticity occurs in the work of d'Alembert (1749) and Euler (1752-5);[2] Lagrange (1760) and Cauchy (1815)[3] were the first to introduce single letters to stand for the vorticity components. All this early work is purely formal and somewhat mystifying. The kinematical significance of the vorticity did not begin to be recognized until MacCullagh (1839)[4] and Cauchy (1841)[5] proved that the components of the curl satisfy the vectorial law of transformation.

Our first duty is to describe the kinematical significance of the vorticity vector.

30. First interpretation: Cauchy's mean value of the rates of rotation of all directions in a plane (1841).[1] Let both the material and the spatial co-ordinates be referred to a single rectangular Cartesian frame. If $d\mathbf{X}$ be an arbitrary material element, at time t it is carried into the spatial element $d\mathbf{x}$ given by $(15.1)_1$. Let $\mathbf{N}$ and $\mathbf{n}$ be unit vectors parallel to these two elements:

$$\mathbf{N} \equiv \frac{d\mathbf{X}}{dX}, \qquad \mathbf{n} \equiv \frac{d\mathbf{x}}{dx}. \tag{30.1}$$

The projections of these unit vectors upon the Y-Z- and y-z- planes are segments of slopes N_z/N_y and n_z/n_y, respectively. The angle K_X subtended by one of these segments upon the other is given by

$$\tan K_X = \frac{\dfrac{n_z}{n_y} - \dfrac{N_z}{N_y}}{1 + \dfrac{n_z}{n_y} \cdot \dfrac{N_z}{N_y}}. \tag{30.2}$$

Similarly, angles K_Y and K_Z may be determined, and these specify completely the rotation of the direction $\mathbf{N}$ into the direction $\mathbf{n}$. Now to measure the local rotation of the medium, it is necessary to take into account the rotations suffered by all possible directions $\mathbf{N}$ for the

[2] [1752, 1, §§45–49][1761, 1, §§46–47, 58–59].

[3] [1762, 3, Ch. XLII][1827, 1, 2nd Part, Sect. 1, §4].

[4] [1848, 3, §II, lemma 2]. MacCullagh's nearest approach to a kinematical interpretation is a single remark in his §III. His line of thought if pursued would have led to what we give in §31 below.

[5] [1841, 1, §II, eq. 12].

[1] Cauchy [1841, 1, Th. IV] analysed static deformation, but only a trivial modification is required in order to adapt his analysis to the case of motion. While this splendid paper is several times mentioned in the literature, its contents has not found its way into the usual expositions of the subject. See also [1896, 2, §§5–6, 11].

particle in question. Consider the set of directions **N** in a plane perpendicular to the X-axis:

$$\mathbf{N} = \mathbf{j} \cos B + \mathbf{k} \sin B, \tag{30.3}$$

where B is the angle subtended by the vector **N** upon some given initial line. For these directions the angles K_Y and K_Z do not exist, so that their rotations are completely specified by K_X. From the foregoing formulae and (15.1) we have

$$\tan K_X(B) = \frac{\cos B\left[\dfrac{\partial z}{\partial Y}\cos B + \dfrac{\partial z}{\partial Z}\sin B\right] - \sin B\left[\dfrac{\partial y}{\partial Y}\cos B + \dfrac{\partial y}{\partial Z}\sin B\right]}{\cos B\left[\dfrac{\partial y}{\partial Y}\cos B + \dfrac{\partial y}{\partial Z}\sin B\right] + \sin B\left[\dfrac{\partial z}{\partial Y}\cos B + \dfrac{\partial z}{\partial Z}\sin B\right]}. \tag{30.4}$$

The *mean local rotation* J_X about the X-axis is then defined as the mean values of $K_X(B)$ over all possible initial directions B:

$$J_X \equiv \frac{1}{2\pi}\int_0^{2\pi} K_X(B)\, dB. \tag{30.5}$$

In a similar way mean rotations J_Y and J_Z about the Y- and Z-axes, respectively, may be defined. These angles do not form components of a vector-field,[2] and indeed it does not seem easy to put the foregoing analysis into invariant form.

By differentiating (30.4) with respect to t and then putting $y = Y$, $z = Z$, *i.e.*, putting $t = 0$, we obtain

$$\begin{aligned}\dot{K}_X(B)\Big|_{t=0} &= \cos^2 B\,\frac{\partial \dot{z}}{\partial y} - \sin^2 B\,\frac{\partial \dot{y}}{\partial z} - \sin B\cos B\left(\frac{\partial \dot{y}}{\partial y} - \frac{\partial \dot{z}}{\partial z}\right), \\ &= \frac{1}{2}\left(\frac{\partial \dot{z}}{\partial y} - \frac{\partial \dot{y}}{\partial z}\right) + \frac{1}{2}\left(\frac{\partial \dot{z}}{\partial y} + \frac{\partial \dot{y}}{\partial z}\right)\cos 2B - \frac{1}{2}\left(\frac{\partial \dot{y}}{\partial y} - \frac{\partial \dot{z}}{\partial z}\right)\sin 2B.\end{aligned} \tag{30.6}$$

Since the sense of the angle B in (30.3) is taken according to the usual convention, the above rate is positive when the direction of rotation is such that a right-handed screw motion with respect to the Y- and Z-co-ordinate axes would advance along the positive X-direction, it being assumed that the X, Y, Z system is right-handed. Putting this result into (30.5) we obtain

$$\dot{J}_X\Big|_{t=0} = \frac{1}{2}\left(\frac{\partial \dot{z}}{\partial y} - \frac{\partial \dot{y}}{\partial z}\right) = \tfrac{1}{2}\, w_x. \tag{30.7}$$

[2] Their law of transformation was given by Cauchy [1841, 1, §I, eq. 37].

Thus the value of w_x at a point is twice the mean value of the rates of right-handed rotation about the x-direction of all directions in the plane through the point and normal to the x-direction.

In view of the vectorial character of **w** we may express the result in the following invariant form: *the projection of* **w** *upon any axis at a point is twice the mean rate of right-handed rotation of all line segments through the point which are perpendicular to that axis.*[3]

31. Second interpretation: Cauchy's mean value of the rates of rotation of perpendicular axes (1841).[1] Still employing a rectangular Cartesian frame, consider the motion of two particles (X, Y, Z) and $(X, Y + \Delta Y, Z)$, which at time $t = 0$ are situate a distance ΔY apart upon a line parallel to the Y-axis. At time Δt we have

$$\begin{aligned} y(X, Y, Z, \Delta t) &= Y + \int_0^{\Delta t} \dot{y}(X, Y, Z, t)\, dt, \\ y(X, Y + \Delta Y, Z, \Delta t) &= Y + \Delta Y + \int_0^{\Delta t} \dot{y}(X, Y + \Delta Y, Z, t)\, dt. \end{aligned} \tag{31.1}$$

Hence

$$\begin{aligned} \Delta y &\equiv y(X, Y + \Delta Y, Z, \Delta t) - y(X, Y, Z, \Delta t), \\ &= \Delta Y + \int_0^{\Delta t} [\dot{y}(X, Y + \Delta Y, Z, t) - \dot{y}(X, Y, Z, t)]\, dt, \\ &= \Delta Y\left[1 + \int_0^{\Delta t} \left\{ \frac{\partial \dot{y}(X, Y', Z, t)}{\partial Y'}\bigg|_{Y' = Y + \Xi} \right\} dt \right], \\ &= \Delta Y\left[1 + \Delta t \frac{\partial \dot{y}(X, Y', Z, \tau)}{\partial Y'}\bigg|_{Y' = Y + \Xi}\right], \end{aligned} \tag{31.2}$$

where we have employed the theorems of mean value of the differential and integral calculus, Ξ and τ being quantities which approach zero with ΔY and with Δt, respectively. Similarly

$$\begin{aligned} \Delta z &\equiv z(X, Y + \Delta Y, Z, \Delta t) - z(X, Y, Z, \Delta t), \\ &= \Delta Y\, \Delta t \frac{\partial \dot{z}(X, Y', Z, \tau')}{\partial Y'}\bigg|_{Y' = Y + \Xi'}, \end{aligned} \tag{31.3}$$

where Ξ' and τ' are quantities which approach zero with ΔY and with Δt, respectively. Now if B_1 be the angle from the Y-axis to the projection

---

[3] [1841, 1, Ths. V, VI, VII].

---

[1] The first result of this section is implied though not actually stated by Cauchy [1841, 1, Th. IX]. The method of proof here used is a modernized analogue to that employed by Euler [1770, 1, §§9–13] is his analysis of **Δ**.

upon the Y-Z-plane of the line connecting the two particles at time Δt we have

$$\tan B_1 = \frac{\Delta z}{\Delta y}. \tag{31.4}$$

From (31.2) and (31.3) it follows that

$$\lim_{\Delta Y \to 0} \lim_{\Delta t \to 0} \frac{B_1}{\Delta t} = \lim_{\Delta Y \to 0} \lim_{\Delta t \to 0} \frac{\tan B_1}{\Delta t} = \frac{\partial \dot{z}}{\partial y}. \tag{31.5}$$

Similarly, if B_2 be the angle from the Z-axis to the line connecting (X, Y, Z) and a third particle $(X, Y, Z + \Delta Z)$ at time Δt, then

$$-\lim_{\Delta Z \to 0} \lim_{\Delta t \to 0} \frac{B_2}{\Delta t} = \frac{\partial \dot{y}}{\partial z}, \tag{31.6}$$

where the negative sign is occasioned by the angle B_2's being taken positive when the increment ratio is negative. Thus we obtain

$$\lim_{\Delta Y, \Delta Z \to 0} \lim_{\Delta t \to 0} \frac{B_1 + B_2}{\Delta t} = w_x. \tag{31.7}$$

That is, *the value of w_x at a point is twice the arithmetic mean of the angular speeds of right-handed rotation of the lines through the point and parallel to the y- and z-axes about the x-direction.* Since the y- and z-axes are simply any two axes forming with the x-axis a triply orthogonal set, the foregoing theorem may be expressed in invariant form as follows: *at a given point, the length of the projection of the vorticity upon a given direction is twice the arithmetic mean of the rates of right-handed rotation about that direction of any two mutually perpendicular line segments lying in a plane through the point and perpendicular to the direction.*

Stated in this way, the result just obtained suggests a certain invariantive character not only for the vorticity itself, but even for its component in a given direction. This invariance attains a precise significance through the theorem on dyadic skew projections stated in §2. By (2.1) it appears that the rates of right-handed rotation $\partial \dot{z}/\partial y$ and $-\partial \dot{y}/\partial z$ of line segments pointing in the y- and z-directions, respectively, about the x-direction, are the diagonal components of the plane *dyadic ${}_x\mathbf{\Omega}$ of right-handed rotations about the x-direction*:

$$\| {}_x\mathbf{\Omega} \| \equiv \left\| \begin{matrix} \dfrac{\partial \dot{z}}{\partial y} & -\dfrac{\partial \dot{y}}{\partial y} \\[2ex] \dfrac{\partial \dot{z}}{\partial z} & -\dfrac{\partial \dot{y}}{\partial z} \end{matrix} \right\|. \tag{31.8}$$

By (2.3) it follows that the *x-component of vorticity w_x is the scalar of the dyadic of rotations ${}_x\mathbf{\Omega}$ about the x-axis*:

$$w_x = {}_x\Omega_{\cdot}\,. \tag{31.9}$$

The dyadic ${}_x\mathbf{\Omega}$ may be represented in the usual way by a quadric, which in this case is an ordinary conic section. Its properties are summarized in the following theorem of Cauchy:[2] *the rate of right-handed rotation of any line through a given point and lying in a plane perpendicular to the x-direction at the point is inversely proportional to the square root of the length of the parallel radius vector to the conic*

$$\bar{\mathbf{r}}\cdot{}_x\mathbf{\Omega}\cdot\bar{\mathbf{r}} = \text{const.}, \tag{31.10}$$

($\bar{\mathbf{r}}$ *being a radius vector in the plane with the given point as origin*). Let the constant be given any non-zero value which renders the locus real. There are then three possibilities. 1°, the conic is an ellipse; the rates of rotation of lines parallel to its major and minor axes are respectively minimal and maximal rates of rotation, and all lines in the plane are instantaneously rotating in the same sense. 2°, the conic is a pair of parallel straight lines; again all lines in the plane are rotating in the same sense, except that the line parallel to the pair of straight lines is not rotating at all. 3°, the conic is a pair of conjugate hyperbolae; then its asymptotes divide the plane into two portions, lines in one of which are rotating in one sense, while lines in the other are rotating in the opposite sense; lines in the directions of the asymptotes are suffering no rotation whatever, while the rates of rotation of lines parallel to the axes are maximal and minimal rates (in absolute value, both are maximal). Since similar results hold for lines in planes perpendicular to the y- and z-axes, in all cases *the component in a given direction of the vorticity at a point is twice the arithmetic mean of the maximum and minimum rates of right-handed rotation about that direction of lines through the point and normal to the direction.*[3]

For the rate of right-handed rotation $\partial\dot{z}'/\partial y'$ of a line which subtends an arbitrary angle B with the y-direction, by (31.8) and the dyadic law of transformation we obtain

$$\frac{\partial\dot{z}'}{\partial y'} = \frac{\partial\dot{z}}{\partial y}\cos^2 B - \frac{\partial\dot{y}}{\partial y}\cos B\sin B + \frac{\partial\dot{z}}{\partial z}\sin B\cos B - \frac{\partial\dot{y}}{\partial z}\sin^2 B. \tag{31.11}$$

By integrating with respect to B we may obtain (30.7) again, this time without having employed the notion of finite rotation.

32. Third interpretation: Stokes's local angular velocity (1845).[1] In the case of a *rigid* motion, by taking the curl of (21.11) we obtain

$$\begin{aligned}\operatorname{curl}\dot{\mathbf{x}} &= \operatorname{curl}(\boldsymbol{\omega}\times\mathbf{r}),\\ &= 2\boldsymbol{\omega}.\end{aligned} \tag{32.1}$$

---

[2] [1841, 1, Th. VIII]. See also [1867, 2].
[3] [1841, 1, Th. IX].

[1] [1845, 1, §2].

Thus *in a rigid motion the* curl *of the velocity is twice the angular velocity.* In the general case when the motion is not rigid, curl $\dot{\mathbf{x}}$ is no longer constant from point to point, but *at each point it may be regarded as twice the angular velocity of a small element of the continuum.* Sometimes it is convenient to speak of the *local angular velocity* $\tilde{\boldsymbol{\omega}}$ of a deformable continuum; in this case $\tilde{\boldsymbol{\omega}}$ is given by the definition

$$\tilde{\boldsymbol{\omega}} \equiv \tfrac{1}{2}\mathbf{w}. \tag{32.2}$$

A dynamical interpretation of this result was suggested by Stokes and generalized by Beltrami (1871).[2] Imagine that a vanishingly small element of mass whose principal axes of inertia coincide with the principal axes of extension at the point and whose center of mass is the point itself should be solidified suddenly into a rigid body, and at the same time all the surrounding material were removed; then *this rigid body would continue to rotate indefinitely with the angular velocity* $\tilde{\boldsymbol{\omega}}$.

A very elegant reformation of Stokes's result is due to Boussinesq.[3] Since the principal axes of extension are instantaneously suffering no rotation with respect to one another, their motion as lines, no account being taken of the motions of the particles along them, is instantaneously rigid; hence *the local angular velocity is the angular velocity of the principal axes of extension* in the ordinary sense of rigid motions.

Further motivation for Stokes's viewpoint is afforded by the nature of the invariance enjoyed by the vorticity, for from (16.9) and (32.1) we have

$$\begin{aligned} \mathbf{w} \equiv \operatorname{curl} \dot{\mathbf{x}} &= \operatorname{curl}\left(\frac{d\mathbf{c}}{dt} + \boldsymbol{\omega} \times \mathbf{r}' + \dot{\mathbf{x}}'\right), \\ &= \operatorname{curl}' (\boldsymbol{\omega} \times \mathbf{r}' + \dot{\mathbf{x}}'), \\ &= 2\boldsymbol{\omega} + \mathbf{w}', \end{aligned} \tag{32.3}$$

or

$$\tilde{\boldsymbol{\omega}} = \boldsymbol{\omega} + \tilde{\boldsymbol{\omega}}': \tag{32.4}$$

the local angular velocity of a motion as apparent to one observer is the vector sum of that apparent to a second observer and the angular velocity of the frame of reference of the second observer with respect to that of the first.

---

[2] [1871, 1, §10]. Stokes considered only a spherical element. See also [1903, 2, ¶63]. Earlier Coriolis [1835, 1, pp. 100–101] had introduced a similar notion for a system of particles.

[3] According to [1921, 2, §706]. The earliest statement of this result which I have found is that of Levy [1890, 1, §5], who does not mention Boussinesq. I am unable to follow the analysis of Valcovici [1946, 3], who claims that the material lines which at a given instant coincide with the principal axes of extension no longer do so after an infinitesimal time and on this ground proposes a new measure of rotation.

33. Fourth interpretation: Hankel (1861),[1] Roch (1863),[2] and Kelvin's (1869)[3] circulation per unit area. By Kelvin's transformation (8.1) we have

$$\oint_{c} d\mathbf{x}\cdot\dot{\mathbf{x}} = \int_{\mathfrak{s}} d\mathbf{s}\cdot\mathbf{w}, \tag{33.1}$$

where $\mathfrak{s}$ is any surface whose bounding circuit is $\mathfrak{c}$, subject to the restrictions stated in §8. The left-hand side of this equation is the mean value of the tangential component of velocity, multiplied by the length of the circuit. Its central importance was revealed by the work of Kelvin,[4] who called it the ***circulation*** around $\mathfrak{c}$. We shall discuss its general significance in §§42 and 46, resting content here to notice that by the mean value theorem follows

$$\oint_{c} d\mathbf{x}\cdot\dot{\mathbf{x}} = \mathcal{S}w_n\Big|_{\mathbf{x}+\mathbf{x}'}, \tag{33.2}$$

where $\mathbf{x}' \to 0$ as $\mathfrak{c}$ is shrunk down to the point $\mathbf{x}$. Thus

$$w_n = \operatorname*{Lim}_{\mathcal{S}\to 0} \frac{\oint_{c} d\mathbf{x}\cdot\dot{\mathbf{x}}}{\mathcal{S}}. \tag{33.3}$$

This formula shows that *if* $\mathcal{S}$ *be the area inclosed by a circuit* $\mathfrak{c}$ *upon a surface* $\mathfrak{s}$, *then the ratio of the circulation around* $\mathfrak{c}$ *to the area* $\mathcal{S}$ *as* $\mathfrak{c}$ *is shrunk down to a point P approaches the magnitude of the component of the vorticity in the direction normal to* $\mathfrak{s}$ *at P*. Hence the direction of $\mathbf{w}$ at P is perpendicular to the plane in which the circulation per unit area is greatest, and the magnitude of $\mathbf{w}$ is the circulation per unit area of this plane.

Bjørgum's formula (9.13) expresses the vorticity in terms of its projections onto the principal directions of the velocity field. Either from it or from (9.3) or by comparing (33.3) with (9.16) we may conclude

$$\Omega = \frac{\mathbf{v}\cdot\mathbf{w}}{v^2} = \frac{\mathbf{t}\cdot\mathbf{w}}{v} = \frac{w_{\mathbf{v}}}{v}: \tag{33.4}$$

the abnormality of the velocity field is the ratio of the component of vorticity in the direction of motion to the speed.

34. The Cauchy-Stokes decomposition. The analysis of foregoing sections has amply demonstrated that the vorticity vector $\mathbf{w}$ represents

---

[1] Hankel [1861, 1, §8] chose to consider only a circular circuit.

[2] [1863, 1, §4].

[3] [1869, 1, §§60(d), 60(e), 60(p)].

[4] [1869, 1, §§59–60].

the *local and instantaneous rate of rotation of the medium.* By combining this fact with the results of our previous analysis of the rate of deformation $\mathbf{\Delta}$(§21) and adding the observation that a uniform translation ($\dot{\mathbf{x}}$ = const.) contributes nothing to grad $\dot{\mathbf{x}}$, we may now fully interpret the basic identity (29.1), obtaining the ***Cauchy-Stokes decomposition theorem*** (1841, 1845):[1] *an arbitrary instantaneous state of continuous motion of a deformable medium is at each point the superposition of a uniform velocity of translation, a motion of extension, a shearing motion, and a rigid rotation.* Another form of the same result is: *an arbitrary instantaneous state of motion may be resolved at each point into a uniform translation, a dilatation along three mutually perpendicular axes, and a rigid rotation of these axes.* This theorem could justly be called the ***fundamental theorem of the kinematics of continua.***[2]

35. Sources and vortices. The result of the last section may be complemented by calculation of the Stokes potentials (16.11) for the velocity field $\dot{\mathbf{x}}$. From the definitions of the expansion and the vorticity and from the assumed solenoidal character of $\boldsymbol{\pi}$ we have at once

$$\nabla^2\phi = -\vartheta, \qquad \nabla^2\boldsymbol{\pi} = -\mathbf{w}. \tag{35.1}$$

Thus ϕ, π_x, π_y, and π_z may be regarded as Newtonian potentials; the "matter" giving rise to the scalar potential is the expansion field, while that giving rise to the vector potential is the vorticity field. The formulae (13.6) put into the present notation become

$$\begin{aligned} 4\pi\phi(\mathbf{x}, t) &= \int_{\mathfrak{v}} \frac{\vartheta(\boldsymbol{\xi}, t)\, dv(\boldsymbol{\xi})}{d(\mathbf{x}, \boldsymbol{\xi})} + h(\mathbf{x}, t), \\ 4\pi\boldsymbol{\pi}(\mathbf{x}, t) &= \int_{\mathfrak{v}} \frac{\mathbf{w}(\boldsymbol{\xi}, t)\, dv(\boldsymbol{\xi})}{d(\mathbf{x}, \boldsymbol{\xi})}, \end{aligned} \tag{35.2}$$

where $\boldsymbol{\xi}$ is a dummy vector variable of integration, $d(\mathbf{x}, \boldsymbol{\xi})$ is the distance from $\mathbf{x}$ to $\boldsymbol{\xi}$, and $h(\mathbf{x}, t)$ is a harmonic function. Putting these formulae into (16.11) yields

$$\begin{aligned} 4\pi\dot{\mathbf{x}}(\mathbf{x}, t) = -\text{grad}\left[\int_{\mathfrak{v}} \frac{\vartheta(\boldsymbol{\xi}, t)\, dv(\boldsymbol{\xi})}{d(\mathbf{x}, \boldsymbol{\xi})} + h(\mathbf{x}, t)\right] \\ + \text{curl} \int_{\mathfrak{v}} \frac{\mathbf{w}(\boldsymbol{\xi}, t)\, dv(\boldsymbol{\xi})}{d(\mathbf{x}, \boldsymbol{\xi})}. \end{aligned} \tag{35.3}$$

[1] Only a trivial modification is required to apply Cauchy's analysis of infinitesimal strain and rotation [1823, 1] [1827, 2] [1828, 1] [1841, 1] to the present subject. The result was explicitly stated by Stokes [1845, 1, §2]. While Euler had analysed $\mathbf{\Delta}$ completely, he did not give a kinematical interpretation for $\mathbf{w}$.

[2] Some other properties of grad $\dot{\mathbf{x}}$ are discussed by Appell [1903, 6].

The harmonic function h may be chosen in such a way that the normal projection $\dot{x}_n$ of $\dot{\mathbf{x}}$ takes on a prescribed value, assumed consistent with the total expansion, upon the boundary $\mathfrak{s}$ of the region $\mathfrak{v}$. For a region $\mathfrak{v}$ such that the solution of the Neumann problem for continuous boundary data exists, we have then the following theorem: *corresponding to a prescribed distribution of the vorticity* $\mathbf{w}$ *and expansion* ϑ *throughout a region there exists one and only one velocity field* $\dot{\mathbf{x}}$ *whose normal projection assumes prescribed continuous values upon the boundary.* The reader will easily formulate a corresponding theorem for infinite regions.

From (35.3) it is easy to construct velocity fields containing singularities of various sorts. If $\vartheta = 0$ except at one point, where it may be supposed to become infinite in a suitable fashion, while $\mathbf{w} = 0$ everywhere without exception, the motion is said to be that induced by a *source*. If $\mathbf{w} = 0$ except upon a certain curve, where it may be supposed to become infinite in a suitable fashion, while $\vartheta = 0$ everywhere without exception, there results a motion said to be induced by a *vortex*. We may interpret (35.3) as a statement that *any motion may be induced by a suitable distribution of sources and vortices.* To give in detail the modern theory of vortices, which was initiated by Helmholtz[1] and forms the subject matter of several treatises, would excessively swell the bulk of the present work, so that our attention must generally be confined to continuous motions.

The formula (35.3) is more important for its implications than for any direct use that is made of it, for it shows that in principle a state of motion is known at any given instant to within the gradient of a harmonic function if one scalar, the expansion, and one vector, the vorticity, be known. Thus the vorticity and the expansion afford a natural invariant description of the motion of a medium, a description particularly appealing because of their simple kinematical significances. For each of the various quantities associated with a motion there exists an expression showing the manner in which vorticity and expansion cooperate to produce it.[2]

---

[1] [1858, 1, §§5–6]. Special cases were considered by Newton [1687, 1, Bk. II, Sect. IX] and other early writers and were treated mathematically by Daniel Bernoulli [1738, 1, Sect. 11], d'Alembert [1744, 1, Bk. III, Ch. IV], Euler [1757, 2, §§XXX–XXXIII] [1770, 1, §§75–87], and Lagrange [1783, 1, §21] [1788, 1, Part II, Sect. 11, ¶19].

[2] In the theory of viscous fluids there are in addition to the general kinematical formulae of the type discussed here special and simpler relations connecting $\mathbf{w}$ and ϑ. These occur because Poisson's dynamical equation (17.1) may be written

$$\mathbf{f} - \rho\ddot{\mathbf{x}} = -\operatorname{grad}\,[-p + (\lambda + 2\mu)\vartheta] + \mu \operatorname{curl} \mathbf{w}:$$

the effective force of viscosity (provided $\lambda = \text{const.}$, $\mu = \text{const.}$*) is expressible in terms of the space derivatives of the expansion and the vorticity.* For such a relation see [1948, 10].

We exemplify the foregoing remarks by calculating the total squared speed. From (16.11) we have, assuming ϕ to be single-valued,

$$\begin{aligned}\dot{x}^2 &= \dot{\mathbf{x}}\cdot[-\text{grad}\,\phi + \text{curl}\,\boldsymbol{\pi}],\\ &= -\text{div}\,(\dot{\mathbf{x}}\phi) + \phi\vartheta + \text{div}\,(\dot{\mathbf{x}} \times \boldsymbol{\pi}) + \boldsymbol{\pi}\cdot\mathbf{w},\end{aligned} \tag{35.4}$$

and hence by (35.2) follows

$$\begin{aligned}\dot{x}^2 = \text{div}\,(\dot{\mathbf{x}} \times \boldsymbol{\pi} - \dot{\mathbf{x}}\phi) &+ \frac{\vartheta(\mathbf{x}, t)}{4\pi}\left[\int_{\mathfrak{v}} \frac{\vartheta(\boldsymbol{\xi}, t)\, dv(\boldsymbol{\xi})}{d(\mathbf{x}, \boldsymbol{\xi})} + h(\mathbf{x}, t)\right]\\ &+ \frac{\mathbf{w}(\mathbf{x}, t)}{4\pi}\cdot\int_{\mathfrak{v}} \frac{\mathbf{w}(\boldsymbol{\xi}, t)\, dv(\boldsymbol{\xi})}{d(\mathbf{x}, \boldsymbol{\xi})}.\end{aligned} \tag{35.5}$$

By integrating this result over a volume and employing Green's transformation we then obtain

$$\begin{aligned}\frac{1}{2}\int_{\mathfrak{v}} \dot{x}^2\, dv = \frac{1}{2}\oint_{\mathfrak{s}} d\mathbf{s}\cdot(\dot{\mathbf{x}} \times \boldsymbol{\pi} - \dot{\mathbf{x}}\phi) + \frac{1}{4\pi}\int_{\mathfrak{v}} \vartheta h\, dv\\ + \frac{1}{8\pi}\iint_{\mathfrak{v}\,\mathfrak{v}} \frac{\vartheta(\boldsymbol{\xi}, t)\vartheta(\mathbf{n}, t) + \mathbf{w}(\boldsymbol{\xi}, t)\cdot\mathbf{w}(\mathbf{n}, t)}{d(\mathbf{n}, \boldsymbol{\xi})}\, dv(\boldsymbol{\xi})\, dv(\mathbf{n}).\end{aligned} \tag{35.6}$$

Under various conditions the first two integrals on the right hand side may vanish, an example being the case when ϑ and $\mathbf{w}$ both vanish outside a finite region and $(\dot{\mathbf{x}} \times \boldsymbol{\pi} - \dot{\mathbf{x}}\phi)_n = \text{o}(r^{-2})$. In this case (35.6) reduces to the elegant formula of A. Föppl (1897):[3]

$$\frac{1}{2}\int_{\mathfrak{v}} \dot{x}^2\, dv = \frac{1}{8\pi}\iint_{\mathfrak{v}\,\mathfrak{v}} \frac{\vartheta(\boldsymbol{\xi}, t)\vartheta(\mathbf{n}, t) + \mathbf{w}(\boldsymbol{\xi}, t)\cdot\mathbf{w}(\mathbf{n}, t)}{d(\mathbf{n}, \boldsymbol{\xi})}\, dv(\boldsymbol{\xi})\, dv(\mathbf{n}). \tag{35.7}$$

36. Irrotational motions. A motion in which the vorticity vanishes,

$$\mathbf{w} = 0, \tag{36.1}$$

is called *irrotational*.[1] To a stationary observer a small object suspended in an irrotational motion would appear to move in such a fashion that the arms of a cross marked upon it remained at all times parallel to their original directions.

In the case of an irrotational motion we have grad $\dot{\mathbf{x}} = \boldsymbol{\Delta}$. Now in §22 we saw that the directions suffering no instantaneous rotation are the directions of the left proper vectors of grad $\dot{\mathbf{x}}$. In the present case

---

[3] [1897, 2, §§32–33].

---

[1] This name was introduced by Kelvin [1869, 1, §§59(e), 59(f), 60(r), 61(c)]. Earlier authors had referred to this class of motions by the more elaborate term *motions in which $\dot{x}dx + \dot{y}dy + \dot{z}dz$ is an exact differential.*

these are the directions of the proper vectors of the symmetric matrix $\mathbf{\Delta}$. Hence in an *irrotational motion there are at each point either three or an infinite number of directions suffering no instantaneous rotation, according as the three principal rates of extension are all distinct or not.*

In the terminology of §12, *a motion is irrotational if and only if its velocity field be lamellar*, and (36.1) is a necessary and sufficient condition that $-\dot{\mathbf{x}}\cdot d\mathbf{x} = -\dot{x}\, dx - \dot{y}\, dy - \dot{z}\, dz$ should be the exact differential of a *velocity-potential* ϕ:

$$\dot{\mathbf{x}} = -\text{grad}\,\phi. \tag{36.2}$$

The velocity-potential is a discovery of Euler (1752-1755).[2]

In simply-connected regions of irrotational motion the velocity-potential is single-valued; its application in multiply-connected regions, where it is generally a cyclic function, was initiated by Helmholtz (1858)[3] and Kelvin (1869).[4] In the former case, we have

$$-\int_{P_1}^{P_2} d\mathbf{x}\cdot\dot{\mathbf{x}} = \int_{P_1}^{P_2} d\phi = \phi(P_2) - \phi(P_1). \tag{36.3}$$

We may call the flow of $\dot{\mathbf{x}}$ along $\mathfrak{c}$ simply *the flow along* $\mathfrak{c}$. *Then the velocity-potential ϕ at a point is the negative of the flow along any curve connecting that point with some arbitrary fixed point at which ϕ is assigned the value zero.* Taking the points P_1 and P_2 as the same and the closed circuit of integration as lying wholly within a simply-connected region, we find the right hand side of (36.3) to be zero, and hence conclude ***Kelvin's kinematical theorem:***[5] *a motion is irrotational if and only if the circulation about every reducible circuit be zero.* In case the region in question be multiply-connected, let it be rendered simply-connected by the imposition of suitable barriers. Then from two reconcileable but not reducible circuits may be formed a single reducible circuit by adding one connecting path, to be traversed in opposite senses, upon each

---

[2] [1761, 1, §§46–48, 50, 55–65] [1757, 2, §§XXVI–XXVII] [1770, 1, §93] [1806, 1, §192]. While d'Alembert [1752, 1, §§59, 86] based his theory of plane and rotationally-symmetric fluid motions on the contention that $\dot{x}dx + \dot{y}dy + \dot{z}dz$ is exact, he did not make direct use of the velocity-potential, whose existence follows from this condition. The term *Geschwindigkeitspotential* was introduced by Helmholtz [1858, 1, Introd.]. The symbol ϕ, though with the opposite sign, was introduced by Earnshaw [1837, 1]. In §14 of this paper Earnshaw considered also the possibility that the quantities $\rho\dot{x}$, $\rho\dot{y}$, $\rho\dot{z}$, $a^2\rho$, where a is a constant speed of wave propagation, be the components of a lamellar 4-vector, somewhat suggestive of the kinematics of special relativity.

[3] [1858, 1, §§5–6].

[4] [1869, 1, §§60(s)–64]. For an elaborate analytical treatment see [1929, 1, Ch. 3, §4].

[5] [1869, 1, §§59(e), 59(f)]. While Hankel did not state this result, it follows immediately from his analysis [1861, 1, §8].

barrier. By applying Kelvin's kinematical theorem to the resulting single reducible circuit and noting that the net flow along the twice traversed paths on the barriers is zero we easily conclude that *a motion is irrotational if and only if the circulation about any two reconcileable circuits be equal.*

Since $(\text{grad}\ \dot{\mathbf{x}})_\times = \text{curl}\ \dot{\mathbf{x}}$, it follows that grad $\dot{\mathbf{x}}$ is symmetric if and only if the motion be irrotational. By applying the algebraic theorems of Cauchy and Kelvin & Tait stated in §2 we then conclude that *a necessary and sufficient condition for a motion to be instantaneously irrotational at a point is that at that point there exist three mutually orthogonal directions suffering no instantaneous rotation.* This result follows also from Cauchy's conics of rotations (§31): *a necessary and sufficient condition that a motion in which* grad $\dot{\mathbf{x}} \neq 0$ *be instantaneously irrotational at a point is that the conics of rotations in three distinct planes through the point be equilateral hyperbolae.* The proof lies in the fact that the asymptotes of a hyperbolic conic of rotations are suffering no rotation, so that if these be orthogonal the mean rate of rotation of directions in the plane of the conic is zero, and hence by (31.7) the component of vorticity normal to that plane is zero.

By (18.3) and (36.2) follows

$$\dot{\phi} = \frac{\partial \phi}{\partial t} + \dot{\mathbf{x}} \cdot \text{grad}\ \phi = \frac{\partial \phi}{\partial t} - \dot{x}^2, \tag{36.4}$$

whence

$$\frac{\partial \phi}{\partial t} - \dot{\phi} = \dot{x}^2 \geqq 0, \qquad \text{or} \qquad -\dot{\mathbf{x}} \cdot \text{grad}\ \phi = -\frac{d\phi}{d\dot{x}} = \dot{x}^2: \tag{36.5}$$

the speed of an irrotational motion is the square root of the difference between the spatial and the material time derivative of the velocity-potential, and consequently the latter can never exceed the former. In particular, *in an irrotational motion the velocity-potential can never increase in the direction of motion along a stream-line,* and *in a steady irrotational motion a particle always moves toward a region of lower velocity-potential. A stagnation point is a stationary point for the velocity-potential along the stream-line on which it is situate.*[6] Thus a closed stream-line lying wholly in a simply-connected region can never exist. In these circumstances also it follows from the continuity of the velocity field that the stream-lines in a simply-connected region cannot return arbitrarily near to themselves as they do generally in a rotational motion.[7]

d'Alembert[8] contended in effect that all motions of inviscid incom-

[6] This theorem is sometimes mistakenly presented as being dynamical.
[7] [1903, 2, ¶67].
[8] [1752, 1, §§46–48].

pressible fluids are irrotational. Euler at first shared this opinion,[9] but soon realized and demonstrated by an example that it was quite false, and both he and Lagrange repeatedly emphasized that irrotational motions constitute only a special case.[10] d'Alembert later admitted rotational motions to be possible,[11] but he still contended that they were unusual. The major part of classical hydrodynamics is the theory of motions which are both irrotational and isochoric. For such motions, by putting (36.2) into (24.7) we have at once

$$\nabla^2\phi = 0, \tag{36.6}$$

a partial differential equation which is universally and erroneously called *Laplace's equation*, although it was given in the present connection by Euler in his first paper on fluid dynamics (1752-1755).[12] Thus the classical theory of irrotational motions of inviscid incompressible fluids is one branch of potential theory. From this analysis we may conclude the measure of indeterminacy represented by the harmonic function h in (35.3): *two motions having the same distributions of vorticity and expansion differ at most by an isochoric irrotational motion.*

As is apparent from the foregoing derivation, classical hydrodynamics can be put in wholly kinematical terms. It is highly simplified in comparison to any truly dynamical theory of deformable bodies because the relation (36.2) expresses the velocity in terms of linear operations upon a single scalar function satisfying the simple linear differential equation (36.6), permitting the analyst to outflank the non-linear convective acceleration in Euler's formula (16.7). In support of irrotational motions stand two powerful theorems of classical hydrodynamics, both enunciated by Lagrange and both perfected in statement and in proof by Cauchy: first,[13] *a rotational flow of an inviscid incompressible fluid cannot be started by impulsive pressures*, and, second (§104), *in a barotropic flow of an inviscid fluid subject to conservative extraneous force a portion of fluid once in irrotational flow must remain ever in irrotational flow.* On the other hand, we shall demonstrate presently that irrotational

---

[9] [1761, 1, §§44–48, 55–65].

[10] [1757, 2, §§XXX–XXXIII] [1757, 3, §LXXX] [1762, 3, §XLIII] [1783, 1, §21] [1788, 1, Part II, sect. 11, ¶19] [1806, 1, §§180–183, 194]. *Cf.* §48[3].

[11] [1761, 2, §§I, XI, XIII]. While denying the importance of rotational motions, d'Alembert characteristically made a great discovery about them (§100), whose significance he apparently did not recognize, so that it lay unnoticed until rediscovered a century later by Helmholtz, to whom it is usually attributed.

[12] [1761, 1, §67] [1770, 1, §93] [1806, 1, §195]. The unacknowledged borrowings of Lagrange and Laplace from Euler are discussed by Hoppe [1926, 1, §9]. *Cf.* §14[2]. In [1761, 1, §§68–74] Euler obtained the homogeneous polynomial harmonics of degrees 0, 1, . . . , 5 and discussed the fluid motions which they yield.

[13] [1783, 1, §20] [1788, 1, 2me Partie, Sect. XI, ¶18] [1827, 1, 1re Partie, Sect. 1].

motion is usually impossible nevertheless. From a dynamical point of view such a conclusion is certainly to be expected. The basic idea of Newtonian mechanics is that applied forces effect changes of momentum, or, equivalently, that the dynamics of a body are specified when its acceleration is specified. A theory of motion in which it is possible to avoid use of the acceleration altogether is quite extraordinary from the Newtonian point of view. When moreover this theory allows motions to be calculated by superposition we must, even if regretfully, confess its inadequacy to represent more than a highly restricted and indeed dynamically degenerate class of phenomena. There are, of course, circumstances in which the vorticity though not zero is negligible; a means of determining these circumstances we shall discuss in Chapter V. In any case our major concern is the vorticity itself. The present work is conceived in the belief that the convective acceleration generates those beautiful, intricate, and perplexing phenomena which make the challenge of the theory of the motion of fluids, whether perfect, viscous, of more complicated in their dynamical response—a challenge for the most part declined by classical hydrodynamics—and that analysis of the basic kinematical properties of vorticity initiates a frontal attack upon the citadel of the non-linear convective acceleration.

37. The impossibility of irrotational motions in general. In two important special cases the kinematical boundary condition (19.8) is incompatible with irrotational motion. Both these cases are included in the following theorem of Kelvin (1849) and Helmholtz (1858):[1] *Consider an irrotational motion in a simply-connected region such that the motion, if any, of any finite boundary be tangential to itself; suppose further that the motion be either isochoric or steady and that in any portion of the region extending to infinity*

$$j\phi\dot{\mathbf{x}}_n = \bar{o}(r^{-2}); \qquad (37.1)$$

then the motion is a state of rest.

Proof. Since the domain is simply-connected, any irrotational motion within it must be possessed of a single-valued velocity-potential ϕ. For an *isochoric* motion we may put $\emptyset = \phi$ in (25.5), thus obtaining

$$\int_{\mathcal{D}} \dot{x}^2\, dv = -\oint_{\mathcal{S}} d\mathbf{s}\cdot\dot{\mathbf{x}}\phi = 0. \qquad (37.2)$$

Since $j = 1$, the theorem follows. For a *steady* motion we may put $\emptyset = \phi$ in (25.6), obtaining

[1] [1849, 1] [1858, 1, §1].

$$\int_v j\dot{x}^2\, dv = -\oint_s d\mathbf{s}\cdot\dot{\mathbf{x}}j\phi = 0. \tag{37.3}$$

Since $j > 0$, the theorem again follows. Q.E.D.

We note the following corollaries. First, for a motion in the interior of a finite region the condition (37.1) becomes superfluous, and the theorem holds unconditionally: *there is no isochoric or steady irrotational motion, other than a state of rest, within a finite simply-connected region with steady boundary.* Second, if the region of motion be such that $\dot{x}^i \to 0$ uniformly at ∞, then for the isochoric case the three functions $\dot{x}^i$ are single-valued harmonic functions, analytic at ∞. This fact together with the condition that $\oint d\mathbf{s}\cdot\dot{\mathbf{x}} = 0$ for a sufficiently large sphere yields $\dot{x}^i = \mathrm{O}(r^{-2})$, whence follows $\phi = \mathrm{O}(r^{-1})$. Thus $\phi\dot{\mathbf{x}}_n = \mathrm{O}(r^{-3}) = \mathrm{o}(r^{-2})$, so (37.1) is satisfied. That is: *in an isochoric irrotational motion in an infinite region such that all bounding surfaces are finite and in motion, if at all, only tangentially to themselves, if the velocity field vanish uniformly at infinity and if no fluid be supplied at infinity then the motion is a state of rest.*

That the foregoing result cannot be extended to multiply-connected regions may be seen from the counter-example of the plane vortex given in cylindrical co-ordinates r, θ, z by

$$\dot{r} = 0, \qquad \dot{\theta} = \frac{K}{r^2}, \qquad \dot{z} = 0, \tag{37.4}$$

which is irrotational in the doubly-connected domain $0 < r_1 \leqq r \leqq r_2$, $|z| \leqq z_0$. The reason the above proof fails to hold for multiply-connected regions lies in the many-valuedness of ϕ, so that the substitution $\emptyset = \phi$ in (25.5) or (25.6) is no longer permissible. That when the motion is not isochoric the restriction to steady motion is essential may be seen from the counter-example of the oscillating motion

$$\begin{aligned} j &= l[1 + q\sin kt \sin x], \qquad 0 < q < 1,\\ \dot{x} &= Jlqk \cos kt \cos x, \qquad \dot{y} = \dot{z} = 0, \end{aligned} \tag{37.5}$$

which is irrotational in the finite simply-connected domain $|x| \leqq \frac{1}{2}\pi$, $|y| \leqq a$, $|z| \leqq b$, upon whose boundaries it satisfies (19.8).

There are, however, objections to be raised against irrotational motion which are more serious than those indicated by the foregoing theorem of Helmholtz and Kelvin. Physical fluids either *adhere* to solid boundaries or very nearly do so, and accordingly in the theory of viscous fluids it is customary to impose the stronger boundary condition

$$\dot{\mathbf{x}} = \dot{\mathbf{x}}^*, \tag{37.6}$$

where $\dot{\mathbf{x}}^*$ is the velocity of the bounding surface. Apart from a few exceptional cases, this boundary condition, which we shall call the *adherence condition*, precludes irrotational motion, as we shall demonstrate in the following paragraphs.

Indeed, it was asserted by Duhem[2] that the adherence condition and irrotational motion are usually incompatible, but his proof is not convincing, and his statement cannot be true in general, since the oscillating motion given in spherical co-ordinates r, θ, ϕ by

$$j = \frac{l}{r^2}[1 + q \sin kt \sin r], \qquad 0 < q < 1,$$

$$\dot{r} = \frac{Jlqk}{r^2} \cos kt \cos r, \qquad \dot{\theta} = 0, \qquad \dot{\phi} = 0, \tag{37.7}$$

is a motion which is continuous and irrotational within the spherical shell $\frac{1}{2}\pi \leqq r \leqq \frac{3}{2}\pi$, to whose stationary boundaries the material adheres. It would be valuable to characterize in full generality all irrotational motions in which there exists a finite or infinite rigid surface to which the material adheres. Following a recent analysis of Supino (1949),[3] we shall perforce rest content with a treatment of the special case of *isochoric* motion, so that the theory of harmonic functions is at our disposal.

First, *the only isochoric irrotational motion in which the material adheres to a finite stationary surface, however small, is a state of rest.*

Proof. The condition (37.6) becomes $\dot{\mathbf{x}}^* = 0$, and thus the velocity-potential ϕ is a harmonic function all of whose first derivatives with respect to the space variables vanish upon the finite surface. Upon that finite surface, consequently, the function itself is constant and its normal derivative vanishes. It was indicated by Kirchhoff[4] that the only such harmonic function is a constant. While the proof of Kirchhoff is not rigorous, the result is true, being in fact but an interpretation of a well-known theorem of potential theory. As an immediate corollary it follows that the only isochoric irrotational motion in which the material adheres to a finite surface in rigid translation is itself a rigid translation. The results just stated and proved refer to the state of motion at any one instant.

Let us now consider the case when the material adheres to surfaces suffering rigid rotation at angular velocity $\boldsymbol{\omega}(t)$. If we refer an irrotational motion whose velocity-potential is ϕ to a co-ordinate frame which is at rest with respect to these surfaces, and whose origin coincides with

[2] [1903, **3**, 5^e partie, Ch. II, §1].

[3] [1949, **5**].

[4] [1876, **1**, Vorl. 16, §6].

the origin of a system with respect to which the motion is irrotational, by (16.9) we obtain

$$\dot{\mathbf{x}}' = -\boldsymbol{\omega} \times \mathbf{r} - \operatorname{grad} \phi, \tag{37.8}$$

or equivalently

$$\dot{x}' = 2\omega_z y - \frac{\partial \chi}{\partial x}, \qquad \dot{y}' = 2\omega_x z - \frac{\partial \chi}{\partial y}, \qquad \dot{z}' = 2\omega_y x - \frac{\partial \chi}{\partial z}, \tag{37.9}$$

where

$$\chi \equiv \phi + x\omega_y z + y\omega_z x + z\omega_x y. \tag{37.10}$$

Evidently χ also is a harmonic function. At any given instant we may orient the axes in such a way that $\omega_x = \omega_y = 0$, so that (37.9) reduces to

$$\dot{x}' = 2\omega y - \frac{\partial \chi}{\partial x}, \qquad \dot{y}' = -\frac{\partial \chi}{\partial y}, \qquad \dot{z}' = -\frac{\partial \chi}{\partial z}. \tag{37.11}$$

Since the adherence condition (37.6) now assumes the form $\dot{\mathbf{x}}' = 0$, the harmonic function χ must satisfy the boundary conditions

$$\frac{\partial \chi}{\partial x} = 2\omega y, \qquad \frac{\partial \chi}{\partial y} = 0, \qquad \frac{\partial \chi}{\partial z} = 0. \tag{37.12}$$

We suppose $\omega \neq 0$ so as to exclude the case treated in the previous paragraph.

Consider first a motion within finite rigid boundary surfaces, on each point of which (37.12) applies. Then $\dot{y}'$ and $\dot{z}'$ are harmonic functions vanishing upon a closed bounding surface, and hence vanishing identically. Thus $\partial\chi/\partial y = 0$, $\partial\chi/\partial z = 0$; since $\nabla^2\chi = 0$ it follows that $\chi = ax + b$. Comparing this result with $(37.12)_1$ yields $2\omega y = a$, but since $\omega \neq 0$ this condition can be satisfied only upon a single plane $y = \text{const.}$, a type of boundary not included in the hypothesis. Hence no such motion exists. Second, consider a motion exterior to finite rigid surfaces, on each point of which (37.12) applies. Let us call *regular* a velocity field which, relative to the surfaces in question, approaches a definite limit at ∞ and possesses a gradient which is $O(r^{-2})$. Then by the uniqueness of solution to the exterior Dirichlet problem[5] the functions $\dot{y}'$ and $\dot{z}'$ must vanish identically, and again there is no such motion. These results constitute the following ***first theorem of Supino***: *no continuous isochoric irrotational motion of a material adhering to boundary surfaces and completely filling a finite domain whose boundaries form a rigid system can exist. No such motion filling an infinite domain exterior*

[5] *E.g.* [1929, 1, Kap. 3, §6].

to a rigid system of surfaces to which the material adheres and possessed of a regular velocity field can exist.

In the statement of the above theorem the term "rigid system" denotes a set of rigid surfaces rigidly attached to one another, so that all are endowed with the same angular velocity. Various irrotational motions of materials adhering to rigid bounding surfaces[6] can indeed exist if these be in relative motion, or if the motion relative to them be unbounded at ∞. A simple example is the plane irrotational vortex already cited, for the stream-lines are the rigid concentric circles $r = r_0$, $z = z_0$ rotating at angular speeds $\omega = Kr_0^{-2}$. Another example is furnished by the rectilinear motion $\dot{x}' = 2\omega y$, $\dot{y}' = 0$, $\dot{z}' = 0$, or, equivalently, $\dot{x} = \omega y$, $\dot{y} = \omega x$, where the material adheres to both the planes $x = 0$ and $y = 0$.

In general, however, as Supino points out, the conditions $(37.12)_2$ and $(37.13)_3$ are themselves sufficient to determine χ, and hence are not likely to be compatible with $(37.12)_1$ except in degenerate cases. Another way of saying this same thing is to say that if there exist a harmonic function χ satisfying (37.12) on any finite surface, however small, then by the theorem of Kirchhoff mentioned above it is unique in all space to within an additive constant. Expressed in kinematical terms, this result becomes the following ***second theorem of Supino***: *suppose there exist an isochoric irrotational motion of a material adhering to a certain rigid boundary surface rotating at a certain angular velocity. Then there is no other such motion adhering to any finite portion of that surface, however small, rotating at the same angular velocity.* Applying this theorem to the second example noted in the preceding paragraph shows that the only such motion of a material adhering to a finite plane area rotating about an axis in its plane is the motion induced by the rigid rotation of the entire plane. Similarly, from the first example we conclude that the only such motion of a material adhering to a finite portion of a circular cylinder rotating about its axis is the irrotational vortex motion induced by the rotation of the entire cylinder.[7]

The foregoing analysis shows that the boundary condition customarily employed in the theory of viscous fluids makes irrotational motion a virtual impossibility. We shall now revert to the analysis of rotational motions.

38. Lagrange's acceleration formula. From the d'Alembert-Euler acceleration formula (16.7) we have

$$\ddot{\mathbf{x}} = \frac{\partial \dot{\mathbf{x}}}{\partial t} + \dot{\mathbf{x}} \cdot \operatorname{grad} \dot{\mathbf{x}} - \operatorname{grad} \dot{\mathbf{x}} \cdot \dot{\mathbf{x}} + \operatorname{grad} \dot{\mathbf{x}} \cdot \dot{\mathbf{x}}, \tag{38.1}$$

---

[6] For a viscous incompressible fluid such motions are studied by Hamel [1941, **2**].

[7] Supino shows further that these two motions are the only possible plane isochoric irrotational motions in which the material adheres to any finite rigid surface.

whence by the identity (3.10) follows ***Lagrange's acceleration formula*** (1781),[1]

$$\ddot{\mathbf{x}} = \frac{\partial \dot{\mathbf{x}}}{\partial t} + \mathbf{w} \times \dot{\mathbf{x}} + \operatorname{grad} \tfrac{1}{2} \dot{x}^2, \tag{38.2}$$

a result of central importance, as will appear from the long sequence of its consequences throughout the rest of this work. While the d'Alembert-Euler formula (16.7) gives the acceleration in terms of two vectors and one dyadic, Lagrange's formula expresses it by means of four vectors: the local acceleration, the velocity, the vorticity, and the gradient of the squared speed. While Euler's acceleration formula is valid in a space of any number of dimensions, for the existence of Lagrange's formula it is requisite that the number of dimensions be three, so that this formula becomes a vehicle especially suited to the presentation of the peculiar features of three-dimensional kinematics.

Lagrange's formula (38.2) separates the convective acceleration $\dot{\mathbf{x}} \cdot \operatorname{grad} \dot{\mathbf{x}}$ into two portions. The second, $\operatorname{grad} \frac{1}{2}\dot{x}^2$, is always lamellar and has relatively slight qualitative effect. The first, the vector $\mathbf{w} \times \dot{\mathbf{x}}$, is of great importance in our subject. As we shall see, its being zero, lamellar, complex-lamellar, or none of these is a determining factor in the nature of the motion. For reasons to appear subsequently, we shall call it the ***Lamb vector.***

In the special case of an irrotational motion, or more generally when $\mathbf{w} \times \dot{\mathbf{x}} = 0$, (38.2) reduces to

$$\ddot{\mathbf{x}} = \frac{\partial \dot{\mathbf{x}}}{\partial t} + \operatorname{grad} \tfrac{1}{2} \dot{x}^2, \tag{38.3}$$

so that the convective acceleration becomes a lamellar field whose normal surfaces are the surfaces of constant speed.

39. The acceleration relative to a rotating frame. Lagrange's acceleration formula maintains its same form when the co-ordinate frame in which the velocity is measured is suffering steady rotation and translation with respect to that in which $\ddot{\mathbf{x}}$ is measured. Putting $d^2\mathbf{c}/dt^2 = 0$, $d\boldsymbol{\omega}/dt = 0$ in (16.10) and comparing the result with (38.2) we obtain

$$\ddot{\mathbf{x}} = \boldsymbol{\omega} \times (\boldsymbol{\omega} \times \mathbf{r}') + 2\boldsymbol{\omega} \times \dot{\mathbf{x}}' + \frac{\partial \dot{\mathbf{x}}'}{\partial t} + \mathbf{w}' \times \dot{\mathbf{x}}' + \operatorname{grad}' \tfrac{1}{2} \dot{x}'^2, \tag{39.1}$$

where primes denote quantities as apparent to an observer in the rotating system. But since

$$\boldsymbol{\omega} \times (\boldsymbol{\omega} \times \mathbf{r}') = \operatorname{grad}' [\tfrac{1}{2}(\boldsymbol{\omega} \cdot \mathbf{r}')^2 - \tfrac{1}{2}\omega^2 r'^2], \tag{39.2}$$

[1] [1783, **1**, §14] [1788, **1**, Part II, ¶12]. The special case for plane motion was implied by d'Alembert [1761, **2**, §XI], the general case by Euler [1757, **3**, §LIV].

we may put (39.1) into the form[1]

$$\ddot{\mathbf{x}} = \frac{\partial \dot{\mathbf{x}}'}{\partial t} + (\mathbf{w}' + 2\boldsymbol{\omega}) \times \dot{\mathbf{x}}' + \text{grad}\ \tfrac{1}{2}\,[\dot{x}'^2 + (\boldsymbol{\omega}\cdot\mathbf{r}')^2 - \omega^2 r'^2]. \tag{39.3}$$

That is, *the acceleration as apparent to an observer in a given frame of reference may be calculated from the Lagrange formula by an observer in a second frame suffering steady translation and rotation with respect to the first, provided the second observer simply increase the apparent local angular velocity* $\frac{1}{2}\mathbf{w}$ *by the angular velocity* $\boldsymbol{\omega}$ *of his co-ordinate frame and increase the apparent squared speed by the quantity* $(\boldsymbol{\omega}\cdot\mathbf{r}')^2 - \omega^2 r'^2$. This result is frequently applied in the study of atmospheric motions. The vector $\mathbf{w}' + 2\boldsymbol{\omega}$, where $\mathbf{w}'$ is the vorticity apparent to a terrestrial observer and $\boldsymbol{\omega}$ is the earth's angular velocity with respect to an approximately inertial frame, is then called the *absolute vorticity,* and the increase in the effective speed, being of the second order in ω, is usually neglected.

40. Duhem's acceleration formula. The acceleration may also be expressed in terms of the Monge potentials of the velocity. By differentiating (16.12) we obtain

$$\begin{aligned} \frac{\partial \dot{\mathbf{x}}}{\partial t} &= \text{grad}\,\frac{\partial h}{\partial t} + \frac{\partial f}{\partial t}\,\text{grad}\ g + f\ \text{grad}\,\frac{\partial g}{\partial t}, \\ &= \text{grad}\left[\frac{\partial h}{\partial t} + f\,\frac{\partial g}{\partial t}\right] - \frac{\partial g}{\partial t}\,\text{grad}\ f + \frac{\partial f}{\partial t}\,\text{grad}\ g. \end{aligned} \tag{40.1}$$

But[1]

$$\mathbf{w} = \text{grad}\ f \times \text{grad}\ g. \tag{40.2}$$

Hence by (18.3) follows

$$\frac{\partial \dot{\mathbf{x}}}{\partial t} + \mathbf{w} \times \dot{\mathbf{x}} = \text{grad}\left[\frac{\partial h}{\partial t} + f\,\frac{\partial g}{\partial t}\right] + \dot{f}\ \text{grad}\ g - \dot{g}\ \text{grad}\ f, \tag{40.3}$$

so that by Lagrange's acceleration formula (38.2) we obtain

$$\ddot{\mathbf{x}} = \text{grad}\left[\tfrac{1}{2}\,\dot{x}^2 + \frac{\partial h}{\partial t} + f\,\frac{\partial g}{\partial t}\right] + \dot{f}\ \text{grad}\ g - \dot{g}\ \text{grad}\ f. \tag{40.4}$$

Analysis equivalent to the foregoing was given by Duhem (1901) and Lamb (1906).[2]

---

[1] An equivalent formula is given by Guglielmi [1922, 1, Appendice].

---

[1] [1889, 2].

[2] [1901, 2, §2] [1906, 1, §166] [1932, 1, §167].

41. Expressions for div $\ddot{\mathbf{x}}$. For frequent later use we shall require expressions for the divergence of the acceleration field. These we shall now systematically calculate.

From the d'Alembert-Euler acceleration formula (16.7) we have

$$\begin{aligned} \operatorname{div} \ddot{\mathbf{x}} &= \frac{\partial \vartheta}{\partial t} + \operatorname{div} (\dot{\mathbf{x}} \cdot \operatorname{grad} \dot{\mathbf{x}}), \\ &= \frac{\partial \vartheta}{\partial t} + (\dot{x}^j \dot{x}^i{}_{,j})_{,i}, \\ &= \frac{\partial \vartheta}{\partial t} + \dot{x}^j \vartheta_{,j} + \dot{x}^j{}_{,i} \dot{x}^i{}_{,j}. \end{aligned} \tag{41.1}$$

But by (29.2) follows

$$\begin{aligned} \dot{x}^i{}_{,j} \dot{x}^j{}_{,i} &= \left(\Delta^{ij} - \tfrac{1}{2} \frac{\epsilon^{ijl}}{\sqrt{g}} w_l\right)\left(\Delta_{ji} - \tfrac{1}{2} \sqrt{g}\, \epsilon_{jik} w^k\right), \\ &= \Delta_{ij}\Delta^{ij} - \tfrac{1}{4}\epsilon_{ijk}\epsilon^{ijl} w^k w_l, \\ &= \Delta_{ij}\Delta^{ij} - \tfrac{1}{2} w^i w_i. \end{aligned} \tag{41.2}$$

By putting (41.2) into (41.1) and using (18.2) we obtain

$$\operatorname{div} \ddot{\mathbf{x}} = \dot{\vartheta} + \Delta_{ij}\Delta^{ij} - \tfrac{1}{2} w^2, \tag{41.3}$$

whence by (21.10) we have

$$\operatorname{div} \ddot{\mathbf{x}} = \dot{\vartheta} + \vartheta^2 - 2II - \tfrac{1}{2} w^2. \tag{41.4}$$

Since $\Delta_{ij}\Delta^{ij} \geqq 0$, it follows from (41.3) that for an isochoric motion

$$\operatorname{div} \ddot{\mathbf{x}} \geqq -\tfrac{1}{2} w^2, \tag{41.5}$$

when the sign of equality can hold only at a point where the motion is rigid. In isochoric irrotational motion, then,

$$\operatorname{div} \ddot{\mathbf{x}} \geqq 0. \tag{41.6}$$

Now by (29.2) follows also

$$\begin{aligned} K \equiv \tfrac{1}{2}\, \delta^{lm}_{jk} \dot{x}^j{}_{,l} \dot{x}^k{}_{,m} &= \tfrac{1}{2}\, \delta^{lm}_{jk}\left(\Delta^j_l - \frac{1}{2} \frac{g^{jq}}{\sqrt{g}} \epsilon_{qlp} w^p\right)\left(\Delta^k_m - \frac{1}{2} \frac{g^{kr}}{\sqrt{g}} \epsilon_{rms} w^s\right), \\ &= \tfrac{1}{2}\, \delta^{lm}_{jk} \Delta^j_l \Delta^k_m + \tfrac{1}{8}\, \delta^{lm}_{jk} \frac{g^{jq} g^{kr}}{g} \epsilon_{qlp}\epsilon_{rms} w^p w^s, \\ &= II + \tfrac{1}{4} w^2, \end{aligned} \tag{41.7}$$

an identity noted by Hamel[1] in the equivalent form

$$2\Delta^{ij}\Delta_{ij} = w^2 + 2\vartheta^2 - 4K. \tag{41.8}$$

Hence from (41.4) follows

$$\operatorname{div} \ddot{\mathbf{x}} = \dot{\vartheta} + \vartheta^2 - 2K. \tag{41.9}$$

From Lagrange's acceleration formula (38.2) we have

$$\begin{aligned} \operatorname{div} \ddot{\mathbf{x}} &= \frac{\partial \vartheta}{\partial t} + \nabla^2 \tfrac{1}{2} \dot{x}^2 + \operatorname{div} (\mathbf{w} \times \dot{\mathbf{x}}), \\ &= \frac{\partial \vartheta}{\partial t} + \nabla^2 \tfrac{1}{2} \dot{x}^2 + \dot{\mathbf{x}} \cdot \operatorname{curl} \mathbf{w} - w^2, \\ &= \frac{\partial \vartheta}{\partial t} + \nabla^2 \tfrac{1}{2} \dot{x}^2 - \dot{\mathbf{x}} \cdot \operatorname{grad} \vartheta - \dot{\mathbf{x}} \cdot \nabla^2 \dot{\mathbf{x}} - w^2, \\ &= \dot{\vartheta} + \nabla^2 \tfrac{1}{2} \dot{x}^2 - \dot{\mathbf{x}} \cdot \nabla^2 \dot{\mathbf{x}} - w^2. \end{aligned} \tag{41.10}$$

If $\partial\vartheta/\partial t = 0$ and $w = 0$, $(41.10)_1$ combined with (41.6) yields

$$\nabla^2 \dot{x}^2 \geqq 0. \tag{41.11}$$

Hence the squared speed is a subharmonic function, and thus *in a region of isochoric irrotational motion the greatest speed cannot occur at an interior point.*

The properties of grad $\ddot{\mathbf{x}}$ have been studied by Appell[2] and Carstoiu.[3]

[1] [1936, 2, §1].
[2] [1903, 1].
[3] [1946, 2].

Chapter IV. THE VORTICITY FIELD

42. Vorticity and circulation. Kelvin's transformation (33.1), *viz.*

$$\oint_{\mathfrak{c}} d\mathbf{x}\cdot\dot{\mathbf{x}} = \int_{\mathfrak{s}} d\mathbf{s}\cdot\mathbf{w}, \tag{33.1}$$

has already been discussed from a local point of view in §33. Its direct significance as an integral theorem is even more important:[1] *let* $\mathfrak{s}$ *be any surface lying upon the boundary of a closed region in which the velocity* $\dot{\mathbf{x}}$ *is continuous and the vorticity* $\mathbf{w}$ *is piecewise continuous; then the flux of vorticity through* $\mathfrak{s}$ *equals the circulation around the boundary circuit* $\mathfrak{c}$. Thus the net rotation of the finite circuit $\mathfrak{c}$ equals the sum of the strengths of all the vortices it embraces. It is important that the vorticity need not be defined on both sides of the surfaces $\mathfrak{s}$: (33.1) can be applied not only to interior surfaces, but also to bounding surfaces.

As a first application of Kelvin's transformation we shall now determine the nature of the vorticity on a rigid surface of *adherence* (§37), where the boundary condition satisfied by the velocity field is (37.6). Let us refer the motion to a co-ordinate frame at rest with respect to the rigid surface, so that the boundary condition becomes $\dot{\mathbf{x}}' = 0$. By applying (33.1) to any closed circuit lying wholly upon the surface we obtain

$$\int_{\mathfrak{s}} d\mathbf{s}\cdot\mathbf{w}' = 0 \tag{42.1}$$

for any area $\mathfrak{s}$ upon the surface, and hence since $\mathbf{w}$ is piecewise continuous and since the normal to $\mathfrak{s}$ is continuous almost everywhere it follows that

$$d\mathbf{s}\cdot\mathbf{w}' = 0 \tag{42.2}$$

almost everywhere; by (32.3) this result may be put into the form

$$d\mathbf{s}\cdot(\mathbf{w} - 2\boldsymbol{\omega}) = 0: \tag{42.3}$$

upon a rigid surface to which the material adheres, the vector difference between the local angular velocity $\frac{1}{2}\mathbf{w}$ *of the motion and the angular velocity*

[1] [1869, 1, §§59–60]. *Cf.* also [1871, 1, §12].

$\boldsymbol{\omega}$ *of the surface must be zero or tangent to the surface.* In particular, for a stationary surface (42.3) becomes[2]

$$d\mathbf{s}\cdot\mathbf{w} = 0: \tag{42.4}$$

the particles rotate, if at all, about axes tangential to the wall.

43. Vortex-lines and vortex-tubes. The vector-lines of the vorticity field are called *vortex-lines*, its vector-surfaces are called *vortex-surfaces*, and its vector-tubes are called *vortex-tubes*. These concepts, which were introduced by Helmholtz (1858),[1] have proved to be of central importance, and they shall be fully analyzed in this work.

Since

$$\text{div}\,\mathbf{w} = 0, \tag{43.1}$$

the vorticity field is solenoidal. Herein lies one reason that the vorticity field is often simpler than the velocity field from which it is derived. From the Helmholtz characterization of solenoidal fields (§10) we now conclude ***Helmholtz's first vorticity theorem*** (1858): *the strength of a vortex-tube is the same at all cross-sections.*[2] By applying (33.1) we may at once put Helmholtz's theorem into a form derived by Kelvin: *the circulations taken in the same sense about any two reconcileable circuits lying upon a given vortex-tube are equal.* In the foregoing proof we have assumed, as mentioned at the outset of this work, that the velocity field is twice continuously differentiable, and in particular that the vorticity field is once continuously differentiable. Stated thus, however, the result holds subject to rather weaker hypotheses,[3] as we shall now discover by following the proof given by Kelvin.[4]

First, *the circulation around any circuit lying wholly upon a vortex-surface and reducible upon it is zero,*[5] for we may choose for the surface in Kelvin's transformation (33.1) the inclosed portion of the vortex-surface, upon which $\mathbf{w}$ is normal to $d\mathbf{s}$, so that

$$0 = \oint_c d\mathbf{x}\cdot\dot{\mathbf{x}}. \tag{43.2}$$

---

[2] This result is well known, but I have not been able to trace its origin. It is proved in [1929, **1**, Ch. 5, §16].

---

[1] [1858, **1**, §2]. Kelvin [1869, **1**, §§60(i), 60(m)] used the terms *axial lines* and *axial sheets* for vortex-lines and vortex-surfaces, respectively. For vortex-surfaces in general, see [1893, **1**, §12].

[2] [1858, **1**, §2]. Helmholtz erroneously concluded that vortex-lines either form closed curves or else end upon a boundary; see §10[2] above. Closed vortex-lines, while uncommon, are of great interest in the theory of vortices.

[3] We owe this observation to Lamb [1895, **1**, §145].

[4] [1869, **1**, §60(l)]. An alternative proof, based on direct analysis of the discontinuity of $\mathbf{w}$ across a surface, is given by Weingarten [1901, **3**].

[5] [1893, **1**, §12] [1921, **2**, §778].

Alternatively, *let two points lying upon the same vortex-surface be connected by two curves* $\mathfrak{c}_1$ *and* $\mathfrak{c}_2$ *reconcileable upon it; then the flow along* $\mathfrak{c}_1$ *equals the flow along* $\mathfrak{c}_2$. Now for the validity of Kelvin's transformation, and hence also for the validity of this result, the vorticity $\mathbf{w}$ need not even be continuous: it suffices that the velocity field $\dot{\mathbf{x}}$ be continuous and that the vortex surface be part of the boundary of a closed region in which $\mathbf{w}$ is piecewise continuous (§8). Thus (43.2) holds for circuits upon vortex-surfaces which suffer sharp but isolated bends. Consider now two simple circuits $\mathfrak{c}_1$ and $\mathfrak{c}_2$ lying upon a vortex-tube, reconcileable upon it but not reducible upon it. The vortex-tube itself need not be reducible. Let the two circuits be connected by a barrier $\mathfrak{k}$ traversed in opposite senses, so that $\mathfrak{c} \equiv \mathfrak{c}_1 + \mathfrak{k} - \mathfrak{c}_2 - \mathfrak{k}$ is a single closed circuit lying upon the surface of the vortex-tube (Fig. 43.1) and inclosing a

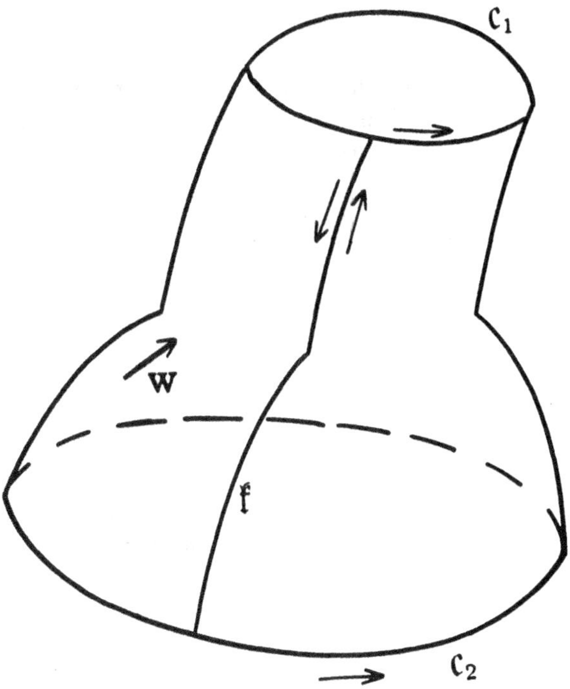

Fig. 43.1

simply-connected portion of it. To this circuit our previous result (43.2) applies:

$$0 = \oint_{\mathfrak{c}} d\mathbf{x}\cdot\dot{\mathbf{x}} = \oint_{\mathfrak{c}_1} + \int_{\mathfrak{k}} - \oint_{\mathfrak{c}_2} - \int_{\mathfrak{k}}, \tag{43.3}$$

whence follows

$$\oint_{c_1} d\mathbf{x}\cdot\dot{\mathbf{x}} = \int_{c_2} d\mathbf{x}\cdot\dot{\mathbf{x}}. \tag{43.4}$$

The result is easily extended to reconcileable circuits which are not simple. We have thus established ***Kelvin's extension*** (1869) of Helmholtz's first vorticity theorem: *let there be given a vortex-tube in a motion where the velocity field is continuous and the vorticity field is piecewise continuous in a closed region, of whose boundary the vortex-tube forms a part; then the circulations about any two circuits lying wholly upon this vortex-tube and reconcileable upon it are equal.*

44. The flexion field. The vector

$$\operatorname{curl}\mathbf{w} = \operatorname{curl}\operatorname{curl}\dot{\mathbf{x}} \tag{44.1}$$

is called the *flexion vector,*[1] and its vector-lines are called the *lines of flexion.* Since a number of the theorems and conditions derived in the present work are valid for the curl of any vector field, corresponding results for the flexion field may be obtained by purely verbal changes, such as the substitution of "line of flexion" for "vortex-line," "vorticity" for "velocity," etc.[2]

The solenoidal character of $\mathbf{w}$ permits us to put $\mathbf{c} = \mathbf{w}$ in the Lamb-Thomson identity (10.11), whence follows[3]

$$\tfrac{1}{2}\int_{\mathfrak{v}} w^2\, dv = \oint_{\mathfrak{s}} d\mathbf{s}\cdot(\tfrac{1}{2}\, w^2\mathbf{r} - \mathbf{w}\mathbf{w}\cdot\mathbf{r}) + \int_{\mathfrak{v}} \mathbf{r}\cdot\operatorname{curl}\mathbf{w}\times\mathbf{w}\, dv, \tag{44.2}$$

from which we may draw the following conclusion: *in a region (not necessarily simply-connected) where the flexion field vanishes or is parallel to the vorticity, the total squared vorticity is determined by the vorticity upon the bounding surface.* In particular, *in a stationary finite domain to whose walls the material adheres* (§37) *or on whose walls the motion is irrotational, if* $\operatorname{curl}\mathbf{w}\times\mathbf{w} = 0$ *then the total squared vorticity is determined by the vorticity magnitude at the boundary*:

$$\int_{\mathfrak{v}} w^2\, dv = \oint_{\mathfrak{s}} d\mathbf{s}\cdot\mathbf{r}w^2, \tag{44.3}$$

and, finally, *in a domain where* $\operatorname{curl}\mathbf{w}\times\mathbf{w} = 0$ *in the interior and* $w = 0$ *upon the boundaries the motion must be irrotational.* Thus if

[1] The importance of the flexion vector lies in its occurrence in Navier's dynamical equation (17.4).

[2] Results of this type are given by Boggio-Lera [1887, 1], who introduced the term *flexion.*

[3] [1951, 2].

$w \neq 0$ but curl $\mathbf{w} = 0$ in a finite region,[4] it is impossible to surround this region with a layer in which $w = 0$ and yet maintain continuity of the partial derivatives of $\mathbf{w}$ throughout the entire region.

Since curl $\mathbf{w}$, curl curl $\mathbf{w}$, . . . , are also solenoidal, similar analysis may be applied to them, yielding analogous results with respect to the vanishing of curl curl $\mathbf{w} \times$ curl $\mathbf{w}$, etc.[5]

There are various interesting rotational motions in which the flexion field vanishes. An example is furnished by the superposition of a rigid motion upon an irrotational motion. A finite motion of this type may be regarded as taking place within a rigidly rotating vessel, the simplest example being the *Couette motion*, which is obtained by superposition of a rigid rotation upon an irrotational vortex with axis parallel to the axis of rotation. Another case is furnished by the *simple Poiseuille motion*: $\dot{x} = ky$, $\dot{y} = 0$, $\dot{z} = 0$. For a *generalized Poiseuille motion*,[6] *viz*.

$$\dot{x} = \dot{x}(y, z), \qquad \dot{y} = 0, \qquad \dot{z} = 0, \tag{44.4}$$

the flexion field vanishes if and only if $\dot{x}$ be harmonic: $\nabla^2\dot{x} = 0$.

It will later appear that a case of particular interest is

$$\text{curl curl}\,\mathbf{w} = 0 = -\nabla^2\mathbf{w}, \tag{44.5}$$

so that the flexion field, assumed continuously differentiable, is lamellar:

$$\text{curl}\,\mathbf{w} = \text{grad}\,\sigma. \tag{44.6}$$

A motion of this type is said to *admit a flexion-potential*[7] σ. Since the flexion field is solenoidal, the flexion-potential must be harmonic:

$$\nabla^2\sigma = 0. \tag{44.7}$$

An example of a motion admitting a flexion-potential is furnished by a generalized Poiseuille motion (44.4) in which $\dot{x}$ satisfies $\nabla^2\dot{x} = $ const.

While in an arbitrary motion we have

$$\oint_{\mathfrak{s}} d\mathbf{s}\cdot\text{curl}\,\mathbf{w} = 0 \tag{44.8}$$

---

[4] The interest of this result lies in the fact that when curl $\mathbf{w} = 0$ Navier's dynamical equation (17.4) for viscous fluids reduces to the same form as Euler's dynamical equation (17.7) for perfect fluids. *Cf.* also §49.

[5] Craig [1881, 2, pp. 15–26] considered motions of viscous fluids in which $\text{curl}^n\,\dot{\mathbf{x}} = 0$ but $\text{curl}^{n-1}\,\dot{\mathbf{x}} \neq 0$.

[6] These motions were introduced and analyzed by Euler [1757, 2, §§XLVIII–XLIX].

[7] The flexion-potential was introduced by Müller [1878, 4, §3].

for any closed surface $\mathfrak{s}$, it follows from (7.2) that *a motion with continuously differentiable flexion field admits a flexion-potential if and only if*

$$\oint_{\mathfrak{s}} d\mathbf{s} \times \operatorname{curl} \mathbf{w} = 0. \tag{44.9}$$

45. Criteria for permanence and constant strength of the vortex-tubes. In order to investigate the possibility that the vortex-lines be material lines, we put $\mathbf{c} = \mathbf{w}$ in the Helmholtz-Zorawski criterion $(28.8)_2$, by (43.1) obtaining

$$\mathbf{w} \times \left[\frac{\partial \mathbf{w}}{\partial t} + \operatorname{curl}(\mathbf{w} \times \dot{\mathbf{x}})\right] = 0. \tag{45.1}$$

But by taking the curl of Lagrange's acceleration formula (38.2) we obtain

$$\frac{\partial \mathbf{w}}{\partial t} + \operatorname{curl}(\mathbf{w} \times \dot{\mathbf{x}}) = \operatorname{curl} \ddot{\mathbf{x}} \tag{45.2}$$

for any motion. Hence (45.1) becomes

$$\mathbf{w} \times \operatorname{curl} \ddot{\mathbf{x}} = 0: \tag{45.3}$$

a necessary and sufficient condition that the vortex-lines be material lines is that the acceleration be lamellar or that its curl *be parallel to the vorticity.*

By putting $\mathbf{c} = \mathbf{w}$ in Zorawski's criterion $(28.2)_2$ we similarly obtain

$$\operatorname{curl} \ddot{\mathbf{x}} = 0: \tag{45.4}$$

a necessary and sufficient condition that the strengths of all vortex-tubes remain constant in time is that the acceleration be lamellar: this condition is also sufficient that the vortex-tubes be material tubes.[1]

From (28.10) follows a possibly more general result: *in order that the flux of vorticity through a material surface F* = const. *be constant in time, it is necessary and sufficient that*

---

[1] Müller [1878, 4] stated in effect that if the vortex-lines be material it follows that curl $\ddot{\mathbf{x}} = 0$; from (45.3) it is plain that this result is false, and indeed Müller's proof rests upon Clebsch's transformation (§101), which is valid only if curl $\ddot{\mathbf{x}} = 0$. The condition (45.3) and the incorrect substitute $\mathbf{w} \cdot \operatorname{curl} \ddot{\mathbf{x}} = 0$ for (45.5) were stated without proof by Levy [1890, 1, §10]; apparently the first published derivations are those of Poincaré [1893, 1, §§5–6, 150–151]. *Cf.* [1900, 1] [1905, 1, §386] [1911, 1, §4] [1921, 2, Ch. XXV, ex. 5] [1950, 1]. By (28.11), Levy's condition $\mathbf{w} \cdot \operatorname{curl} \ddot{\mathbf{x}} = 0$ is necessary and sufficient that the strengths of the vortex-tubes at material normal cross-sections be constant in time. In general, no such cross-sections exist. If they do, suppose Levy's condition satisfied. Then the strength of the vortex-tubes at these cross-sections is constant. But the strength of a vortex-tube is the same at all cross-sections (§43); hence, by the theorem in the text above, curl $\ddot{\mathbf{x}} = 0$. Thus in the restricted case when Levy's condition is relevant it reduces to (45.4).

$$\operatorname{grad} F \cdot \operatorname{curl} \ddot{\mathbf{x}} = 0. \tag{45.5}$$

In general, no such material surfaces exist, since F must satisfy both (45.5) and (19.6), both of which are linear first order partial differential equations. Alternatively, we may eliminate (19.6) altogether by using a material formulation: $F = F(\mathbf{X})$; but, when (45.5) is converted to material form, the coefficients involve not only $\mathbf{X}$ but also t, while the solution F must be independent of t—in general, impossible. Of course there are special cases in which solutions exist. For example, the flux of vorticity through a vortex-surface is always zero, so these, if they be material, furnish a trivial example. In this case (45.3) obviously implies (45.5) when $F =$ const. is a vortex-surface.

46. Circulation-preserving motions. The Helmholtz theorems. Now the condition just derived, *viz.*

$$\operatorname{curl} \ddot{\mathbf{x}} = 0, \tag{45.4}$$

is of great importance. Both d'Alembert (1749)[1] and Euler (1752-1755)[2] derived it as a consequence of the dynamical equations for barotropic motions of perfect[3] fluids subject to conservative extraneous force, and we shall call it the ***d'Alembert-Euler condition.*** Its meaning emerges when we put $\emptyset = \dot{\mathbf{x}}$ in (23.2), or, alternatively, interpret the second theorem of §45 in the light of Kelvin's transformation: *the d'Alembert-Euler condition is necessary and sufficient that the circulation of every reducible material circuit remain constant in time.*[4] Classical hydrodynamics is characterized by this one basic statement, and all the main theorems of that subject are consequences of it.[5] Motions in which the circulation of reducible material circuits does not change we shall call ***circulation-preserving motions.*** They will be frequent subject of remark and illustration in this work, and the last chapter presents their peculiar properties.

For a simple example of the use of the d'Alembert-Euler condition, we may note that by (45.2) it is equivalent to

$$\frac{\partial \mathbf{w}}{\partial t} + \operatorname{curl} (\mathbf{w} \times \dot{\mathbf{x}}) = 0. \tag{46.1}$$

---

[1] [1752, 1, §86] [1761, 2, §§X, XV].

[2] [1761, 1, §58] [1757, 2, §XXXV] [1770, 1, §§42–43] [1806, 1, §§154–155]. Euler noted the connection of this condition with Bernoulli's equation.

[3] It was remarked by Brandes [1806, 1, §150, footnote] that in a theory where friction is taken into account it will generally follow that curl $\ddot{\mathbf{x}} \neq 0$.

[4] The sufficiency was proved by Hankel [1861, 1, §8] and Kelvin [1869, 1, §59(e)].

[5] We owe this observation to Levy [1890, 1, §7].

Hence *a motion with steady vorticity* ($\partial \mathbf{w}/\partial t = 0$) *is circulation-preserving if and only if its Lamb vector be lamellar:*[6]

$$\text{curl}\,(\mathbf{w} \times \dot{\mathbf{x}}) = 0. \tag{46.2}$$

In particular, any slow motion with steady vorticity is circulation-preserving,[7] since in a slow motion $\mathbf{w} \times \dot{\mathbf{x}} \approx 0$.

We may now reformulate the results of §45 in a fashion so closely connected with the hydrodynamical theorems of Helmholtz (1858)[8] that although they are purely kinematical, we shall call them the ***second and third vorticity theorems of Helmholtz***: (2) *in a circulation-preserving motion the vortex-lines are material lines, and* (3) *in a motion such that the vortex-lines are material, in order that the strengths of all vortex-tubes remain constant in time it is necessary and sufficient that the motion be circulation-preserving.* The Helmholtz theorems yield a full visualization of the general character of circulation-preserving motions, whose theory may thus be said to be, in a certain sense, closed. Most hydrodynamical motions are not circulation-preserving, however, and one of the major aims of the present work is to clarify the manner in which a general motion departs from a circulation-preserving character, or, equivalently, to discover the manner in which the vortex-tubes depart from material character and change their strength—that is to say, to generalize the Helmholtz theorems. This investigation is presented in Chapter VIII.

47. Invariance of the circulation-preserving property. The circulation-preserving quality is evidently a Galilean invariant. More generally, by forming the curl of (16.10) we obtain

---

[6] This condition is the real starting point of many investigations which are deceptively phrased in dynamical terms. For example, the completion of Castoldi's analysis of complex-lamellar velocity fields by Prim [1948, **6**, §§4–5] [1952, **4**, Ch. V, Sect. B] is equivalent to combining (46.2) with $\dot{\mathbf{x}} = \dot{x}\mathbf{n}$, $\mathbf{n} \equiv \text{grad}\,\chi/|\,\text{grad}\,\chi\,|$, and (see §16²) div $\mathbf{n} = 0$ to conclude that in order for a circulation-preserving complex-lamellar velocity field to be such that the velocity magnitude is constant along each stream-line at any instant it is necessary and sufficient that it be possible to parametrize the minimal surfaces $\chi = \text{const.}$ in such a way that $\nabla^2\chi = 0$, and hence that the geometric variety of these flow fields is exactly that of irrotational fields satisfying the same condition, which were analysed by Hamel [1937, **1**]. *Cf.* also [1953, **6**].

[7] In particular, the slow motions of viscous fluids, to which much hydrodynamical literature has been devoted, are circulation-preserving; an equivalent remark was made by Craig [1880, **7**, pp. 272–273].

[8] [1858, **1**, §2]. The validity of the Helmholtz theorems was extended by Kelvin [1869, **1**, §§59(d), 60(f)–60(i)]. *Cf.* [1874, **1**]. That these theorems are essentially kinematical was remarked by Schütz [1895, **2**]. Larmor (quoted in [1895, **1**, §143] [1932, **1**, §146]) observed that Helmholtz's own proofs of the second and third theorems are open to the same objection as that raised by Stokes against certain faulty proofs of the velocity-potential theorem (see §104).

$$\text{curl}'\,\ddot{\mathbf{x}}' = \text{curl}'\left[\ddot{\mathbf{x}} - \frac{d^2\mathbf{c}}{dt^2} - \frac{d\boldsymbol{\omega}}{dt}\times\mathbf{r}' - \boldsymbol{\omega}\times(\boldsymbol{\omega}\times\mathbf{r}') - 2\boldsymbol{\omega}\times\dot{\mathbf{x}}'\right]. \tag{47.1}$$

Now by (39.2) the centripetal acceleration field is lamellar:

$$\text{curl}'\,[\boldsymbol{\omega}\times(\boldsymbol{\omega}\times\mathbf{r}')] = 0. \tag{47.2}$$

But

$$\text{curl}'\left[\frac{d\boldsymbol{\omega}}{dt}\times\mathbf{r}'\right] = 2\,\frac{d\boldsymbol{\omega}}{dt} \tag{47.3}$$

and

$$\text{curl}'\,[\boldsymbol{\omega}\times\dot{\mathbf{x}}'] = \boldsymbol{\omega}\,\text{div}'\,\dot{\mathbf{x}}' - \boldsymbol{\omega}\cdot\text{grad}'\,\dot{\mathbf{x}}'. \tag{47.4}$$

Also

$$\begin{aligned}\text{curl}'\,\ddot{\mathbf{x}} &= [\text{grad}'\,\ddot{\mathbf{x}}]_\times = [\text{grad}'\,(\mathbf{r}' - \mathbf{c})\cdot\text{grad}\,\ddot{\mathbf{x}}]_\times \\ &= [\mathbf{I}\cdot\text{grad}\,\ddot{\mathbf{x}}]_\times = \text{curl}\,\ddot{\mathbf{x}}.\end{aligned} \tag{47.5}$$

Hence (47.1) becomes

$$\text{curl}'\,\ddot{\mathbf{x}}' = \text{curl}\,\ddot{\mathbf{x}} - 2\left[\frac{d\boldsymbol{\omega}}{dt} + \boldsymbol{\omega}\,\text{div}'\,\dot{\mathbf{x}}' - \boldsymbol{\omega}\cdot\text{grad}'\,\dot{\mathbf{x}}'\right]. \tag{47.6}$$

From this identity and from the d'Alembert-Euler condition (45.4) we conclude that *a motion which is circulation-preserving for an observer in a certain frame is also circulation-preserving for an observer in any other frame suffering translation only with respect to the first*, since then $\boldsymbol{\omega} = 0$. For the validity of this result the rate of translation need not be constant in time.

In the case when the second frame is rotating with respect to the first, however, the circulation-preserving property will generally fail to be preserved. To find those circulation-preserving motions which are rotational invariants, we must first find the class of motions $\dot{\mathbf{x}}$ such that

$$\boldsymbol{\omega}\cdot\text{grad}\,\dot{\mathbf{x}} - \boldsymbol{\omega}\,\text{div}\,\dot{\mathbf{x}} = \frac{d\boldsymbol{\omega}}{dt}, \tag{47.7}$$

$\boldsymbol{\omega}$ being a given vector function of time only.

For ease of analysis[1] choose the Cartesian axes such that at the instant considered we have

$$\boldsymbol{\omega} = \omega\mathbf{i}, \qquad \frac{d\boldsymbol{\omega}}{dt} = \alpha\mathbf{i} + \beta\mathbf{j}. \tag{47.8}$$

Then (47.7) becomes

$$\frac{\partial\dot{y}}{\partial y} + \frac{\partial\dot{z}}{\partial z} = -\frac{\alpha}{\omega}, \qquad \frac{\partial\dot{y}}{\partial x} = \frac{\beta}{\omega}, \qquad \frac{\partial\dot{z}}{\partial x} = 0. \tag{47.9}$$

[1] I am obliged to my colleague Professor Hlavatý for help on this problem.

We first seek motions satisfying (47.9) for arbitrary $\boldsymbol{\omega}$ and $d\boldsymbol{\omega}/dt$. The x-direction then becomes arbitrary, and the system (47.9) incompatible unless $\alpha = \beta = 0$, in which case its only solution is $\dot{x}^i = \text{const.}$ Thus we have proved that *there exists no motion which is circulation-preserving to all observers; the only motion which is circulation-preserving to all observers in steady relative rotation is a uniform translation.*

One may take up instead the interesting problem of finding that class of motions which are circulation-preserving in *two* preferred frames, supposed to be in relative rotation at angular velocity $\pm\boldsymbol{\omega}(t)$. Then α, β, and ω in (47.9) become given constants at any one instant, and the system is compatible. We have in fact

$$\begin{aligned} \dot{x} &= f(x, y, z, t), \\ \dot{y} &= -\int \frac{\partial g(z, y, t)}{\partial z}\, dy + \frac{\beta x - \alpha y}{\omega} + h(z, t), \\ \dot{z} &= g(y, z, t), \end{aligned} \tag{47.10}$$

the functions f, g, h being arbitrary. For this class of motions, and only for this class, $\boldsymbol{\omega}$ and $d\boldsymbol{\omega}/dt$ being given, we shall have curl $\ddot{\mathbf{x}} = \text{curl}'\ \ddot{\mathbf{x}}'$. The solution of our problem is then obtained by finding the subclass of (47.10) for which curl $\ddot{\mathbf{x}} = 0$. The system which must be solved is rather elaborate and will not be written out. We must rest content with a partial discussion of the very special case when $\boldsymbol{\omega} = \text{const.}$ and the motion itself is required to be steady in one of the frames. We have then from (47.10)

$$\begin{aligned} \ddot{x} &= f\frac{\partial f}{\partial x} + \left[-\int \frac{\partial g}{\partial z}\, dy + h\right]\frac{\partial f}{\partial y} + g\frac{\partial f}{\partial z}, \\ \ddot{y} &= -\left[-\int \frac{\partial g}{\partial z}\, dy + h\right]\frac{\partial g}{\partial z} + \left[-\int \frac{\partial^2 g}{\partial z^2}\, dy + h'\right]g, \\ \ddot{z} &= \left[-\int \frac{\partial g}{\partial z}\, dy + h\right]\frac{\partial g}{\partial y} + g\frac{\partial g}{\partial z}. \end{aligned} \tag{47.11}$$

The condition curl $\ddot{\mathbf{x}} = 0$ may then be put into the form

$$\begin{aligned} &f\frac{\partial f}{\partial x} + \left[-\int \frac{\partial g}{\partial z}\, dy + h\right]\frac{\partial f}{\partial y} + g\frac{\partial f}{\partial z} = k(x), \\ &\left[\frac{\partial}{\partial z} g^2 \frac{\partial}{\partial z} + \frac{\partial}{\partial y} g^2 \frac{\partial}{\partial y}\right] \frac{-\int \frac{\partial g}{\partial z}\, dy + h}{g} = 0. \end{aligned} \tag{47.12}$$

These conditions involve the angular velocity $\boldsymbol{\omega}$ only through its direction, selected as the axis of x, but are independent of its magnitude ω.

If f, g, h, k be selected so as to satisfy the conditions (47.12) and then put back into (47.10), there results a motion which is circulation-preserving when viewed in the x, y, z frame or in any other frame whose angular velocity, assumed constant in time, is parallel to the x-axis. Since only the two conditions (47.12) are to be satisfied by the four otherwise arbitrary functions occurring, it is plain that, given any two observers, there is a great variety of motions which are circulation-preserving for both. A simple example of a solution of (47.12) is

$$f = f(x), \qquad g = g(y \pm iz), \qquad h = 0. \tag{47.13}$$

The general effect of rotation of the frame of reference upon the apparent value of the circulation will be discussed in full generality in §81.

48. The acceleration-potential. A function ϕ^* such that

$$\ddot{\mathbf{x}} = -\operatorname{grad} \phi^* \tag{48.1}$$

is called an ***acceleration-potential.***[1] The function ϕ^* may be single-valued or many-valued. By the first result in §12, we may reformulate the d'Alembert-Euler condition (45.4) in the following way: *a motion is circulation-preserving if and only if it possesses an acceleration-potential.*

An evident example of a circulation-preserving motion is an irrotational motion. By putting (36.2) into (38.3) we obtain

$$\ddot{\mathbf{x}} = -\operatorname{grad}\left[\frac{\partial \phi}{\partial t} - \tfrac{1}{2}(\operatorname{grad} \phi)^2\right], \tag{48.2}$$

whence comparison with (48.1) shows that if a meaningless function of time only be absorbed into ϕ we shall have

$$\phi^* = \frac{\partial \phi}{\partial t} - \tfrac{1}{2}(\operatorname{grad} \phi)^2, \tag{48.3}$$

a formula given by Euler.[2]

Another example is furnished by the *simple vortices*, given in cylindrical co-ordinates by the equations

$$\dot{r} = 0, \qquad \dot{\theta} = \omega(r, t), \qquad \dot{z} = 0 \tag{48.4}$$

(note that $\dot{x} = \dot{x}^{\theta} = r\dot{\theta} = r\omega$). While in some respects these flows are rather degenerate specimens, in others they furnish valuable and typical illustrations for the general propositions of our subject. Their theory

[1] The acceleration-potential is a discovery of Euler [1757, 2, §XXXV] [1770, 1 §42]. Its importance was stressed by Levy [1890, 1, §7]. *Cf.* [1893, 1, §6] [1921, 2, §§729, 753, *et passim*].

[2] [1761, 1, §§79–80]. *Cf.* [1911, 1, §4] [1946, 2, §3].

is due to Euler.[3] Their stream-lines are concentric circles in the z = const. planes. Their vorticity is $(2\omega + r\omega')\mathbf{k}$; the circulation around a stream-line of radius r is

$$\oint d\mathbf{x}\cdot\dot{\mathbf{x}} = 2\pi r^2\omega = K(r, t), \qquad \text{say}. \tag{48.5}$$

Thus it is evident that these flows are potential flows[4] if and only if $r^2\omega = f(t)$, or, equivalently, $\dot{x} = f(t)/r$; that they are circulation-preserving if and only if they be steady. In the latter case

$$\phi^* = \int r\omega^2\, dr, \tag{48.6}$$

a result due to Euler.[5]

49. Parenthesis: circulation-preserving motions in fluid dynamics. 1. *Perfect fluids.* By taking the curl of Euler's dynamical equation (17.7) we obtain

$$\text{curl}\,\ddot{\mathbf{x}} = \text{grad}\, p \times \text{grad}\,\frac{1}{\rho}. \tag{49.1}$$

Hence curl $\ddot{\mathbf{x}} = 0$ if and only if

$$f(p, \rho, t) = 0. \tag{49.2}$$

Flows satisfying (49.2), as already mentioned in §17, are called *barotropic*; in such a flow at each instant either (1) the density is uniform, (2) the pressure is uniform, (3) both pressure and density are uniform, or (4) the surfaces p = const. coincide with the surfaces ρ = const. By comparing (49.1) with the d'Alembert-Euler condition (45.4) we thus obtain the ***Hankel-Kelvin circulation theorem*** (1861, 1869):[1] *a flow of an inviscid fluid subject to conservative extraneous force is circulation-preserving if and only if it be barotropic.* Classical hydrodynamics is the

---

[3] [1757, 2, §§XXX–XXXIII] [1757, 3, §§LVII–LXI]. In the steady case these flows, being circulation-preserving, are dynamically possible for an inviscid incompressible fluid subject to any conservative extraneous force. It is this fact, of course differently worded, which Euler observed and used to disprove d'Alembert's contention that only potential flows can occur in nature. See my history [1954, 1, parts VIII, XII and XVI]. Certain unsteady simple vortices are dynamically possible flows of viscous liquids (*cf.* §104[13]).

[4] Among the simple vortices it is only these "potential vortices" (37.4) which fall into the class of vortices defined in §35.

[5] [1757, 2, §XXXI].

---

[1] Hankel [1861, 1, §8] proved the sufficiency of (49.2) only incidentally and apparently did not recognize the central importance of this theorem in fluid dynamics, which was revealed by the discussion of Kelvin [1869, 1, §59(d)], who discovered it independently.

science of these barotropic flows, and the foregoing result is its fundamental theorem. Flows in which (49.2) does not hold are called *baroclinic.* The fluid motions of principal interest in meteorology and aerodynamics fall into this second and more interesting class, to which the Hankel-Kelvin theorem does not apply.

When (49.2) holds we have

$$\frac{1}{\rho}\,\mathrm{grad}\,p = \mathrm{grad}\int\frac{dp}{\rho}, \tag{49.3}$$

and hence by comparing (17.7) with (48.1) we obtain the following expression for the acceleration-potential:

$$\phi^* = \upsilon + \int\frac{dp}{\rho} + f(t). \tag{49.4}$$

In order for a flow to be physically possible it is necessary that p be non-negative and single-valued. For the barotropic relations occurring in physical problems this property is shared by the function $\int dp/\rho$. From (49.4) we then obtain the following characterization: *given a motion with acceleration-potential ϕ^* and a conservative extraneous force field with potential υ, the motion is physically possible for an inviscid fluid subject to this force-field if and only if $\phi^* - \upsilon$ be single-valued and bounded below.* In particular, a motion with a many-valued acceleration-potential is not physically realizable in a force field with a single-valued potential.

2. *Viscous liquids.* By taking the curl of Navier's dynamical equation (17.4) we obtain

$$\mathrm{curl}\,\ddot{\mathbf{x}} = -\frac{\mu}{\rho}\,\mathrm{curl}\,\mathrm{curl}\,\mathbf{w}. \tag{49.5}$$

Hence curl $\ddot{\mathbf{x}} = 0$ if and only if curl curl $\mathbf{w} = 0$. By comparing this result with (44.5) and the d'Alembert-Euler condition (45.4) we obtain a theorem implied by Craig (1880):[2] *a flow of a fluid of uniform density and viscosity, subject to conservative extraneous force, is circulation-preserving if and only if it admits a flexion-potential.*[3] *Thus all the* kinematical

---

[2] [1880, **7**, pp. 272–273]. *Cf.* [1913, **1**] [1918, **1**, §III] [1927, **3**, §§1, 4] [1932, **2**].

[3] This class includes not only the physically improbable irrotational flows of viscous liquids, which are discussed by Hamel [1941, **2**], but also the important Couette and Poiseuille flows. *Cf.* [1913, **1**]. Görtler & Wieghart [1942, **5**] show that the only *steady plane* flows of a viscous fluid of uniform density and viscosity for which an acceleration-potential exists are flows obtainable by superposition from irrotational flows, Couette flows, Poiseuille flows, and flows whose stream-function is of the form $\psi = r^2\,(C\log r + D)$; the last type of flow leads to a many-valued pressure if the region under consideration contains a curve encircling the origin. The corresponding question for truly spatial flows and for unsteady flows remains unanswered.

theorems of classical hydrodynamics remain valid, though perhaps with a a different dynamical *interpretation, for this class of flows of viscous liquids.*

By putting (44.6) into (17.4) and comparing the result with (48.1) we obtain the following expression for the acceleration-potential in this class of flows:

$$\phi^* = v + \frac{p}{\rho} + \frac{\mu}{\rho}\sigma + f(t). \tag{49.6}$$

When $\sigma = 0$, *i.e.* when the flexion field vanishes, the flow has the same pressure field as a possible flow of a perfect incompressible fluid; when $\sigma \neq 0$ the pressure field of the flow of the viscous liquid differs from that of the kinematically identical flow of a perfect liquid by an apparent pressure $\mu\sigma$.

50. Complex-lamellar motions. For a complex-lamellar motion application of (12.4) and (12.5) combined with (16.12) yields

$$\dot{\mathbf{x}} = f \operatorname{grad} g, \tag{50.1}$$

$$\mathbf{w} = \operatorname{grad} f \times \operatorname{grad} g, \tag{50.2}$$

$$\mathbf{w}\cdot\dot{\mathbf{x}} = 0. \tag{50.3}$$

From the last of these we conclude that *a rotational motion is complex-lamellar if and only if its vortex-lines and its stream-lines form orthogonal families of curves.*

In the special case of a plane motion the vortex-lines are the straight lines normal to the x-y-plane, and in the special case of a rotationally-symmetric motion the vortex-lines are the circles whose centres are upon the axis of rotational symmetry and whose planes are normal to it. In both these cases the vorticity field is thus also complex-lamellar. It would be valuable to find necessary and sufficient conditions that a complex-lamellar velocity field be endowed with a complex-lamellar vorticity field, and conversely. The solution of this problem in a special case is presented in §72.

By differentiating (50.3) we obtain

$$\mathbf{w}\cdot\frac{\partial\dot{\mathbf{x}}}{\partial t} + \dot{\mathbf{x}}\cdot\frac{\partial\mathbf{w}}{\partial t} = 0. \tag{50.4}$$

When the stream-lines are steady, taking the curl of (28.9) yields

$$\mathbf{w}\cdot\frac{\partial\dot{\mathbf{x}}}{\partial t} - \dot{\mathbf{x}}\cdot\frac{\partial\mathbf{w}}{\partial t} = 0. \tag{50.5}$$

Hence in a complex-lamellar motion with steady stream-lines we have

$$\mathbf{w}\cdot\frac{\partial\dot{\mathbf{x}}}{\partial t} = \dot{\mathbf{x}}\cdot\frac{\partial\mathbf{w}}{\partial t} = 0. \tag{50.6}$$

In the general case (50.3) is to be replaced by

$$\mathbf{w}\cdot\dot{\mathbf{x}} = \frac{\partial(h, f, g)}{\partial(x, y, z)}, \tag{50.7}$$

where f, g, and h are the Monge potentials (16.12).

Complex-lamellar motions possess some of the distinguishing properties of irrotational motions. The function $-g$ is closely analogous to the potential function ϕ, since we have

$$-\dot{g} = -\frac{\partial g}{\partial t} + \dot{\mathbf{x}}\cdot\text{grad}\,(-g) = -\frac{\partial g}{\partial t} - f\dot{x}^2, \quad \text{or} \quad -\frac{dg}{d\dot{x}} = f\dot{x}^2, \tag{50.8}$$

a result similar to (36.5). Thus the conclusions regarding the stream-lines of an irrotational motion given in §36 carry over to a region of complex-lamellar motion where f is of one sign. Now by the continuity of the velocity field, f can be of opposite sign at two points upon a stream-line only if there be an intermediate point where $f = 0$, and by (50.1) this point is a stagnation point. *Consequently in a region of complex-lamellar motion without stagnation points, the conclusions of* §36 *regarding the speed and the stream-line pattern of an irrotational motion continue to hold, provided* $-g$ *be substituted for* ϕ *if* $f > 0$, g *for* ϕ *if* $f < 0$.

51. Steady vortex-lines. Steady vorticity. *The vortex-lines are steady* if and only if

$$\frac{\partial \mathbf{w}}{\partial t} \times \mathbf{w} = 0. \tag{51.1}$$

For the vortex-lines to be steady it is of course sufficient, but not necessary, that vorticity itself be steady:

$$\frac{\partial \mathbf{w}}{\partial t} = 0. \tag{51.2}$$

From (51.1) follows

$$\frac{\partial \mathbf{w}}{\partial t} = \lambda \mathbf{w}, \tag{51.3}$$

where λ is a scalar quantity. Thus we have

$$\text{curl}\,\frac{\partial \dot{\mathbf{x}}}{\partial t} = \text{curl}\,(\lambda\dot{\mathbf{x}}) - \text{grad}\,\lambda \times \dot{\mathbf{x}}. \tag{51.4}$$

Taking the divergence of this equation yields

$$\begin{aligned} 0 &= \text{div}\,(\text{grad}\,\lambda \times \dot{\mathbf{x}}),\\ &= \text{grad}\,\lambda\cdot\mathbf{w}. \end{aligned} \tag{51.5}$$

Hence it follows that *in a motion where the vortex-lines are steady but the vorticity is not steady, the surfaces upon which* $\partial \mathbf{w}/\partial t$ *bears a constant ratio to* $\mathbf{w}$ *are vortex-surfaces.*

In any plane or rotationally-symmetric motion the vortex-lines are steady. In any complex-lamellar motion in which the vortex-lines are normal to a family of steady stream-surfaces the vortex-lines are steady. More generally, it would be valuable to delimit that class of complex-lamellar motions which are endowed with steady vortex-lines.

A type of motion in which both the stream-lines and the vortex-lines are steady is given by

$$\dot{\mathbf{x}}(\mathbf{x}, t) = T(t)\mathbf{u}(\mathbf{x}), \tag{51.6}$$

a class introduced by d'Alembert[1] and analysed by Lagrange and Cauchy.[2] It is convenient to call these motions *d'Alembert motions.*

Following the analysis of Masotti,[3] we may characterize flows with steady vorticity by noting that (51.2) is equivalent to

$$\operatorname{curl} \frac{\partial \dot{\mathbf{x}}}{\partial t} = 0. \tag{51.7}$$

Hence (51.2) is satisfied if and only if there exist a scalar $\chi(\mathbf{x}, t)$ such that

$$\frac{\partial \dot{\mathbf{x}}}{\partial t} = \operatorname{grad} \chi; \tag{51.8}$$

that is, the local acceleration is lamellar, and hence by integrating we obtain

$$\dot{\mathbf{x}}(\mathbf{x}, t) = \operatorname{grad}\left[\int \chi(\mathbf{x}, t)\, dt\right] + \mathbf{u}(\mathbf{x}): \tag{51.9}$$

a necessary and sufficient condition for a motion with steady vorticity is that the velocity be the sum of a lamellar field and a steady field. An irrotational motion furnishes a special case, with $\chi = -\partial\phi/\partial t$, $\mathbf{u} = 0$. Henceforward the symbol χ will be employed only in the sense indicated by (51.8).

For any steady motion with steady vorticity we may put (51.8) into Lagrange's acceleration formula (38.2), obtaining

$$\ddot{\mathbf{x}} = \mathbf{w} \times \dot{\mathbf{x}} + \operatorname{grad}\left(\tfrac{1}{2}\dot{x}^2 + \chi\right), \tag{51.10}$$

a result whose significance will appear in Chapter VII.

The condition (51.8) for steady vorticity may be put into a different

---

[1] [1752, 1, §148] [1761, 2, §§I, XV]. As usual, d'Alembert claimed that only this type of motion can occur in nature.

[2] [1762, 3, §XLIII] [1823, 2]. Only isochoric motions are considered.

[3] [1927, 3, §2].

form by considering the circulation around a stationary closed spatial curve or *control circuit* c:

$$\frac{\partial}{\partial t}\oint_{c} d\mathbf{x}\cdot\dot{\mathbf{x}} = \oint_{c} d\mathbf{x}\cdot\frac{\partial \dot{\mathbf{x}}}{\partial t}. \tag{51.11}$$

Since the integral on the right vanishes for all reducible circuits if and only if (51.8) hold, we conclude that *a necessary and sufficient condition that the vorticity be steady is that the circulation about every reducible control circuit be constant in time.*

The type of motion treated in the classical linearized theory of acoustics is always a motion of steady vorticity. In fact, from (16.16) we have

$$\operatorname{curl}\ddot{\mathbf{x}} \approx \frac{\partial \mathbf{w}}{\partial t}, \tag{51.12}$$

and hence by the d'Alembert-Euler condition (45.4) *a slow motion is a circulation-preserving motion if and only if the vorticity be steady.*[4] In the circulation-preserving case we have $\chi = -\phi^* + \text{const.}$

52. Beltrami motion. A motion whose velocity field is a Beltrami field,

$$\mathbf{w} \times \dot{\mathbf{x}} = 0, \qquad \mathbf{w} \neq 0, \tag{52.1}$$

is called a *Beltrami motion.*[1] While the possibility of such motions was known to Stokes (by 1880)[2] and Craig (1880),[3] their importance in hydrodynamics became widely realized only through the work of Beltrami (1889),[4] who constructed some examples. Nearly all of Beltrami's results, however, had been obtained earlier by Gromeka (1881).[5]

Beltrami motions and complex-lamellar motions are mutually exclusive types, and each shares some of the properties of irrotational motions, which, by the definitions employed in the present work, are included as special cases of the latter. In an irrotational motion there are no vortex-lines, the stream-lines are normal to the equipotential surfaces, and the convective acceleration may be determined from the speed alone. In a complex-lamellar motion the vortex-lines are normal to the stream-lines,

---

[4] Substantially this same result is stated by Lamb [1879, **1**, §24].

---

[1] This name was originated by Cisotti [1923, **1**].

[2] While at one time Stokes [1842, **2**, p. 3] stated that $w = 0$ follows from $(52.1)_1$, he later realized his error [footnote, p. 3, 1880 reprint of [1842, **2**]].

[3] Craig [1880, **5**, p. 225] [1880, **6**, p. 276] [1881, **2**, pp. 5–6] casually referred to them as *screw motions*.

[4] [1889, **1**]. Beltrami used the term *helicoidal* for such motions. See also [1921, **2**, §763].

[5] [1881, **4**, Gl. 2].

the stream-lines are endowed with normal surfaces, but the acceleration possesses no particularly simple quality. In a Beltrami motion the vortex-lines coincide with the stream-lines, the acceleration is given by the same simple formula (38.3) as for an irrotational motion, but a congruence of surfaces normal to the stream-lines cannot exist.

For a steady Beltrami motion we have from (38.3)

$$\ddot{\mathbf{x}} = \text{grad}\, \tfrac{1}{2}\dot{x}^2. \tag{52.2}$$

Hence follows the ***first Gromeka-Beltrami theorem*** (1881, 1889): *any steady Beltrami motion is circulation-preserving,*[6] *its acceleration-potential being*

$$\phi^* = -\tfrac{1}{2}\dot{x}^2 + \text{const.} \tag{52.3}$$

For any steady Beltrami motion it follows from (24.6) that we may put $\sigma = j$ in the theorem of §12, *sub fine*, and thus we obtain the ***second Gromeka-Beltrami theorem*** (1881-1889): *in a Beltrami motion in which j is steady the surfaces*

$$\frac{\mathbf{w}}{j\dot{\mathbf{x}}} = \text{const.} \tag{52.4}$$

are stream-surfaces, and in particular (52.4) *holds on each stream-line.*[7]

Finally, Beltrami showed in effect that *a circulation-preserving Beltrami motion is necessarily steady.* The following elegant proof is due to Bjørgum.[8] First, it follows immediately from (46.1) and (45.4) that in *a circulation-preserving Beltrami motion, the vorticity is steady.* Since the abnormality Ω of a Beltrami field is determined by the vorticity field alone (§12), Ω is steady if $\mathbf{w}$ is steady. But $\dot{\mathbf{x}} = \mathbf{w}/\Omega$, and hence *a Beltrami motion is steady if and only if its vorticity be steady.* Several of the foregoing statements may be combined in a single ***theorem of Beltrami*** (1889): *a Beltrami motion is circulation-preserving if and only if it satisfy the following two fully equivalent conditions: it is steady, or its vorticity is steady.*

For an isochoric motion we may put $\mathbf{c} = \dot{\mathbf{x}}$ in (10.11), obtaining the kinetic energy formula of Lamb and J. J. Thomson:[9]

---

[6] Consequently any such motion is dynamically possible for an inviscid fluid obeying any prescribed relation of barotropy $\rho = f(p)$ and subject to any conservative extraneous force. (*Cf.* §49.)

[7] Dr. Van Tuyl has pointed out to me that Gromeka and Beltrami's second theorem is also an immediate consequence of the facts that (1) $\mathbf{w}$ and $j\dot{\mathbf{x}}$ are solenoidal and (2) any two solenoidal vector fields with common vector-lines are proportional along these lines.

[8] [1951, 12, §2.5]. *Cf.* also [1940, 2, §6] [1948, 12, §4]. In [1940, 3] it is shown that for a viscous incompressible fluid the assumption $\Omega = \Omega(z, t)$ implies $\Omega = \text{const.}$

[9] [1879, 1, §136] [1883, 1, §6] [1932, 1, §153].

$$\tfrac{1}{2}\int_{\mathfrak{v}} \dot{x}^2\, dv = \oint_{\mathfrak{s}} d\mathbf{s}\cdot(\tfrac{1}{2}\dot{x}^2\mathbf{r} - \dot{\mathbf{x}}\dot{\mathbf{x}}\cdot\mathbf{r}) + \int_{\mathfrak{v}} \mathbf{r}\cdot\mathbf{w} \times \dot{\mathbf{x}}\, dv. \tag{52.5}$$

In particular, by considering the case when $\mathbf{w} \times \dot{\mathbf{x}} = 0$ throughout $\mathfrak{v}$ and $\dot{\mathbf{x}} = 0$ on $\mathfrak{s}$, as in §10 we may conclude[10] that *in a domain whose finite boundaries are stationary walls to which the material adheres, if in any portion extending to* ∞ *the conditions*

$$\dot{x}^2 = \bar{o}(r^{-3}), \qquad \mathbf{r}\cdot\dot{\mathbf{x}}\dot{\mathbf{x}}_n = \bar{o}(r^{-2}), \tag{52.6}$$

be satisfied, then the only possible continuous isochoric Beltrami or irrotational motion is a state of rest. This result shows that one of the theorems of impossibility of irrotational motion given in §37 can be extended to the case of Beltrami motion. A sufficient condition stronger than (52.6) is

$$\dot{x} = o(r^{-\frac{3}{2}}). \tag{52.7}$$

It would be valuable to investigate in general the question of the consistency of Beltrami motion with the adherence condition (37.6) at a rigid boundary.

There are certain restrictions on the stream-line patterns which can appertain to Beltrami motions.[11] These must follow, of course, from the intrinsic formulae of §9 and §12, which in the present notation become

$$\dot{x}\Omega \neq 0, \qquad \frac{d\dot{x}}{db} = 0, \qquad \dot{x} = \dot{x}_0 \exp \int_0^n \kappa\, dn,$$
$$\Omega = \frac{\dot{\mathbf{x}}\cdot\mathbf{w}}{\dot{x}^2} = \frac{\mathbf{w}\cdot\operatorname{curl}\mathbf{w}}{w^2}. \tag{52.8}$$

In particular, a restatement of a result derived in §12 is: *in a Beltrami motion the vorticity field cannot be lamellar, nor can the lines of flexion be normal to the stream-lines.* But the full consequences of (52.8) are not yet known.

---

10 [1951, 2].

11 Morera [1889, 2] noticed that certain general properties of vortex-tubes imply properties of the stream-tubes in a Beltrami motion. For example, the circulation about any reducible circuit on a stream-tube is zero, and the circulations about any two reconcileable circuits on a stream-tube are equal. Hence if the stream-lines possess a single orthogonal trajectory which is a closed circuit $\mathfrak{c}_0$, since $d\mathbf{x}\cdot\dot{\mathbf{x}} = 0$ upon it, the circulations about all circuits lying upon the stream-tube through $\mathfrak{c}_0$ are zero. However, Morera's conclusion that such a condition is incompatible with Beltrami motion is at best unproved, if not false.

Chapter V. VORTICITY MEASURES

53. The theorem of average intensity balance.[1] The decomposition theorem of Cauchy and Stokes (§34) resolves the local and instantaneous motion into deformative and rotatory parts. We begin our analysis of the significance of these two portions relative to one another by considering their space averages.

From (29.2) we have[2]

$$\dot{x}^j{}_{,i}\dot{x}^i{}_{,j} = \left(\Delta_{ji} - \tfrac{1}{2}\sqrt{g}\,\epsilon_{jik}w^k\right)\left(\Delta^{ij} - \tfrac{1}{2}\frac{\epsilon^{ijl}}{\sqrt{g}}w_l\right),$$

$$= \Delta_{ij}\Delta^{ij} - \tfrac{1}{4}\epsilon_{ijk}\epsilon^{ijl}w^k w_l \qquad (53.1)$$

$$= \Delta_{ij}\Delta^{ij} - \tfrac{1}{2}w^i w_i.$$

By (21.10) follows then[3]

$$\dot{x}^i{}_{,i}\dot{x}^i{}_{,j} = \vartheta^2 - 2II - \tfrac{1}{2}w^2. \qquad (53.2)$$

But by (24.4) we have

$$(\dot{x}^j\dot{x}^i{}_{,j} - \vartheta\dot{x}^i)_{,i} = \dot{x}^j{}_{,i}\dot{x}^i{}_{,j} + \dot{x}^j\dot{x}^i{}_{,ji} - \vartheta_{,i}\dot{x}^i - \vartheta^2,$$
$$= -\vartheta^2 + \dot{x}^j{}_{,i}\dot{x}^i{}_{,j}. \qquad (53.3)$$

Combination of this identity with (53.2) yields a result which can be put into the form

$$\operatorname{div}(\dot{\mathbf{x}}\cdot\operatorname{grad}\dot{\mathbf{x}} - \vartheta\dot{\mathbf{x}}) = -2II - \tfrac{1}{2}w^2. \qquad (53.4)$$

By integrating this equation over a volume and then employing (7.2) and (16.7) we obtain

$$\int_v (4II + w^2)\,dv = -2\oint_s d\mathbf{s}\cdot(\dot{\mathbf{x}}\cdot\operatorname{grad}\dot{\mathbf{x}} - \vartheta\dot{\mathbf{x}}),$$
$$= -2\oint_s d\mathbf{s}\cdot\left(\ddot{\mathbf{x}} - \frac{\partial\dot{\mathbf{x}}}{\partial t} - \vartheta\dot{\mathbf{x}}\right). \qquad (53.5)$$

By formulating conditions under which the surface integral on the right vanishes, we obtain the following remarkable ***theorem of average***

---

[1] The analysis of this and the succeeding section is extracted from [1950, **3**].

[2] [1948, **10**].

[3] This identity was used by Hamel [1936, **2**, §1].

intensity balance between vorticity and deformation: *if all finite boundaries be stationary, and if upon them the material adhere without slipping, while in any portion of the material extending to ∞ the condition*

$$(\dot{\mathbf{x}}\cdot\text{grad}\ \dot{\mathbf{x}} - \vartheta\dot{\mathbf{x}})_{\mathsf{n}} = \bar{o}(r^{-2}) \tag{53.6}$$

be satisfied, then the average value of $4K = 4II + w^2$ *over the entire motion is zero*:

$$\int_{\mathfrak{v}} (4II + w^2)\, dv = 0. \tag{53.7}$$

Corollary 1. *For a motion of the type described in the theorem to be rotational it is necessary that there exist within it a region where*

$$II < 0. \tag{53.8}$$

Corollary 2. *In an irrotational motion satisfying the hypotheses of the theorem the average value of* II *is zero*:

$$\int_{\mathfrak{v}} II\, dv = 0. \tag{53.9}$$

Corollary 3. *In an isochoric motion of the type described in the theorem the average value of the squared vorticity must equal twice the average value of the squared intensity of deformation*:

$$\int_{\mathfrak{v}} w^2\, dv = 2\int_{\mathfrak{v}} \Delta_{ij}\Delta^{ij}\, dv. \tag{53.10}$$

The proofs of corollaries 1 and 2 are immediate, and corollary 3 follows from (21.10). We notice in passing that from corollary 3 and the fact that $\Delta_{ij}\Delta^{ij} > 0$ if $\mathbf{\Delta} \neq 0$ we may derive anew a part of the theorem on the impossibility of irrotational motions in a finite domain (§§37, 52).

54. The Pompeiu-Bilimovitch theorem. The theorem of average intensity balance states that in a broad class of motions the average value of the invariant $4K = 4II + w^2$ is zero. We may now seek to characterize those motions in which this average balance between deformation and vorticity is satisfied by the exact balance $K = 0$ at every point. The condition $K = 0$ occurs in a work of Bilimovitch (1948)[1], who investigated the possible extension of an earlier and apparently unrelated result of Pompeiu.[2] Guided by the outcome of

---

[1] [1948, **7**]. *Note added in proof:* In a new study [1953, **7**] Bilimovitch has considered the dimensions of (54.9).

[2] Pompeiu [1929, **4**] considers only the case of plane isochoric motion. His result is that the analogous areas are related by $\mathfrak{S}_{\mathsf{V}} = \mathfrak{S}_{\mathsf{S}} + \mathfrak{S}_{\mathsf{H}}$. (*Cf.* (54.13).)

this study, we shall present a somewhat more general analysis, which will yield the theorem of Bilimovitch as a special case.

In the present section we shall suppose the spatial frame to be rectangular Cartesian, so that the components of $\mathbf{r}$ are also the components of $\mathbf{x}$.

Let a closed surface $\mathfrak{S}$ be given by the parametric equation $\mathbf{x} = \mathbf{x}(l, m)$, which we may write equivalently as $\mathbf{r} = \mathbf{r}(l, m)$. Then from (7.2) and (15.3) it follows that the volume $\mathcal{V}_{\mathsf{S}}$ of the region inclosed by $\mathfrak{S}$ is given by

$$3\mathcal{V}_{\mathsf{S}} = \oint_{\mathfrak{S}} d\mathbf{s}\cdot\mathbf{r} = \oint_{\mathfrak{S}} \frac{\partial \mathbf{r}}{\partial l} \times \frac{\partial \mathbf{r}}{\partial m}\, dl\, dm, \tag{54.1}$$

where a plus or minus sign is attached quite arbitrarily to $\mathcal{V}_{\mathsf{S}}$ as a consequence of the manner of parametrization. Now consider the velocity vectors $\dot{\mathbf{x}}(\mathbf{r}, t) = \dot{\mathbf{x}}(\mathbf{r}(l, m), t) = \dot{\mathbf{x}}(l, m, t)$ at the points $\mathbf{r}$ upon $\mathfrak{S}$. The termini of these vectors sweep out a surface, whose inclosed volume $\mathcal{V}_{\mathsf{V}}$ is obtained by replacing $\mathbf{r}$ by $\mathbf{r} + \dot{\mathbf{x}}$ in (54.1), *viz.*

$$3\mathcal{V}_{\mathsf{V}} = \oint_{\mathfrak{S}} \frac{\partial(\mathbf{r} + \dot{\mathbf{x}})}{\partial l} \times \frac{\partial(\mathbf{r} + \dot{\mathbf{x}})}{\partial m}\cdot(\mathbf{r} + \dot{\mathbf{x}})\, dl\, dm. \tag{54.2}$$

We shall call $\mathcal{V}_{\mathsf{V}}$ the *velocital volume* of the surface. If the origins of these same velocity vectors be put at the origin of co-ordinates, their termini sweep out a third surface, whose inclosed volume $\mathcal{V}_{\mathsf{H}}$ is given by

$$3\mathcal{V}_{\mathsf{H}} = \oint_{\mathfrak{S}} \frac{\partial \dot{\mathbf{x}}}{\partial l} \times \frac{\partial \dot{\mathbf{x}}}{\partial m}\, dl\, dm. \tag{54.3}$$

We shall call $\mathcal{V}_{\mathsf{H}}$ the *hodographic volume* of the surface. The parameters l and m are selected once for all in the parametrization of $\mathfrak{S}$, so that the parametrization in (54.2) and (54.3) is induced by this initial choice; in particular, while the sign of $\mathcal{V}_{\mathsf{S}}$ is arbitrary, the signs of $\mathcal{V}_{\mathsf{V}}$ and of $\mathcal{V}_{\mathsf{H}}$ are induced by it. By multiplying out in (54.2) we obtain

$$\begin{aligned} 3\mathcal{V}_{\mathsf{V}} = \oint_{\mathfrak{S}} \Big\{ & \frac{\partial \mathbf{r}}{\partial l} \times \frac{\partial \mathbf{r}}{\partial m}\cdot\mathbf{r} + \frac{\partial \dot{\mathbf{x}}}{\partial l} \times \frac{\partial \dot{\mathbf{x}}}{\partial m}\cdot\dot{\mathbf{x}} + \frac{\partial \mathbf{r}}{\partial l} \times \frac{\partial \mathbf{r}}{\partial m}\cdot\dot{\mathbf{x}} \\ & + \frac{\partial \dot{\mathbf{x}}}{\partial l} \times \frac{\partial \dot{\mathbf{x}}}{\partial m}\cdot\mathbf{r} + \left[\frac{\partial \dot{\mathbf{x}}}{\partial l} \times \frac{\partial \mathbf{r}}{\partial m} + \frac{\partial \mathbf{r}}{\partial l} \times \frac{\partial \dot{\mathbf{x}}}{\partial m}\right]\cdot(\mathbf{r} + \dot{\mathbf{x}}) \Big\}\, dl\, dm, \\ = 3\mathcal{V}_{\mathsf{S}} + 3\mathcal{V}_{\mathsf{H}} & + \oint_{\mathfrak{S}} d\mathbf{s}\cdot\dot{\mathbf{x}} + \oint_{\mathfrak{S}} \frac{\partial \dot{\mathbf{x}}}{\partial l} \times \frac{\partial \dot{\mathbf{x}}}{\partial m}\cdot\mathbf{r}\, dl\, dm \\ & + \oint_{\mathfrak{S}} \left[\frac{\partial \dot{\mathbf{x}}}{\partial l} \times \frac{\partial \mathbf{r}}{\partial m} + \frac{\partial \mathbf{r}}{\partial l} \times \frac{\partial \dot{\mathbf{x}}}{\partial m}\right]\cdot(\mathbf{r} + \dot{\mathbf{x}})\, dl\, dm. \end{aligned} \tag{54.4}$$

Now the second integral on the right in this identity may be obtained from the first by interchanging the roles of $\mathbf{r}$ and $\dot{\mathbf{x}}$. Thus by (7.2) we have

$$\int_{\mathfrak{s}} \frac{\partial \dot{\mathbf{x}}}{\partial l} \times \frac{\partial \dot{\mathbf{x}}}{\partial m} \cdot \mathbf{r} \, dl \, dm = \int_{\mathfrak{v}} \operatorname{div}' \mathbf{r}' \, dv', \tag{54.5}$$

where primes indicate that the components of $\dot{\mathbf{x}}$ are to be taken as independent variables. By an evident analogue of (15.7) we have $dv' = j' dv$, where $j' \equiv \partial(\dot{x}, \dot{y}, \dot{z})/\partial(x, y, z)$, while from an analogue of (14.12) and from $(22.5)_1$ we have[3]

$$j' \operatorname{div}' r = \frac{\partial(\dot{y}, \dot{z})}{\partial(y, z)} + \frac{\partial(\dot{z}, \dot{x})}{\partial(z, x)} + \frac{\partial(\dot{x}, \dot{y})}{\partial(x, y)} = K. \tag{54.6}$$

Thus (54.5) becomes

$$\oint_{\mathfrak{s}} \frac{\partial \dot{\mathbf{x}}}{\partial l} \times \frac{\partial \dot{\mathbf{x}}}{\partial m} \cdot \mathbf{r} \, dl \, dm = \int_{\mathfrak{v}} K \, dv. \tag{54.7}$$

To evaluate the last integral in (54.4) it is convenient to replace $\mathfrak{v}$ by the material volume $\mathfrak{V}$ with which it instantaneously coincides, and to regard the parameters l and m as material also, so that l, m, and t are independent. Then

$$\begin{aligned}
\oint_{\mathfrak{s}} \left[\frac{\partial \dot{\mathbf{x}}}{\partial l} \times \frac{\partial \mathbf{r}}{\partial m} + \frac{\partial \mathbf{r}}{\partial l} \times \frac{\partial \dot{\mathbf{x}}}{\partial m} \right] \cdot (\mathbf{r} + \dot{\mathbf{x}}) \, dl \, dm &= \oint_{\mathfrak{s}} \frac{\delta \, d\mathbf{s}}{\delta t} \cdot (\mathbf{r} + \dot{\mathbf{x}}), \\
&= \frac{\delta}{\delta t} \oint_{\mathfrak{S}} d\mathbf{s} \cdot \mathbf{r} - \oint_{\mathfrak{S}} d\mathbf{s} \cdot \dot{\mathbf{x}} + \frac{\delta}{\delta t} \oint_{\mathfrak{S}} d\mathbf{s} \cdot \dot{\mathbf{x}} - \oint_{\mathfrak{S}} d\mathbf{s} \cdot \ddot{\mathbf{x}}, \\
&= 3 \frac{\delta}{\delta t} \int_{\mathfrak{V}} dv - \oint_{\mathfrak{S}} d\mathbf{s} \cdot \dot{\mathbf{x}} + \frac{\delta}{\delta t} \int_{\mathfrak{V}} \vartheta \, dv - \int_{\mathfrak{V}} \operatorname{div} \ddot{\mathbf{x}} \, dv, \qquad (54.8) \\
&= 2 \oint_{\mathfrak{S}} d\mathbf{s} \cdot \dot{\mathbf{x}} + \int_{\mathfrak{V}} [\dot{\vartheta} + \vartheta^2 - \operatorname{div} \ddot{\mathbf{x}}] \, dv, \\
&= 2 \oint_{\mathfrak{s}} d\mathbf{s} \cdot \dot{\mathbf{x}} + 2 \int_{\mathfrak{v}} K \, dv,
\end{aligned}$$

where we have employed (25.2) and (41.9).

By putting (54.7) and (54.8) into (54.4) we finally obtain the identity

$$\mathcal{V}_{\mathsf{V}} = \mathcal{V}_{\mathsf{S}} + \mathcal{V}_{\mathsf{H}} + \oint_{\mathfrak{s}} d\mathbf{s} \cdot \dot{\mathbf{x}} + \int_{\mathfrak{v}} K \, dv. \tag{54.9}$$

[3] Hamel [1936, **2**, Satz 6] proves that if $\vartheta = 0$, $K = 0$, and $j' \neq 0$ then $\operatorname{div}' \mathbf{r}' = 0$. The identity (54.6) shows that his condition $\vartheta = 0$ is superfluous.

Now all the foregoing analysis possesses the rather unusual quality of lacking dimensional homogeneity. Since $\dot{\mathbf{x}}$ and $\mathbf{r}$ are of different dimensions, we may hold $\mathcal{V}_{\mathsf{S}}$ fixed and give $\mathcal{V}_{\mathsf{V}}$ a perfectly arbitrary numerical value by simply keeping the unit of length constant and varying the unit of time. During this process $\mathcal{V}_{\mathsf{H}}$ also will vary. All this is perfectly natural: the change of the time unit changes the geometrical shape of the surfaces bounding $\mathcal{V}_{\mathsf{V}}$ and $\mathcal{V}_{\mathsf{H}}$. Now the two integrals on the right in (54.9) are of different dimensions: for a particular choice of units it is possible that numerically they may have a cancelling effect upon each other, but in order for the relation

$$\mathcal{V}_{\mathsf{V}} = \mathcal{V}_{\mathsf{S}} + \mathcal{V}_{\mathsf{H}} \qquad (54.10)$$

to hold *irrespective of the choice of units* it is necessary and sufficient that the two integrals vanish *separately*:

$$\oint_{\mathfrak{s}} d\mathbf{s}\cdot\dot{\mathbf{x}} = 0, \qquad \int_{\mathfrak{v}} K\, dv = 0. \qquad (54.11)$$

Hence we obtain the ***generalized Pompeiu-Bilimovitch theorem:***[4] *in order that the velocital volume of an arbitrary region equal the sum of its volume and its hodographic volume, for any choice of units of length and time, it is necessary and sufficient that the motion be isochoric and satisfy the further requirement.*

$$4K = 4II + w^2 = 0. \qquad (54.12)$$

The result which we desire here is slightly different. From (54.9) we may conclude also the ***local intensity balance theorem***: *in order that*[5]

$$4II + w^2 = 0 \qquad (54.13)$$

throughout the motion it is necessary and sufficient that the total flux of velocity out of any closed surface shall equal the excess of the velocital

---

[4] In the treatment of Bilimovitch it is assumed at the outset that the motion is isochoric. The following related theorem is proved by Jacob [1944, 1, §3]. Let $\mathbf{u} \equiv \dot{\overline{j}}\, j\dot{\mathbf{x}}$, and in the definitions of $\mathcal{V}_{\mathsf{V}}$ and $\mathcal{V}_{\mathsf{H}}$ replace $\dot{\mathbf{x}}$ by $j\dot{\mathbf{x}}$; then in steady motion

$$\mathcal{V}_{\mathsf{V}} = \mathcal{V}_{\mathsf{S}} + \mathcal{V}_{\mathsf{H}} - \tfrac{1}{2}\oint_{\mathfrak{s}} d\mathbf{s}\cdot\mathbf{u},$$

$$= \mathcal{V}_{\mathsf{S}} + \mathcal{V}_{\mathsf{H}} - \tfrac{1}{4}\oint_{\mathfrak{s}} d\mathbf{s}\cdot \operatorname{grad}\,(j\dot{x})^2 - \tfrac{1}{2}\oint_{\mathfrak{s}} d\mathbf{s}\, j\dot{\mathbf{x}}\cdot \operatorname{curl}\,(j\dot{\mathbf{x}}).$$

For isochoric motion this relation and (54.9) can be shown to reduce to the same form.

[5] The class of motions in which $K = 0$ is studied, first kinematically and then for the case of a viscous liquid, by Hamel [1936, 2, §§2–3].

volume over the sum of the volume and the hodographic volume, whatever the choice of units of length and time:

$$\oint_{s} d\mathbf{s}\cdot\dot{\mathbf{x}} = \mathcal{V}_{\mathsf{V}} - (\mathcal{V}_{\mathsf{S}} + \mathcal{V}_{\mathsf{H}}); \qquad (54.14)$$

in particular, for an isochoric motion the condition becomes

$$\mathcal{V}_{\mathsf{V}} = \mathcal{V}_{\mathsf{S}} + \mathcal{V}_{\mathsf{H}}. \qquad (54.15)$$

In order for the local intensity balance theorem to hold for an isochoric motion it is necessary, except in the trivial case of a rigid translation, that the motion be rotational. For the assumption $w = 0$ in (54.13) yields $II = 0$, putting the which along with $I = 0$ into (21.10) yields $\Delta^{ij}\Delta_{ij} = 0$, a result equivalent to $\mathbf{\Delta} = 0$, so that an isochoric irrotational motion in which the local intensity balance theorem holds is also rigid. An example of an isochoric motion in which the local intensity balance theorem holds is evidently furnished by the generalized Poiseuille motions (44.4).

55. The need for a measure of rotationality.[1] The vorticity $\mathbf{w}$ is a rate, its dimensions being T^{-1}, where T is a unit of time. If $w = 0$ the motion is not rotational; if $w \neq 0$ the motion is rotational, but the numerical magnitude of w is a perfectly arbitrary quantity, being dependent on the choice of the unit of time, and thus can give no indication of *how* rotational a particular motion is. We may therefore set the following problem: *to find a quantitative measure of the amount of rotation in a motion.* We might of course select some particular motion as a standard and rate all others in terms of it. This procedure is used for the choice of physical units: when we say a time is "very short" we mean usually that it is smaller than the 51,400th part of the period of the earth's rotation about its axis. When we say a motion is "slightly rotational," however, we do not really mean merely that it is much less rotational than some other perfectly arbitrarily selected motion. We desire rather some absolute criterion of rotationality.[2] A particular

[1] The analysis of the remainder of this chapter is abridged from [1953, 1].

[2] This point may perhaps be clarified by noting the example which gave rise to the problem. The Gerstner waves, while exact solutions of the hydrodynamical equations, are rotational and thus cannot be generated by the action of impulsive pressures or gravity. For physical usefulness it is of course meaningless to require that $w = 0$ rigorously: it is sufficient that the rotation be negligibly small. There can be no question of smallness with respect to an arbitrary standard, for then whether or not these waves were physically meaningful would become a matter of mere convention. An absolute measure of degree of rotation, much as the strain intensity is an absolute measure of strain, is what is required. For the solution of this problem in the case of the Gerstner waves, see [1953, 1, §12].

class of motions will then indeed appear as a unit or standard, but the precision of this class should be an object of research rather than a matter of mere definition.

In taking up this question Levi-Civita[3] introduced the measure

$$\mathbf{\Omega}(t) \equiv \tfrac{1}{2} \int_0^t \mathbf{w}\, dt, \tag{55.1}$$

the space co-ordinates being held constant during the integration. This quantity is indeed dimensionless, but as a vorticity measure it is quite unsatisfactory nevertheless. For a rigid rotation at constant angular velocity $\frac{1}{2}\mathbf{w}$ we obtain $2\mathbf{\Omega} = \mathbf{w}t$, and thus the *measure* of the rotation increases with the time without limit, while the *character* of the motion is absolutely unchanged: surely it is not more rotational at $t = 10$ than at $t = 0$. If $\mathbf{\Omega}$ be calculated for two different rigid rotations, its magnitude for that at greater angular speed exceeds its magnitude for that at lesser angular speed, but surely any two rigid rotations are equally rotational: the measure of rotation should indicate not the relative speeds but the rotational *quality* or *degree*, and thus for steady rigid motions should be constant in time and independent of the angular speed. $\mathbf{\Omega}$ is a vector, but surely a measure of the rotational quality should be a simple number: in particular, we should expect all rigid rotations to be equally rotational in quality and thus to have the same measure, but $\mathbf{\Omega}$ depends upon the axis of rotation. Finally, according to Levi-Civita's criterion, at some perfectly arbitrary initial instant $t = 0$, every motion has the measure $\mathbf{\Omega}(0) = 0$, the same as that for an irrotational motion.

We shall therefore search for some other dimensionless measure of vorticity.

56. The kinematical vorticity number. The direction in which we should seek a solution to the problem of measuring the rotational quality of a motion is really implied by the Cauchy-Stokes decomposition theorem (§34). The amount of rotation relative to the amount of deformation is

$$\frac{w}{\sqrt{k\Delta^{ij}\Delta_{ij}}} = \sqrt{\frac{w_x^2 + w_y^2 + w_z^2}{k[(\Delta_1)^2 + (\Delta_2)^2 + (\Delta_3)^2]}}, \tag{56.1}$$

where k is a dimensionless constant. The numerator vanishes if and only if every component of $\mathbf{w}$ vanish, and is otherwise positive and increases if any component of $\mathbf{w}$ increase in magnitude; the denominator is endowed with analogous properties with respect to the components of the rate of deformation $\mathbf{\Delta}$, and the ratio is dimensionless.

[3] [1940, 1].

In order to find a suitable value for k, let us consider the special case when the deformation reduces to a pure shearing $\dot{x} = f(y)$, $\dot{y} = 0$, $\dot{z} = 0$. The x = const. planes are instantaneously rotating at angular speed $\partial\dot{x}/\partial y$ about the z-direction, the y = const. planes are instantaneously stationary. The rotations of these two sets of planes define the shearing, and the rate of shearing is the mean $\psi = \Delta_{xy}$ of the two angular rates. The angular speed of rotation, as always, is given by $\omega = \frac{1}{2}w$. Thus (56.1) becomes

$$\frac{w}{\sqrt{2k(\Delta_{xy})^2}} = \frac{2\omega}{\sqrt{2k}\psi}. \tag{56.2}$$

Now it would seem most natural if our measure of the proportion of rotation in a motion should in this special case reduce to the ratio ω/ψ of the one angular speed to the other. From (56.2) it follows that (56.1) will indeed reduce to ω/ψ if and only if $k = 2$. Accordingly we define the ***kinematical vorticity number*** as the dimensionless invariant

$$\mathfrak{W}_{\mathrm{K}} \equiv \frac{w}{\sqrt{2\Delta^{ij}\Delta_{ij}}}. \tag{56.3}$$

This measure may be calculated from the instantaneous velocity field at a point. It is generally a function both of location and of time.

Since a rigid rotation is characterized by $\mathbf{\Delta} = 0$, $w \neq 0$, while an irrotational non-rigid motion is characterized by $w = 0$, $\Delta^{ij}\Delta_{ij} \neq 0$, from (56.3) it follows that *at a given point, a rotational motion is instantaneously rigid if and only if*

$$\mathfrak{W}_{\mathrm{K}} = \infty, \tag{56.4}$$

while a non-rigid motion is instantaneously irrotational if and only if

$$\mathfrak{W}_{\mathrm{K}} = 0. \tag{56.5}$$

Only when the velocity gradient vanishes does the measure (56.3) fail to exist. Thus all possible motions with the sole exception of rigid translations are assigned a numerical degree of rotationality on a scale from 0 to ∞, a rigid motion being the most rotational type of motion possible.

If a reduction in numerical scale were considered desirable, we might replace $\mathfrak{W}_{\mathrm{K}}$ by

$$\mathfrak{V}_{\mathrm{K}} \equiv \frac{1}{\pi} \operatorname{Arc\,tan} \mathfrak{W}_{K}. \tag{56.6}$$

This quantity assumes values in the interval $0 \leqq \mathfrak{V}_{\mathrm{K}} \leqq 1$, with the very interesting class of motions in which $\mathfrak{W}_{\mathrm{K}} = 1$ (to be determined in §58) being assigned a rating $\mathfrak{V}_{\mathrm{K}} = \frac{1}{2}$ midway between the irrotational case ($\mathfrak{V}_{\mathrm{K}} = 0$) and the rigid rotation ($\mathfrak{V}_{\mathrm{K}} = 1$).

A *nearly irrotational motion* is one in which

$$\mathfrak{W}_K \ll 1. \tag{56.7}$$

57. Alternative formulae for the kinematical vorticity number. We shall now establish five alternative expressions for $\mathfrak{W}_K$.

First, by putting (53.1) into (56.3) we obtain

$$\mathfrak{W}_K = \left(1 + \frac{2\dot{x}^i{}_{,j}\dot{x}^j{}_{,i}}{w^2}\right)^{-\frac{1}{2}}. \tag{57.1}$$

Second, by putting (21.10) into (56.3) we obtain

$$\mathfrak{W}_K = \frac{w}{\sqrt{2\vartheta^2 - 4II}}. \tag{57.2}$$

Third, by putting (41.4) into (56.3) we obtain

$$\mathfrak{W}_K = \left[1 + 2\frac{\operatorname{div}\ddot{\mathbf{x}} - \dot{\vartheta}}{w^2}\right]^{-\frac{1}{2}}. \tag{57.3}$$

Fourth, by putting (41.8) into (56.3) we obtain

$$\mathfrak{W}_K = \left[\frac{\nabla^2\dot{x}^2 - 2\dot{\mathbf{x}}\cdot\nabla^2\dot{\mathbf{x}}}{w^2} - 1\right]^{-\frac{1}{2}}. \tag{57.4}$$

Fifth, by putting (16.13) into (56.3) we obtain

$$\mathfrak{W}_K = \left[1 - 2\frac{\nabla^2\phi^* + \dot{\vartheta}}{w^2}\right]^{-\frac{1}{2}}, \tag{57.5}$$

a result remarkable because it does not depend explicitly[1] upon the vector potential $\boldsymbol{\pi}^*$. For a circulation-preserving motion ϕ^* is to be taken as the acceleration-potential.

58. Motions of unit rotationality. Our next problem is to determine the unit of rotationality; that is, to characterize the class of all motions in which $\mathfrak{W}_K = 1$. From the reasons motivating our choice $k = 2$ in the fundamental definition (56.3) it follows that a simple shearing motion is a member of this class, and indeed any generalized Poiseuille motion (44.4) may easily be seen to have $\mathfrak{W}_K = 1$ everywhere. More generally, however, from the formulae (56.3) and (57.1-5) we conclude at once that *in a rotational motion, necessary and sufficient that*

$$\mathfrak{W}_K = 1 \tag{58.1}$$

---

[1] Thus it follows from (17.5) that for fluids of uniform density and viscosity, subject to conservative solenoidal extraneous force, we have

$$\mathfrak{W}_K = \left[1 - \frac{2\nabla^2 p}{\rho w^2}\right]^{-\frac{1}{2}},$$

whatever the value of the viscosity, the inviscid case $\mu = 0$ included.

is any one of the six equivalent conditions

$$w^2 = 2\Delta^{ij}\Delta_{ij}, \tag{58.2}$$

$$x^i{}_{,j}x^j{}_{,i} = 0, \tag{58.3}$$

$$4II + w^2 = 2\vartheta^2, \tag{58.4}$$

$$\operatorname{div}\ddot{\mathbf{x}} = \dot{\vartheta}, \tag{58.5}$$

$$\nabla^2\dot{\mathbf{x}}^2 - 2\dot{\mathbf{x}}\cdot\nabla^2\dot{\mathbf{x}} = 2w^2, \tag{58.6}$$

$$\nabla^2\phi^* = -\dot{\vartheta}. \tag{58.7}$$

The generalized Poiseuille flows appear as the simplest special cases satisfying these conditions.

59. The vorticity number of an isochoric motion. In an isochoric motion the formulae (57.2-5) reduce to the simpler forms

$$\begin{aligned}\mathfrak{W}_K &= \frac{w}{\sqrt{-4II}},\\ &= \left[1 + \frac{2\operatorname{div}\ddot{\mathbf{x}}}{w^2}\right]^{-\frac{1}{2}},\\ &= \left[\frac{\nabla^2\dot{x}^2 - 2\dot{\mathbf{x}}\cdot\operatorname{curl}\mathbf{w}}{w^2} - 1\right]^{-\frac{1}{2}},\\ &= \left[1 - \frac{2\nabla^2\phi^*}{w^2}\right]^{-\frac{1}{2}}.\end{aligned} \tag{59.1}$$

From $(59.1)_2$ or (58.5) it follows that *in a rotational isochoric motion* $\mathfrak{W}_K = 1$ *if and only if the acceleration field be solenoidal*:

$$\operatorname{div}\ddot{\mathbf{x}} = 0; \tag{59.2}$$

in particular, $\mathfrak{W}_K = 1$ *in an isochoric motion without acceleration.* In §24 we noticed that (59.2) is satisfied in any slow isochoric motion; hence *in any slow rotational isochoric motion* $\mathfrak{W}_K = 1$ *at all points.*[1]

From $(59.1)_1$ or (58.4) it follows that *in a rotational isochoric motion* $\mathfrak{W}_K = 1$ *if and only if the local intensity balance theorem holds* (§54).

From $(59.1)_4$ or (58.7) it follows that *in a rotational isochoric motion* $\mathfrak{W}_K = 1$ *if and only if the scalar potential* ϕ^* *of the acceleration be harmonic:*[2]

$$\nabla^2\phi^* = 0. \tag{59.3}$$

---

[1] Thus in particular in a motion consisting of the propagation of infinitesimal shear waves $\mathfrak{W}_K = 1$.

[2] From this result and (17.5) it follows that *for a fluid of uniform density and viscosity, subject to conservative solenoidal extraneous force,* $\mathfrak{W}_K = 1$ *if and only if the pressure be harmonic:* $\nabla^2 p = 0$. Hence $\mathfrak{W}_K = 1$ at all points in the theory of slow motions of viscous fluids, both in the first approximation of Stokes and in the second approximation of Oseen.

60. Further simple cases. If curl $\mathbf{w} = 0$ the formula $(59.1)_3$ reduces to

$$\mathfrak{W}_K = \left[\frac{\nabla^2\dot{x}^2}{w^2} - 1\right]^{-\frac{1}{2}}. \tag{60.1}$$

A plane isochoric motion of uniform vorticity 2ω, given by $\dot{\mathbf{x}} = \mathbf{k} \times \operatorname{grad} \psi(x, y)$, $\nabla^2\psi = 2\omega = w$, may serve as an example. We easily obtain

$$\mathfrak{W}_K = \left[1 + \frac{\left\{\left(\frac{\partial^2\psi}{\partial x\,\partial y}\right)^2 - \frac{\partial^2\psi}{\partial x^2}\frac{\partial^2\psi}{\partial y^2}\right\}}{\omega^2}\right]^{-\frac{1}{2}}. \tag{60.2}$$

Such a motion may be regarded as taking place within the rigid cylindrical boundary $\psi = \text{const.}$, which is supposed to rotate about the z-axis at angular speed ω. For a vessel whose cross-section is a conic, $\psi = (\frac{1}{2}\omega + A)x^2 + (\frac{1}{2}\omega - A)y^2 + Kx + Ly + M$, and (60.2) yields

$$\mathfrak{W}_K = \left|\frac{\omega}{2A}\right|. \tag{60.3}$$

Thus $\mathfrak{W}_K$ is constant throughout the motion, and for different types of vessels assumes all possible values from 0 to ∞. For a circular vessel $\mathfrak{W}_K = \infty$ and the motion is rigid; for an elliptic vessel $\mathfrak{W}_K > 1$; for a parabolic vessel $\mathfrak{W}_K = 1$; for a hyperbolic vessel $\mathfrak{W}_K < 1$. More specifically, for the elliptical vessel $x^2/a^2 + y^2/b^2 = 1$ we obtain

$$\mathfrak{W}_K = \left|\frac{a^2 + b^2}{a^2 - b^2}\right|, \tag{60.4}$$

while for the hyperbolic vessel $x^2/a^2 - y^2/b^2 = 1$ we obtain

$$\mathfrak{W}_K = \left|\frac{a^2 - b^2}{a^2 + b^2}\right|, \tag{60.5}$$

whence it follows that the vorticity number depends only upon the shape of the vessel and is the same at all speeds of rotation, for vessels whose cross-section is a conic.

Another case when the expression for $\mathfrak{W}_K$ is simple is that of an isochoric Beltrami motion, since by (57.3) and (38.3) we then obtain

$$\mathfrak{W}_K = \left[1 + \frac{\nabla^2\dot{x}^2}{w^2}\right]^{-\frac{1}{2}}. \tag{60.6}$$

Hence in an isochoric Beltrami motion for which $\nabla^2\dot{x}^2 = 0$ we have $\mathfrak{W}_K = 1$.

Finally, we note that for unsteady motions of the d'Alembert type (51.6) the vorticity number $\mathfrak{W}_K$ is steady.

61. The range of the kinematical vorticity number. From (57.2) it follows that the identity (53.5) may be put into the form

$$\int_{\mathfrak{v}} w^2\left(\frac{1}{\mathfrak{W}_K^2} - 1\right) dv = 2 \int_{\mathfrak{v}} \vartheta^2\, dv + 2 \oint_{\mathfrak{s}} d\mathbf{s}\cdot(\dot{\mathbf{x}}\cdot\operatorname{grad}\dot{\mathbf{x}} - \vartheta\dot{\mathbf{x}}). \tag{61.1}$$

In particular, the average intensity balance theorem (§53) is equivalent to

$$\int_{\mathfrak{v}} w^2\left(\frac{1}{\mathfrak{W}_K^2} - 1\right) dv = 2 \int_{\mathfrak{v}} \vartheta^2\, dv. \tag{61.2}$$

Since the right hand side is essentially positive, we may conclude that *under the conditions sufficient for the validity of the average balance theorem, it is necessary that in some region*

$$\mathfrak{W}_K \leqq 1; \tag{61.3}$$

if there be any volume changes whatever, there must be a region where

$$\mathfrak{W}_K < 1. \tag{61.4}$$

62. The dynamical vorticity number. The magnitude of $\mathfrak{W}_K$ indicates the amount of vorticity relative to deformation, and thus measures the rotational quality of a known motion. When one is trying to solve hydrodynamical equations by approximate methods one needs a different measure, for the question is then, not how rotational a motion is in the absolute or kinematical sense, but how important the rotation is in determining the flow. In particular, methods of solution based upon the existence of a velocity-potential can be valid only if the *effect* (rather than the amount) of the vorticity is negligible.

Now for any medium (*cf.* §17) the dynamical equation is

$$\rho\ddot{\mathbf{x}} = \rho\left[\frac{\partial\dot{\mathbf{x}}}{\partial t} + \mathbf{w}\times\dot{\mathbf{x}} + \operatorname{grad}\tfrac{1}{2}\dot{x}^2\right] = \cdots, \tag{62.1}$$

where the dots indicate a term which varies according to whether or not gravity, viscosity, or other effects are taken into account. The equation resulting when the vorticity term is neglected is

$$\rho\ddot{\mathbf{x}} \approx \rho\left[\frac{\partial\dot{\mathbf{x}}}{\partial t} + \operatorname{grad}\tfrac{1}{2}\dot{x}^2\right] = \cdots. \tag{62.2}$$

Thus the number

$$\mathfrak{W}_D \equiv \frac{|\mathbf{w}\times\dot{\mathbf{x}}|}{\left|\dfrac{\partial\dot{\mathbf{x}}}{\partial t} + \operatorname{grad}\tfrac{1}{2}\dot{x}^2\right|} \tag{62.3}$$

is a proper measure of the dynamical importance of vorticity, and we shall call it the ***dynamical vorticity number,*** although of course it is determined by purely kinematical quantities. Unlike the kinematical vorticity number $\mathfrak{W}_K$, it depends not only upon the spatial configuration of the velocity field but also upon its local time rate of change. The criterion for *dynamically negligible rotation* is

$$\mathfrak{W}_D \ll 1. \qquad (62.4)$$

From (62.3) it follows at once that *supposing* $\partial\dot{\mathbf{x}}/\partial t + \text{grad}\,\frac{1}{2}\dot{x}^2 \neq 0$, *then*

$$\mathfrak{W}_D = 0 \qquad (62.5)$$

if and only if the motion be an irrotational or a Beltrami motion.

In a steady rigid motion from (16.9) and (16.10) we have $\dot{\mathbf{x}} = \boldsymbol{\omega} \times \mathbf{r}$, where $\boldsymbol{\omega}$ is a constant, $\ddot{\mathbf{x}} = \boldsymbol{\omega} \times (\boldsymbol{\omega} \times \mathbf{r})$, $\mathbf{w} \times \dot{\mathbf{x}} = 2\boldsymbol{\omega} \times (\boldsymbol{\omega} \times \mathbf{r})$, and hence from (62.3) it follows that *in a steady rigid rotation*

$$\mathfrak{W}_D = 2. \qquad (62.6)$$

It is equally evident that *in a steady motion which is neither an irrotational nor a Beltrami motion, if the speed be uniform then*

$$\mathfrak{W}_D = \infty. \qquad (62.7)$$

In these motions the rotation term $\mathbf{w} \times \dot{\mathbf{x}}$ is the entire acceleration, and to neglect it would be to neglect the inertia of the material altogether.

Directly from (62.3) follows that *in order that*

$$\mathfrak{W}_D = 1 \qquad (62.8)$$

it is necessary and sufficient that the acceleration $\ddot{\mathbf{x}}$ *and the Lamb vector* $\mathbf{w} \times \dot{\mathbf{x}}$ *form the base and one leg, respectively, of an isosceles triangle.* In particular, *in any motion without acceleration* $\mathfrak{W}_D = 1$. The simplest case included by this corollary is that of a Poiseuille motion. That this type of motion again appears as strongly rotational is to be expected: if one attempted to neglect its rotation in solving the appropriate dynamical equations one would write $\ddot{\mathbf{x}} \approx \text{grad}\,\frac{1}{2}\dot{x}^2$ instead of the correct result $\ddot{\mathbf{x}} = 0$, and the resulting solution would be completely wrong.

The vorticity numbers $\mathfrak{W}_K$ and $\mathfrak{W}_D$ are easily calculated for any given motion. For the simple vortices (48.4), for example, if we put $n \equiv 2 + r\omega'/r$, then[1]

$$\mathfrak{W}_K = \left|\frac{n}{n-2}\right|, \qquad \mathfrak{W}_D = \left|\frac{n}{n-1}\right|. \qquad (62.9)$$

[1] In [1953, 1] I discuss in greater detail this and the following examples: Couette flow, Gerstner waves, flow behind a curved shock wave.

Chapter VI. Vorticity Averages

63. The linear balance theorems. The previous chapter opened with a demonstration that in a broad class of motions an average balance between the second deformation invariant and the squared magnitude of the vorticity is maintained. We shall now establish two simpler but kinematically less informative relations of balance connecting the vorticity vector $\mathbf{w}$ and the expansion ϑ.

Let h be any single-valued harmonic field: $\mathbf{h} = \operatorname{grad} \chi$, $\nabla^2\chi = 0$. In (7.4) put $n = 0$, $\mathbf{b} = \dot{\mathbf{x}}$, $\mathbf{c} = \mathbf{h}$. Then there results

$$\oint_{\mathfrak{s}} [d\mathbf{s}\cdot(\dot{\mathbf{x}}\mathbf{h} + \mathbf{h}\dot{\mathbf{x}}) - d\mathbf{s}\;\dot{\mathbf{x}}\cdot\mathbf{h}] = \int_{\mathfrak{v}} [\mathbf{h}\vartheta - \mathbf{h} \times \mathbf{w}]\, dv. \tag{63.1}$$

By formulating conditions sufficient for the vanishing of the surface integral we obtain the ***first linear balance theorem***:[1] *let* $\mathbf{h}$ *be a single-valued harmonic vector and let* ϑ *be the expansion and* $\mathbf{w}$ *the vorticity of a motion in a region* $\mathfrak{v}$ *such that*

1. *Each finite boundary is stationary, and upon it the material adheres without slipping;*
2. *In any portion of* $\mathfrak{v}$ *which extends to* ∞,

$$(\dot{\mathbf{x}}\mathbf{h} + \mathbf{h}\dot{\mathbf{x}})_n = \bar{o}(r^{-2}), \qquad \dot{\mathbf{x}}\cdot\mathbf{h} = \bar{o}(r^{-2}); \tag{63.2}$$

then

$$\int_{\mathfrak{v}} [\mathbf{h}\vartheta - \mathbf{h} \times \mathbf{w}]\, dv = 0. \tag{63.3}$$

A sufficient limit condition weaker than (63.2) is

$$h\dot{x} = o(r^{-2}). \tag{63.4}$$

Let $\mathbf{f}$ be a field whose curl is a harmonic field: $\operatorname{curl}\mathbf{f} = \operatorname{grad}\psi$, $\nabla^2\psi = 0$, the harmonic function ψ being single-valued. Then we have

$$\begin{aligned} \operatorname{div}(\psi\dot{\mathbf{x}}) &= \psi\vartheta + \dot{\mathbf{x}}\cdot\operatorname{curl}\mathbf{f}, \\ &= \psi\vartheta + \operatorname{div}(\mathbf{f} \times \dot{\mathbf{x}}) + \mathbf{f} \times \mathbf{w}. \end{aligned} \tag{63.5}$$

Hence by Green's transformation follows

$$\oint_{\mathfrak{s}} d\mathbf{s}\cdot[\psi\dot{\mathbf{x}} + \dot{\mathbf{x}} \times \mathbf{f}] = \int_{\mathfrak{v}} [\psi\vartheta + \mathbf{f}\cdot\mathbf{w}]\, dv. \tag{63.6}$$

---

[1] [1951, 3]. The special case $\vartheta = 0$ for a finite domain is given by Berker [1949, 7, Th. IV].

By formulating conditions sufficient for the vanishing of the surface integral we obtain the ***second linear balance theorem***:[2] *let* ψ *be a single-valued harmonic function, and let* curl $\mathbf{f} = \text{grad } \psi$; *let* ϑ *be the expansion and* $\mathbf{w}$ *the vorticity of a motion in a region* $\mathfrak{v}$ *such that*

1. *Each finite boundary is stationary, and upon it the material adheres without slipping;*
2. *In any portion of* $\mathfrak{v}$ *which extends to* ∞,

$$(\psi \dot{\mathbf{x}} + \dot{\mathbf{x}} \times \mathbf{f})_n = \bar{o}(r^{-2}); \tag{63.7}$$

then

$$\int_{\mathfrak{v}} [\psi\vartheta + \mathbf{f}\cdot\mathbf{w}]\, dv = 0. \tag{63.8}$$

A sufficient limit condition weaker than (63.7) is

$$\psi \dot{x}_n = o(r^{-2}), \qquad (\dot{\mathbf{x}} \times \mathbf{f})_n = o(r^{-2}). \tag{63.9}$$

An immediate corollary of (63.8), following from the choice $\mathbf{f} = 0$, $\psi = 1$, is the vanishing of the total expansion:

$$\int_{\mathfrak{v}} \vartheta\, dv = 0. \tag{63.10}$$

Putting $\mathbf{h}$ = const. in (63.3) and employing (63.10) yields

$$\mathbf{h} \times \int_{\mathfrak{v}} \mathbf{w}\, dv = \mathbf{h} \times \mathcal{W} = 0, \tag{63.11}$$

whence, since $\mathbf{h}$ is arbitrary, follows

$$\mathcal{W} = 0: \tag{63.12}$$

in a motion such that both the linear balance theorems hold, the total vorticity vanishes. We shall discover a broad generalization of this result in §65.

For the case of a motion in a finite domain, from each of the two linear balance formulae it is possible to form sufficient as well as necessary conditions for an adhering motion. That (63.3) is not by itself sufficient is plain from the example $\vartheta = 0$, $\mathbf{w} = \text{grad } r^{-1}$ when the bounding surface $\mathfrak{s}$ is a pair of concentric spheres, since then we have

---

[2] [1951, 4]. The special case $\vartheta = 0$ for a finite domain is given by Berker [1949, 7, Th. I].

$$\int_{\mathfrak{v}} \mathbf{h} \times \mathbf{w}\, dv = \int_{\mathfrak{v}} \mathbf{h} \times \operatorname{grad} \frac{1}{r}\, dv,$$

$$= \int_{\mathfrak{v}} \operatorname{curl}\left(\chi \operatorname{grad} \frac{1}{r}\right) dv, \tag{63.13}$$

$$= \oint_{\mathfrak{s}} d\mathbf{s} \times \chi \operatorname{grad} \frac{1}{r} = 0.$$

For this special case, then, the condition (63.3) is satisfied, but upon the spherical boundaries we have

$$\mathbf{w_n} = \frac{\partial\left(\frac{1}{r}\right)}{\partial r} = -\frac{1}{r^2} \neq 0, \tag{63.14}$$

whence by (42.4) it follows that the material cannot adhere to the boundary. Guided by this particular example, but restricting our attention for simplicity to simply-connected regions, we may now formulate the ***first characterization of a motion of a material which adheres to the bounding surface:***[3] *let* $\mathfrak{v}$ *be a finite simply-connected domain, sufficiently smooth that the transformation of Green is valid for a certain class* C' *of vector fields, that, further, the problem of Dirichlet admits a solution for continuous boundary data, and that for fields* $v \in C'$ *the transformation of Kelvin is valid for each reducible area upon the boundary* $\mathfrak{s}$ *of* $\mathfrak{v}$; *let* C_0' *be the class of fields* $v \in C'$ *which* vanish upon $\mathfrak{s}$; *let* H *be the class of fields harmonic throughout* $\mathfrak{v}$; *let* ϑ *be a function which satisfies a Hölder condition in* $\mathfrak{v}$, *and let* $\mathbf{w}$ *be a solenoidal field whose first partial derivatives exist and satisfy a Hölder condition in* $\mathfrak{v}$; *then in order that there exist a field* $\mathbf{v} \in C_0'$ *such that* $\operatorname{div} \mathbf{v} = \vartheta$, $\operatorname{curl} \mathbf{v} = \mathbf{w}$, *it is both necessary and sufficient that* ϑ *and* $\mathbf{w}$ *satisfy the two further conditions*

$$\mathbf{w_n} = 0, \tag{63.15}$$

$$\int_{\mathfrak{v}} [\mathbf{h}\vartheta - \mathbf{h} \times \mathbf{w}]\, dv = 0, \tag{63.3}$$

for every $\mathbf{h} \in H$. If these conditions be satisfied, then, there exists a field $\mathbf{v}$ which may be regarded as the velocity field $\dot{\mathbf{x}}$ of a motion of a material adhering to the stationary boundary $\mathfrak{s}$.

The part of the theorem which refers to the necessity of (63.3) is included as a special case of the first linear balance theorem, while the necessity of (63.15) was proved in §42. To prove the sufficiency of these

[3] [1951, **3**]. Regularity assumptions for this theorem and the following are stated in detail because these are results of pure potential theory.

two conditions we require the following lemma: *suppose* $\mathbf{u} \in C'$, $\mathbf{w}' \equiv \text{curl } \mathbf{u}$, $\mathbf{w}_n' = 0$; *then there exists a function f of position upon $\mathfrak{s}$, such that the tangential component $\mathbf{u}_t$ of $\mathbf{u}$ is the gradient of f upon $\mathfrak{s}$.* To prove this lemma we observe that by the transformation of Kelvin we have

$$0 = \int_{\mathfrak{s}} ds\, \mathbf{w}_n' = \oint_{\mathfrak{c}} d\mathbf{x}\cdot\mathbf{u} = \oint_{\mathfrak{c}} d\mathbf{x}\cdot\mathbf{u}_t \tag{63.16}$$

for each reducible circuit $\mathfrak{c}$ upon $\mathfrak{s}$. From this fact and from our hypotheses regarding $\mathfrak{s}$ it follows that the integral $\int d\mathbf{x}\cdot\mathbf{u}_t$ with variable upper limit defines a single-valued continuous function f upon $\mathfrak{s}$, and the surface gradient of f is $\mathbf{u}_t$.

Next we require another lemma: *from the hypotheses of the theorem concerning $\mathbf{w}$ and ϑ, it follows that there exists a field $\mathbf{v}' \in C'$ such that* $\text{div } \mathbf{v}' = \vartheta$, $\text{curl } \mathbf{v}' = \mathbf{w}$, $\mathbf{v}_t' = 0$. To prove this lemma, guided by Stokes's formula (35.3) we define

$$4\pi\mathbf{u} \equiv -\text{grad} \int \frac{\vartheta\, dv}{d} + \text{curl} \int \frac{\mathbf{w}\, dv}{d}. \tag{63.17}$$

From the assumptions concerning ϑ and $\mathbf{w}$ it is then possible to show[4] that $\text{div } \mathbf{u} = \vartheta$, $\text{curl } \mathbf{u} = \mathbf{w}$, and $\mathbf{u}$ is continuous upon $\mathfrak{s}$. By the hypothesis (63.15) and the previous lemma, $\mathbf{u}_t$ is the gradient of a certain function f upon $\mathfrak{s}$. The field $\mathbf{v}'$ whose existence is asserted by the lemma is then given by

$$\mathbf{v}' \equiv \mathbf{u} - \text{grad } \lambda \tag{63.18}$$

where λ is the solution of the Dirichlet problem for $\mathfrak{v}$ which assumes the boundary value $\lambda = f$ upon $\mathfrak{s}$.

To prove the theorem itself, finally, we insert the field $\mathbf{v}'$ in place of $\dot{\mathbf{x}}$ in the identity (63.1), a substitution permissible because $\mathbf{v}' \in C'$ and $\text{div } \mathbf{v}' = \vartheta$, $\text{curl } \mathbf{v}' = \mathbf{w}$. By the hypothesis (63.3) concerning the functions ϑ and $\mathbf{w}$ we have then, for any $\mathbf{h} \in H$,

$$\oint_{\mathfrak{s}} [d\mathbf{s}\cdot(\mathbf{v}'\mathbf{h} + \mathbf{h}\mathbf{v}') - d\mathbf{s}\mathbf{v}'\cdot\mathbf{h}] = 0. \tag{63.19}$$

But since $\mathbf{v}_t' = 0$, we have $\mathbf{h}\cdot\mathbf{v}' = \mathbf{h}_n v'$, $ds\, v' = ds\, \mathbf{v}'$, and hence

$$d\mathbf{s}\cdot\mathbf{h}\mathbf{v}' - ds\, \mathbf{v}'\cdot\mathbf{h} = ds\, \mathbf{v}'\cdot\mathbf{h} - ds\, \mathbf{v}'\cdot\mathbf{h} = 0. \tag{63.20}$$

Since $d\mathbf{s}\cdot\mathbf{v}' = ds\, v'$, the result (63.19) becomes

$$\oint_{\mathfrak{s}} ds\, v'\mathbf{h} = 0. \tag{63.21}$$

---

[4] To this end one may use the results of Lichtenstein [1929, 1, Kap. III, §§8, 13–14].

Hence, in particular,

$$\oint_{\mathfrak{s}} ds\, v' h_x = 0. \tag{63.22}$$

But h_x, being one component of a harmonic vector, is itself a harmonic function, and since the problem of Dirichlet admits a solution for $\mathfrak{v}$, h_x may be assigned an arbitrary continuous value upon $\mathfrak{s}$. From (63.22) we have then $\mathbf{v}' = 0$ upon $\mathfrak{s}$, so that the field $\mathbf{v}'$ furnished by the second lemma must belong to C_0', thus being itself the field $\mathbf{v}$ whose existence is asserted by the theorem.

The second linear balance condition (63.8) is as it stands sufficient for the existence of a velocity field of a motion of adherence, as is stated in the ***second characterization of a motion of a material which adheres to the bounding surface:***[5] *let* $\mathbf{w}$, ϑ, C_0', *and* $\mathfrak{v}$ *be defined by the conditions stated in the first characterization theorem; let F be the class of twice continuously differentiable fields such that* curl $\mathbf{f}$ = grad ψ, *where ψ is a single-valued harmonic function; then in order that there exist a field* $\mathbf{v} \in C_0'$ *such that* div $\mathbf{v} = \vartheta$, curl $\mathbf{v} = \mathbf{w}$, *it is necessary and sufficient that*

$$\int_{\mathfrak{v}} [\psi\vartheta + \mathbf{f}\cdot\mathbf{w}]\, dv = 0 \tag{63.8}$$

for each $\mathbf{f} \in F$. The necessity of the condition stated is included as a special case of the second linear balance theorem. To prove its sufficiency, we shall show that both (63.15) and (63.3) are consequences of (63.8), whence the result follows by the first characterization theorem. In fact, by putting $\psi = 1$, $\mathbf{f} = \operatorname{grad} \chi$, where χ is an arbitrary twice continuously differentiable function, from (63.8) and (63.10) we obtain

$$0 = \int_{\mathfrak{v}} \operatorname{grad} \chi\cdot\mathbf{w}\, dv = \int_{\mathfrak{v}} \operatorname{div}(\chi\mathbf{w})\, dv = \oint_{\mathfrak{s}} d\mathbf{s}\cdot\mathbf{w}\chi. \tag{62.23}$$

[5] I discovered this theorem in October 1950; my proof is published in [1951, **4**]. The theorem was published shortly thereafter by Van den Dungen [1951, **11**] and Synge [1951, **10**]; the former's proof of sufficiency is void, while the latter's is admittedly incomplete. The theorem itself is suggested by a result of Synge [1950, **6**] for plane motions, a result which in its turn generalizes an earlier one of Hamel [1911, **2**, p. 266] for plane isochoric motions. To obtain the characterization of Synge, let ψ be a plane harmonic function and let χ be its harmonic conjugate; then if $\mathbf{f} \equiv -\mathbf{k}\chi$, we have curl $\mathbf{f}$ = grad ψ, and (63.8) becomes equivalent to

$$\int_{\mathfrak{s}} ds\, (\psi\vartheta - \chi w) = 0.$$

Since the value of χ upon $\mathfrak{s}$ is arbitrary, (63.15) follows. Second, let $\mathbf{h} \equiv \text{grad}\,\phi$, where ϕ is an arbitrary harmonic function, and let $\mathbf{f}_1 \equiv \mathbf{j}\phi_z - \mathbf{k}\phi_y$. Then

$$\begin{aligned}\text{curl}\,\mathbf{f}_1 &= -\mathbf{i}(\phi_{yy} + \phi_{zz}) + \mathbf{j}\phi_{xy} + \mathbf{k}\phi_{xz},\\ &= \mathbf{i}\phi_{xx} + \mathbf{j}\phi_{xy} + \mathbf{k}\phi_{xz},\\ &= \text{grad}\,\phi_x.\end{aligned} \tag{63.24}$$

Hence $\mathbf{f}_1 \in F$, whence and by the hypothesis (63.8) follows

$$\int_{\mathfrak{v}} [\phi_x\vartheta - (\phi_y w_z - \phi_z w_y)]\,dv = 0. \tag{63.25}$$

By cyclic permutation we may obtain two analogous equations whose integrands are $\phi_y\theta \ldots, \phi_z\theta \ldots$, and by adding the three such equations we obtain the condition (63.3), thus completing the proof of the second characterization theorem.

64. The vorticity average theorems of Lamb, Poincaré, J. J. Thomson, and Bjørgum. Turning aside from the connections between the vorticity and the expansion, we shall now learn the regularities inherent in the distribution of rotation alone by extablishing theorems concerning the average value of the vorticity itself. The first such results to be discovered concern only isochoric motions, upon which we now fix our attention. Since in an isochoric motion the velocity field is solenoidal, we may put $\mathbf{c} = \dot{\mathbf{x}}$ in (10.8), obtaining

$$\int_{\mathfrak{v}} \mathbf{w} \times \dot{\mathbf{x}}\,dv = \oint_{\mathfrak{s}} [d\mathbf{s}\cdot\dot{\mathbf{x}}\dot{\mathbf{x}} - \tfrac{1}{2}\dot{x}^2\,d\mathbf{s}]. \tag{64.1}$$

For a motion within finite stationary boundaries this formula becomes

$$\int_{\mathfrak{v}} \mathbf{w} \times \dot{\mathbf{x}}\,dv = \tfrac{1}{2}\oint_{\mathfrak{s}} \dot{x}^2\,d\mathbf{s}: \tag{64.2}$$

for a continuous isochoric motion in a finite stationary domain, the average value of $\mathbf{w} \times \dot{\mathbf{x}}$ *is determined by the speed on the bounding surfaces; in the case when the speed is constant on each closed boundary surface, the average value of* $\mathbf{w} \times \dot{\mathbf{x}}$ *is zero.* More generally, by formulating conditions under which the surface integral in (64.1) vanishes we obtain the following generalization of ***Lamb's vorticity average theorem*** (1879):[1] *in any continuous isochoric motion, if all finite boundaries be stationary and the material adhere to them, while in any portion of the material extending to infinity the conditions*

$$\dot{\mathbf{x}}\dot{x}_n = \bar{o}(r^{-2}), \qquad \dot{x}^2 = \bar{o}(r^{-2}) \tag{64.3}$$

[1] [1879, 1, §136].

be satisfied, then

$$\int_{\mathfrak{v}} \mathbf{w} \times \dot{\mathbf{x}}\, dv = 0; \tag{64.4}$$

that is, the average value of the Lamb vector is zero. A sufficient condition weaker than (64.3) is

$$\dot{x} = \mathrm{o}(r^{-1}). \tag{64.5}$$

The property which defines the class of irrotational and Beltrami motions, *i.e.* $\mathbf{w} \times \dot{\mathbf{x}} = 0$, thus has been shown to hold *on the average* for a much greater class of motions.

By putting $\mathbf{c} = \dot{\mathbf{x}}$ in (10.12) we similarly obtain

$$\int_{\mathfrak{v}} \mathbf{r} \times (\mathbf{w} \times \dot{\mathbf{x}})\, dv = \oint_{\mathfrak{s}} [d\mathbf{s}\cdot\dot{\mathbf{x}}\mathbf{r} \times \dot{\mathbf{x}} + \tfrac{1}{2}\dot{x}^2\, d\mathbf{s} \times \mathbf{r}]. \tag{64.6}$$

By formulating conditions sufficient for the vanishing of the surface integral we obtain the following generalization of ***Poincaré's vorticity average theorem*** (1893):[2] *in a continuous isochoric motion, if all finite boundaries be stationary and the material adhere to them, while in any portion of the material extending to infinity the conditions*

$$\mathbf{r} \times \dot{\mathbf{x}}\dot{\mathbf{x}}_n = \bar{\mathrm{o}}(r^{-2}), \qquad \dot{x}^2\mathbf{r}_t = \bar{\mathrm{o}}(r^{-2}) \tag{64.7}$$

be satisfied, then

$$\int_{\mathfrak{v}} \mathbf{r} \times (\mathbf{w} \times \dot{\mathbf{x}})\, dv = 0. \tag{64.8}$$

A sufficient condition weaker than (64.7) is

$$\dot{x} = \mathrm{o}(r^{-\frac{3}{2}}), \tag{64.9}$$

sufficient also for the simultaneous validity of Lamb's theorem.

(Two vorticity average theorems for isochoric motions are obtained by putting $\vartheta = 0$ in the two linear balance theorems (63.3) and (63.8).)

J. J. Thomson[3] has studied two other averages connected with the vorticity. First, in (7.4) put $n = 0$, $\mathbf{b} = \dot{\mathbf{x}}$, $\mathbf{c} = \mathbf{r}$:

$$\begin{aligned}\int_{\mathfrak{v}} (\mathbf{w} \times \mathbf{r} + 3\dot{\mathbf{x}} + \mathbf{r}\vartheta)\, dv &= \oint_{\mathfrak{s}} [d\mathbf{s}\cdot(\dot{\mathbf{x}}\mathbf{r} + \mathbf{r}\dot{\mathbf{x}}) - ds\, \dot{\mathbf{x}}\cdot\mathbf{r}], \\ &= -\oint_{\mathfrak{s}} [\mathbf{r} \times (d\mathbf{s} \times \dot{\mathbf{x}}) + d\mathbf{s}\cdot\dot{\mathbf{x}}\mathbf{r}].\end{aligned} \tag{64.10}$$

---

[2] [1893, 1, §115].

[3] [1883, 1, §§4–5].

Since div $(\dot{\mathbf{x}}\mathbf{r}) = \mathbf{r}\vartheta + \dot{\mathbf{x}}$, application of (7.2) to the last term yields

$$\int_{\mathfrak{v}} \mathbf{r} \times \mathbf{w}\, dv = 2 \int_{\mathfrak{v}} \dot{\mathbf{x}}\, dv + \oint_{\mathfrak{s}} \mathbf{r} \times (d\mathbf{s} \times \dot{\mathbf{x}}). \tag{64.11}$$

Second, if we integrate the identity

$$\text{curl}\,(r^2\mathbf{c}) = r^2 \,\text{curl}\,\mathbf{c} + 2\mathbf{r} \times \mathbf{c} \tag{64.12}$$

and then apply (7.3), we obtain

$$-\int_{\mathfrak{v}} r^2 \,\text{curl}\,\mathbf{c}\, dv = 2 \int_{\mathfrak{v}} \mathbf{r} \times \mathbf{c}\, dv - \oint_{\mathfrak{s}} d\mathbf{s} \times r^2\mathbf{c}, \tag{64.13}$$

a result which becomes a vorticity theorem when we put $\mathbf{c} = \dot{\mathbf{x}}$. The surface integrals in (64.11) and (64.13) vanish when the velocity is normal to $\mathfrak{s}$, or, in the case of an infinite region, when $\dot{\mathbf{x}}_r = o(r^{-3})$, $o(r^{-4})$, respectively. The interest in these formulae lies in the fact that for an isochoric motion they express the momentum and the moment of momentum of the material in $\mathfrak{v}$ in terms of the vorticity and the normal component of velocity upon the boundary.[4]

Finally, consider the case when the Monge representation (16.12) is valid over an entire motion. By $(9.3)_3$ and (50.7) we get for the abnormality

$$\begin{aligned} \dot{x}^2\Omega &= \text{grad}\, h \cdot \text{grad}\, f \times \text{grad}\, g, \\ &= \text{div}\,(h\, \text{grad}\, f \times \text{grad}\, g), \\ &= \text{div}\,(h\mathbf{w}), \end{aligned} \tag{64.14}$$

where the second step requires that h be single-valued and the last follows by (40.2). Integration over a volume yields the ***vorticity average theorem of Bjørgum:***[5]

$$\int_{\mathfrak{v}} \dot{x}^2\Omega\, dv = \oint_{\mathfrak{s}} d\mathbf{s} \cdot \mathbf{w}h; \tag{64.15}$$

[4] Moreau [1948, **14**] [1949, **14**] [1950, **9**] [1952, **3**, §§14–22] has studied the rates of change of these integrals in an isochoric motion, expressing them in terms of the resultant force and moment of force of any non-conservative extraneous forces. Proofs of essentially equivalent but purely kinematical formulae based on (74.1) are given in my paper [1951, **1**, §8]. Moreau notes also some alternative forms of (64.13), which are easily obtained by contracting (65.1) in the case $n = 2$. He emphasizes the application of his results to a limitless fluid, all but a finite interior part of which is in irrotational or circulation-preserving motion. In this connection we should beware of the extremely strong order conditions at ∞ required in order to get simple results, order conditions, indeed, which possibly may never be satisfied.

[5] [1951, **12**, §6.6].

hence in a region where (16.12) *with single-valued h is valid and on all whose finite boundaries* **w** *is tangential, while in any portion extending to* ∞ *we have*

$$h\mathbf{w_r} = \bar{o}(r^{-2}), \tag{64.16}$$

then

$$\int_{\mathfrak{v}} \dot{x}^2 \Omega \, dv = 0. \tag{64.17}$$

In such a region, then, if the motion be not complex-lamellar the abnormality Ω *must assume both positive and negative values.* In particular, the theorem holds for a motion in a finite domain to whose walls the material adheres without slipping.[6]

65. The vorticity moment theorem. Since the vorticity field is solenoidal, we may put $\mathbf{c} = \mathbf{w}$ in (10.3), thus obtaining[1]

$$\mathfrak{W}_n \equiv \int_{\mathfrak{v}} \{\mathbf{r}^{(n)}\mathbf{w}\} \, dv = \oint_{\mathfrak{s}} d\mathbf{s}\cdot\mathbf{w}\mathbf{r}^{(n+1)}: \tag{65.1}$$

all the moments $\mathfrak{W}_n$ *of the vorticity field over any region* $\mathfrak{v}$ *are independent of conditions at interior points, being completely determined by the normal component of vorticity upon the boundary* $\mathfrak{s}$. This purely kinematical statement indicates the predominant effect of boundaries upon vorticity. It may be regarded as implying that the familiar hydrodynamical theorem that vorticity cannot be generated in the interior of a homogeneous viscous liquid subject to conservative extraneous force, but must be diffused inward from the boundaries, continues to hold for arbitrary continuous media, provided it be expressed in terms of the *average* rather than the local vorticity.

Now upon a stationary boundary to which the material adheres, by (42.4) the normal component of vorticity vanishes, and thus *in any motion bounded by finite stationary walls, to which the material adheres, all moments* $\mathfrak{W}_n$ *vanish.* By formulating more general conditions sufficient for the vanishing of the surface integral on the right in (65.1), we obtain the following ***vorticity moment theorem:***[2] *if upon any finite*

---

[6] It is worth noting that on the right hand side in (64.15) h may be replaced by f, which is necessarily single-valued, or by g, assumed single-valued. Thus in case any one of the functions f, g, h vanish on a closed surface, the result (64.17) follows independently of any further hypothesis regarding **w**.

---

[1] The case $n = 0$ was apparently known to A. Föppl [1897, **2**, §§4, 32] and is stated in [1941, **1**, eq. 7].

[2] [1949, **3**] [1951, **1**, §11].

boundary of a continuous motion the normal component $\mathbf{w}_n$ *of the vorticity be zero, while in any portion extending to infinity the condition*

$$\mathbf{w}_n = \bar{o}(r^{-n-3}) \qquad (65.2)$$

be satisfied, then the first $n + 1$ *moments of vorticity vanish:*[3]

$$\mathcal{W}_0 = \mathcal{W}_1 = \ldots = \mathcal{W}_n = 0. \qquad (65.3)$$

As a special case we may notice that the result holds *for all n* for the material within a closed vortex-tube of any continuous motion; for the entire material of any continuous motion in a bounded domain upon whose boundaries $w = 0$; and for any such motion within finite stationary boundaries to which the material adheres without slipping. The special case $n = 0$ is a result obtained in another way in §63.

The origin with respect to which $\mathbf{r}$ is taken is arbitrary. To verify the invariance of the result obtained, select a new origin: $\mathbf{r}' = \mathbf{r} + \mathbf{r}_0$, where $\mathbf{r}_0$ is a constant vector, and distinguish by primes moments taken with respect to the new origin. Then by (65.1)

$$\mathcal{W}_n' = \int_{\mathfrak{v}} \{(\mathbf{r} + \mathbf{r}_0)^{(n)}\mathbf{w}\}\, dv = \oint_{\mathfrak{s}} d\mathbf{s}\cdot\mathbf{w}(\mathbf{r} + \mathbf{r}_0)^{(n+1)}. \qquad (65.4)$$

Conditions sufficient to insure the vanishing of $\mathcal{W}_0, \mathcal{W}_1, \ldots, \mathcal{W}_n$ insure also the vanishing of the right-hand side of (65.4), and hence $\mathcal{W}_0'$, $\mathcal{W}_1', \ldots, \mathcal{W}_n'$ vanish.

For the special case of *a motion enclosed by finite stationary boundaries to which the material adheres*, our several investigations (§§37, 53, 65) have revealed a high degree of regularity: *the motion is almost certainly rotational (if isochoric, certainly rotational), the intensity balance theorem holds, the linear balance theorems hold, and all the moments* $\mathcal{W}_n$ *are zero.* It is to be borne in mind that no restriction regarding the connectivity has been presupposed, and indeed the main interest here is in the case of multiply-connected regions.

66. A generalization. The vorticity moment theorem is included as a special case of a result analogous to the general convection theorem of §25 for isochoric motions, since the solenoidal character of $\mathbf{w}$ permits us to put $\mathbf{c} = \mathbf{w}$ in (10.5):

$$\oint_{\mathfrak{v}} \mathbf{w}\cdot\operatorname{grad}\emptyset\, dv = \oint_{\mathfrak{s}} d\mathbf{s}\cdot\mathbf{w}\emptyset. \qquad (66.1)$$

Thus *for any twice continuously differentiable quantity* $\emptyset$, *the total* $\mathbf{w}\cdot\operatorname{grad}\emptyset$ *in a region is completely determined by the values of* $\emptyset$ *and of the normal*

[3] A special case of the case $n = 0$ was given by A. Föppl [1897, 2, §§4, 32].

component of vorcitity $\mathbf{w_n}$ *upon the boundary* $\mathfrak{s}$. Our vorticity moment theorem (65.1) is the special $\emptyset = \mathbf{r}^{(n+1)}$ in (66.1).

Another special case has been obtained by Berker.[1] First, put $\emptyset = b$ and write $\mathbf{f} = \text{grad } b$ in (66.1):

$$\int_{\mathfrak{v}} \mathbf{w} \cdot \mathbf{f} \, dv = \oint_{\mathfrak{s}} d\mathbf{s} \cdot \mathbf{w} b. \tag{66.2}$$

By formulating conditions sufficient for the vanishing of the surface integral we obtain the following vorticity average theorem: *given a motion in a region* $\mathfrak{v}$ *such that upon any finite boundaries the normal component of vorticity vanish, let* b *be a twice continuously differentiable scalar and let* $\mathbf{f} \equiv \text{grad } b$; *if in any portion of* $\mathfrak{v}$ *extending to* ∞ *the condition*

$$b\mathbf{w_n} = \bar{o}(r^{-2}) \tag{66.3}$$

be satisfied, then

$$\int_{\mathfrak{v}} \mathbf{w} \cdot \mathbf{f} \, dv = 0. \tag{66.4}$$

The conditions of this theorem are satisfied by the material within a closed vortex-tube of any continuous motion; by the entire material of any continuous motion in a bounded domain upon whose boundaries $w = 0$; and for any such motion within finite stationary boundaries to which the material adheres without slipping. Putting $\mathbf{f} = \mathbf{r} = \text{grad } \frac{1}{2}r^2$, we obtain

$$\int_{\mathfrak{v}} \mathbf{r} \cdot \mathbf{w} \, dv = 0, \tag{66.5}$$

a result which follows equally by taking the scalar of the equation $\mathfrak{W}_1 = 0$.

The form of the result (66.4) coincides with that of the special case $\vartheta = 0$ of (63.8), but the conditions under which the two statements hold are somewhat different. For the validity of (66.4) it is required that $\text{curl } \mathbf{f} = 0$, but the motion need not be isochoric, while for the special case of (63.8) valid in an isochoric motion it is sufficient that $\text{curl curl } \mathbf{f} = 0$.

67. An essential difference between three-dimensional and plane motions. A hasty judgment might form the opinion that results proved in general for three-dimensional motions necessarily have counterparts in plane motions, but such an opinion is quite false. Consider, for example, the vorticity moment theorem proved in §65.

[1] [1949, **7**, Th. II].

Like any other meaningful statement, it rests upon certain assumptions. In the simplest case, these are that the motion should vanish at ∞ to a certain order, and that the material adhere to finite stationary boundaries—assumptions natural enough and satisfied by a large class of three-dimensional motions. However, a plane motion whose generating plane is that of x and y never vanishes at ∞ in the z direction, and thus the whole motion can never satisfy the conditions of our theorem. No end is served by applying the theorem to the finite region bounded by the planes $z = 0$ and $z = 1$ and a right cylinder with generators parallel to the z-axis and base a certain area $\mathfrak{S}$ in the x-y plane: the integrands in (65.1) fail to become independent of z, and thus the vanishing of $\mathcal{W}_n$ yields no information whatever concerning the average value of $\mathbf{w}$ over the plane area $\mathfrak{S}$. The uselessness of the result is to be expected from its triviality, since $\mathcal{W}_n = 0$ for the cylindrical region may be seen to be a consequence of mere symmetry.

In general, as the preceding example indicates, the theorems of average value deduced in the foregoing three sections fail to reduce to a non-trivial special case when applied to a plane motion. The underlying reason for this difference lies in the fact that *for the local rotation to be describable by a vector it is necessary and sufficient that the number of dimensions be three.* Our whole analysis rests upon the possibility of forming Gibbsian cross-operators. Not only can it not be generalized to spaces of higher dimension, but also it generally fails to carry down even to the Euclidean plane.

Chapter VII. BERNOULLIAN THEOREMS

68. The classical Bernoulli-Euler theorem. So as to anchor to the unshakeable bed rock of classical tradition the rather general treatment which it is the place of this chapter to present, we shall begin by restating the simple theorem of D. Bernoulli (1730),[1] as generalized by Euler (1755).[2] Let an inviscid fluid subject to conservative extraneous force be in irrotational motion over a period of time. Since any persistently irrotational motion of any medium is circulation-preserving, by the Hankel-Kelvin theorem (§49) it follows that if the medium be an inviscid fluid subject to conservative extraneous force the flow must be barotropic and the formula (49.4) for the acceleration-potential must be valid. In an irrotational motion of any medium, however, the acceleration-potential is given by (48.3). Comparing these two results yields the ***Bernoulli-Euler theorem for potential flow***:

$$-\frac{\partial \phi}{\partial t} + \tfrac{1}{2}\dot{x}^2 + \int \frac{dp}{\rho} + \upsilon = f(t). \tag{68.1}$$

In a steady motion (68.1) reduces to

$$\tfrac{1}{2}\dot{x}^2 + \int \frac{dp}{\rho} + \upsilon = \text{const.} \tag{68.2}$$

As it stands, the Bernoulli-Euler theorem may naturally be regarded in one of two ways: *it is a formula for the squared speed* $\dot{x}^2$ in terms of dynamical variables, or *it is a formula for the pressure* p in terms of kinematical variables. Dynamical considerations, however, are not really needed to derive it, and the dynamical aspects of the result are entirely a matter of interpretation. It is rather Euler's formula (48.3), *viz.*

$$\phi^* + \tfrac{1}{2}\dot{x}^2 - \frac{\partial \phi}{\partial t} = 0, \tag{48.3}$$

[1] The "Bernoulli theorem" antedates hydrodynamics, having originated in the early hydraulics. The result of D. Bernoulli [1738, 1, Sect. XII, §10], valid only in steady flow of an incompressible fluid, follows from a daring but dubious use of the principle of conservation of energy. The extension to unsteady flow was obtained by John Bernoulli [1743, 1, Part I, §IX; Part II, §VI], to whom is due both the first derivation from the momentum principle and also the now usual derivation from the energy principle. The now familiar hydraulic form was first published in Euler's papers on hydraulic machines (see my history [1954, 1, Parts IV & VI]).

[2] [1761, 1, §§51–52, 62–64] [1757, 2, §XXVII]. See also §75[3].

which from the present viewpoint we shall prefer to regard as the essential statement of Bernoulli's theorem. Put in this way, the Bernoulli-Euler theorem becomes *a theorem of pure kinematics, valid in any irrotational motion of any medium whatever.* For the case of an inviscid incompressible fluid we have also the formula (49.4), *viz.*

$$\phi^* = \int \frac{dp}{\rho} + \upsilon + f(t), \tag{49.4}$$

furnishing *a dynamical interpretation for* ϕ^*, *and hence for the Bernoulli-Euler theorem* (48.3). The classical formula (68.1) results from elimination of ϕ^* between (48.3) and (49.4).

This method of looking upon Bernoulli's theorem was given in essence by Euler (1755),[3] and the equation (49.4) is his. Euler himself made no use of this easy conquest, and like many another outpost seized in passage during his effulgent campaigns, the vantage-point of the acceleration-potential has been but intermittently held by succeeding investigators, though it possesses two tactical advantages. The first and lesser is that in the form (48.3) the Bernoulli equation holds for steady irrotational motions of various other media, for which ϕ^* has a dynamical interpretation[4] different from (49.4). The second is that a whole vista of different generalizations of (48.3) itself opens here before our eyes. It is to these latter that the present chapter is devoted. Considerations of the detailed conclusions which can be drawn from the classical theorem (68.1) it is more convenient to defer until after the more general results of the same type have been obtained.

69. The nature of Bernoullian theorems. In any continuous motion the acceleration $\ddot{\mathbf{x}}$ may be expressed in terms of its Stokes potentials (16.13). The scalar potential ϕ^* we shall regard as a generalization of the acceleration-potential ϕ^* occurring in the classical Bernoulli-Euler theorem (48.3). *By a Bernoullian theorem we shall mean a formula for the squared speed* $\dot{x}^2$, *the scalar potential* ϕ^*, *or the* ***Bernoulli function*** *or flow energy* $\phi^* + \frac{1}{2}\dot{x}^2$.

The Bernoullian theorems we shall obtain are of five types, according as $\dot{x}^2$, ϕ^*, or $\phi^* + \frac{1}{2}\dot{x}^2$ equals

1. A volume integral.
2. A line integral.

---

[3] [1761, 1, §§79–81] [1757, 2, §XXVIII] [1770, 1, §§42, 50–51].

[4] Consider, for example, a viscous compressible fluid. From Poisson's dynamical equation (17.1) it follows that an irrotational motion is possible only if $p + (\lambda + 2\mu)\vartheta = f(\rho)$, the extraneous force being supposed conservative, and in this case

$$\phi^* = \int \frac{df}{\rho} + \upsilon.$$

3. A scalar function of position, time, and a particular curve.
4. A scalar function of position, time, and a particular surface.
5. A scalar function of position and time.

The classical theorem of §68 is a special case in every instance.

70. Poisson equations for $\phi^* + \frac{1}{2}\dot{x}^2$ or for ϕ^*. From the expression (16.13) for the acceleration in terms of its Stokes potentials we have immediately

$$-\nabla^2\phi^* = \operatorname{div} \ddot{\mathbf{x}}. \tag{70.1}$$

Substitution for div $\ddot{\mathbf{x}}$ of the various equivalents derived in §41 then yields a sequence of Poisson equations for ϕ^* or for $\phi^* + \frac{1}{2}\dot{x}^2$, whose significance lies in the fact that the *vector potential* $\boldsymbol{\pi}^*$ *has been eliminated.*[1] All these results are here presented for the first time in full generality, but we shall name them after the discoverers of the special cases occurring in the theory of viscous fluids.

First, from $(41.10)_1$ and $(41.10)_2$ we have the ***differential equations of Bobylew*** (1873):[2]

$$\begin{aligned}\nabla^2\left(\phi^* + \tfrac{1}{2}\dot{x}^2\right) &= -\frac{\partial\vartheta}{\partial t} + \operatorname{div}(\dot{\mathbf{x}} \times \mathbf{w}),\\ &= -\frac{\partial\vartheta}{\partial t} + w^2 - \dot{\mathbf{x}}\cdot\operatorname{curl}\mathbf{w}.\end{aligned} \tag{70.2}$$

As a corollary it follows that *in any motion with steady expansion, in order that the Bernoulli function be harmonic*:

$$\nabla^2(\phi^* + \tfrac{1}{2}\dot{x}^2) = 0, \tag{70.3}$$

it is necessary and sufficient that the Lamb vector be solenoidal. In particular, (70.3) *holds in any irrotational or Beltrami motion in which the expansion is steady.* As a further corollary it follows that *in an irrotational or*

---

[1] From (17.5) it follows that for a viscous incompressible fluid of uniform viscosity *the viscosity* μ *does not appear in* ϕ^*, and hence when the theorems of this and the next section are applied to incompressible fluids *their form is independent of* μ *and thus is precisely the same for viscous and for inviscid liquids.* This fact has been remarked in special cases by Hamel [1936, **2**, §1] and by Carstoiu [1947, **8**, Ch. V, §§7–8].

[2] Bobylew [1873, **1**, §4] considered the case of a viscous incompressible fluid; it was noted (in effect) by Forsyth [1879, **3**, pp. 139–139] that from the remark in the preceding footnote it follows that *if a given velocity field* $\dot{\mathbf{x}}(\mathbf{x}, t)$ *be dynamically possible both for a viscous and for a perfect incompressible fluid of the same uniform density, if the pressure assumes the same boundary values in the two cases it also assumes the same values at interior points.* A special case of the equation (70.3) is given also in [1880, **5**, pp. 223–225] [1880, **7**, p. 274] [1880, **8**, p. 268].

Beltrami motion filling all space, if the expansion be steady and if the Bernoulli function $\phi^* + \frac{1}{2}\dot{x}^2$ *be uniformly bounded then the classical Bernoulli theorem holds:*[3]

$$\phi^* + \tfrac{1}{2}\dot{x}^2 = f(t). \tag{70.4}$$

From (41.3) follows the ***differential equation of Rowland*** (1880):[4]

$$-\nabla^2\phi^* = \dot{\vartheta} + \Delta^{ij}\Delta_{ij} - \tfrac{1}{2}w^2. \tag{70.5}$$

For the special case $\dot{\vartheta} = 0$ Hamel[5] has remarked that since $w^2 \geqq 0$, $\Delta^{ij}\Delta_{ij} \geqq 0$, we may obtain bounds for $\nabla^2\phi^*$, *viz.*, in a motion in which *the expansion remains constant for each particle we have*

$$-\Delta^{ij}\Delta_{ij} \leqq \nabla^2\phi^* \leqq \tfrac{1}{2}w^2. \tag{70.6}$$

This result, which holds *a fortiori* for isochoric motions, may be expressed in terms of the vorticity number $\mathfrak{W}_K$ of §56:

$$\frac{\nabla^2\phi^*}{\Delta^{ij}\Delta_{ij}} \leqq \mathfrak{W}_K^{\,2}, \qquad \frac{\nabla^2\phi^*}{\frac{1}{2}w^2} \geqq -\frac{1}{\mathfrak{W}_K^{\,2}}. \tag{70.7}$$

From (70.5) it is plain that the function ϕ^* may be either subharmonic ($\nabla^2\phi^* \geqq 0$), harmonic ($\nabla^2\phi^* = 0$), or superharmonic ($\nabla^2\phi^* \leqq 0$). Circumstances leading to these cases become apparent as soon as we write Rowland's equation (70.5) in the form

$$-\nabla^2\phi^* = \dot{\vartheta} + \Delta^{ij}\Delta_{ij}[1 - \mathfrak{W}_K^{\,2}], \tag{70.8}$$

inspection of which yields the following results: *in a non-rigid motion,*

if $\dot{\vartheta} \geqq 0$ *and* $\mathfrak{W}_K \leqq 1$, *then* ϕ^* *is superharmonic;*
if $\dot{\vartheta} = 0$ *and* $\mathfrak{W}_K = 1$, *then* ϕ^* *is harmonic;*
if $\dot{\vartheta} \leqq 0$ *and* $\mathfrak{W}_K \geqq 1$, *then* ϕ^* *is subharmonic.*

In the special case when $\dot{\vartheta} = 0$ the value of $\mathfrak{W}_K$ becomes the sole criterion of the character of ϕ^*. From analytical theorems it follows that ϕ^* cannot attain a maximum in the interior of a region where it is subharmonic, nor a minimum in the interior of a region where it is superharmonic. Hence, for example, from (70.8) follow quite specific conclusions regarding various types motions: first, the ***maximum theorem for***

---

[3] Consequently in any such Beltrami motion, steady or not, of a viscous liquid we have $p/\rho + \frac{1}{2}\dot{x}^2 + \upsilon = f(t)$. The result is of interest only for the unsteady case; in a steady motion of this type the stronger result of §76 applies.

[4] Rowland [1880, 8, p. 267] gives the special case valid in an incompressible inviscid fluid, subject to conservative extraneous force.

[5] [1936, 2, §1].

motions in which $\mathfrak{W}_K \geqq 1$: *given a motion in which the expansion experienced by a particle does not increase* ($\dot{\vartheta} \leqq 0$), *the greatest value of the scalar potential* ϕ^* *in a region where* $\mathfrak{W}_K \geqq 1$ *cannot be attained in the interior but must be attained in the boundary*; and, second, the ***minimum theorem for motions in which*** $\mathfrak{W}_K \leqq 1$: *given a motion in which the expansion experienced by a particle does not decrease* ($\dot{\vartheta} \geqq 0$), *the least value of* ϕ^* *in a region where* $\mathfrak{W}_K \leqq 1$ *cannot be attained in the interior, but must be attained on the boundary.*[6] The former theorem includes the case of rigid motion ($\mathfrak{W}_K = \infty$, $\vartheta = 0$); the latter the case of irrotational motion ($\mathfrak{W}_K = 0$); both include the case $\vartheta = 0$, $\mathfrak{W}_K = 1$, which was mentioned from another point of view in §59. Both theorems have immediate and important application in hydrodynamics.[7]

For irrotational or Beltrami motions, an analogous result can be obtained for the speed $\dot{x}$, broadly generalizing that obtained in §41. Write (70.8) in the form

$$\operatorname{div} \ddot{\mathbf{x}} = \dot{\vartheta} + \Delta^{ij}\Delta_{ij}[1 - \mathfrak{W}_K{}^2] \tag{70.9}$$

---

[6] These results are broad generalizations of those of Hamel [1936, **2**, §1]; they were first published in [1953, **1**, §10].

[7] For the special case of a homogeneous viscous incompressible fluid subject to conservative extraneous force we have the simple expression (17.5) for ϕ^*, and hence interpretation of the general theorems just stated yields the ***Hamel theorem of pressure extremes for an incompressible fluid*** (1936): *given a flow of an incompressible fluid, whether viscous or not, subject to conservative extraneous force, then*

1. *In a region where* $\mathfrak{W}_K \geqq 1$ *the greatest value of* $p/\rho + v$ *must occur on the boundary and cannot occur in the interior, while*
2. *In a region where* $\mathfrak{W}_K \leqq 1$ *the least value of* $p/\rho + v$ *must occur on the boundary and cannot occur in the interior;*

in case the potential v *of the extraneous forces be harmonic, the pressure itself may be substituted for* $p/\rho + v$ *in the statements* 1 *and* 2. (The result actually given by Hamel [1936, **2**, §1] is not put in terms of $\mathfrak{W}_K$.)

In the case of irrotational motion ($\mathfrak{W}_K = 0$) we shall call the statement 2 ***Bouligand's theorem,*** since it was Bouligand [1927, **6** & **7**] who discovered it for inviscid fluids and deduced from it some important dynamical consequences. For example, in the free fall, starting from rest and subject to gravity, of a mass of incompressible inviscid fluid in a vacuum, so long as the motion be continuous cavitation can never result; for, the motion being initially irrotational, by the second Lagrange-Cauchy theorem (§§36, 104) it must remain so, so that the Bouligand theorem is applicable, from which it follows that, since the pressure is zero upon the entire boundary, it can never be negative at an interior point.

Since the hypotheses leading to the Bouligand theorem insure the simultaneous validity of the Bernoulli theorem (68.1), by making additional assumptions we can derive a statement regarding the greatest speed. The result concerning irrotational motions which is given in the text above is more general, however. Extension of the celebrated theorem of Poincaré regarding the greatest possible angular speed of a rotating mass of fluid is given in [1953, **1**, §10].

and note that by $(41.10)_1$ we get consequently whenever $\mathbf{w} \times \dot{\mathbf{x}} = 0$ simply

$$\nabla^2 \tfrac{1}{2}\dot{x}^2 = \dot{\vartheta} - \frac{\partial \vartheta}{\partial t} + \Delta^{ij}\Delta_{ij}[1 - \mathfrak{W}_K{}^2], \qquad (70.10)$$
$$= \dot{\mathbf{x}}\cdot\text{grad}\,\vartheta + \Delta^{ij}\Delta_{ij}[1 - \mathfrak{W}_K{}^2].$$

By formulating conditions sufficient that the right hand side be non-negative (zero) (non-positive), we get the following ***theorem of maximum speed for irrotational or Beltrami motions:***[8] *in a region of irrotational or Beltrami motion such that* $\mathfrak{W}_K \leqq 1$ $(\mathfrak{W}_K = 1)$ $(\mathfrak{W}_K \geqq 1)$, *while the expansion* ϑ *does not decrease (change) (increase) in the direction of motion along a stream-line, then at an interior point the speed cannot experience a maximum (maximum or minimum) (minimum).* In the irrotational case, of course, $\mathfrak{W}_K = 0$, so the results concerning minimum speed do not apply (*cf.* §36). A corollary is that in a region of isochoric Beltrami motion where $\mathfrak{W}_K \geqq 1$ there can be no stagnation point.

From (41.9) and (70.1) we readily obtain the ***differential equation of Lichtenstein and Lagally*** (1929, 1937):[9]

$$-\nabla^2\phi^* = \dot{\vartheta} + \vartheta^2 - 2K, \qquad (70.11)$$

or, equivalently,

$$-\nabla^2\phi^* = \dot{\vartheta} + \vartheta^2 - 2(II + \tfrac{1}{4}w^2). \qquad (70.12)$$

Comparing this result with the Bilimovitch theorem (§54) and the next to last result of §59 yields the following ***theorem of the harmonic scalar potential of acceleration***: *in an isochoric motion such that the velocital volume of any region equals the sum of its volume and its hodographic volume, or, equivalently,* $\mathfrak{W}_K = 1$, *it follows that the scalar potential* ϕ^* *of the acceleration is harmonic; consequently, the greatest and least values of* ϕ^* *occur upon the boundary and not in the interior of the region, and in the case when the region is the whole of space, then if* ϕ^* *be uniformly bounded it must be constant.* For incompressible fluids this theorem, like the theorems of maximum and minimum stated before, admits an interpretation in terms of the pressure.

The equation (70.2) shows that $\phi^* + \frac{1}{2}\dot{x}^2$ is subharmonic if and only if $-\partial\vartheta/\partial t + w^2 - \dot{\mathbf{x}}\cdot\text{curl}\,\mathbf{w} \geqq 0$. Thus, in particular, *in a motion in which the expansion nowhere increases in time and in which the velocity*

---

[8] [1953, 5]. In the isochoric case the result reduces to that of §41, which was noted by Jacob [1944, 1, §5].

[9] Lichtenstein [1929, 1, Ch. 10, §6] gives the special case for inviscid incompressible fluids, subject to conservative extraneous force. The more general result for viscous fluids is given by Lagally [1937, 2].

and the flexion vectors nowhere subtend an acute angle, the function $\phi^* + \frac{1}{2}\dot{x}^2$ *cannot experience a maximum at an interior point.*

71. Volume integral Bernoulli theorems. Each of the Poisson equations just derived may be integrated by Poisson's formula, and in each case there results an expression for ϕ^* or for $\phi^* + \frac{1}{2}\dot{x}^2$ as a volume integral.

Directly from $(70.2)_1$ follows the ***Bernoulli theorem of Bobylew:***[1]

$$\phi^* + \tfrac{1}{2}\dot{x}^2 = \frac{1}{4\pi}\int_{\mathfrak{v}} \frac{\operatorname{div}\left[\dfrac{\partial \dot{\mathbf{x}}(\boldsymbol{\xi}, t)}{\partial t} + \mathbf{w}(\boldsymbol{\xi}, t) \times \dot{\mathbf{x}}(\boldsymbol{\xi}, t)\right]}{d(\mathbf{x}, \boldsymbol{\xi})}\, dv(\boldsymbol{\xi}) + h_1, \quad (71.1)$$

where $\mathfrak{v}$ is the entire volume occupied by the motion and h_1 is a suitably chosen harmonic function of $\mathbf{x}$, containing the time t as a parameter. This result is a generalization of (70.3), since it shows the manner in which the Bernoulli function is determined up to a harmonic function by the distribution of the expansion ϑ and the Lamb vector $\mathbf{w} \times \dot{\mathbf{x}}$, in any continuous motion of any medium.

Now in general we have, employing rectangular Cartesian co-ordinates,

$$\frac{\operatorname{div} \mathbf{b}(\boldsymbol{\xi})}{d(\mathbf{x}, \boldsymbol{\xi})} = \frac{\operatorname{div} \mathbf{b}(\boldsymbol{\xi})}{|\mathbf{x} - \boldsymbol{\xi}|} = \operatorname{div} \frac{\mathbf{b}(\boldsymbol{\xi})}{|\mathbf{x} - \boldsymbol{\xi}|} - \frac{\mathbf{b}(\boldsymbol{\xi})\cdot(\mathbf{x} - \boldsymbol{\xi})}{|\mathbf{x} - \boldsymbol{\xi}|^3}. \quad (71.2)$$

Substitution of this result into (71.1) and application of Green's transformation then yields the ***Bernoulli theorem of Forsyth:***[2]

$$\begin{aligned} \phi^* + \tfrac{1}{2}\dot{x}^2 &= \frac{1}{4\pi}\int_{\mathfrak{v}} \frac{\dfrac{\partial \dot{\mathbf{x}}(\boldsymbol{\xi}, t)}{\partial t} + \mathbf{w}(\boldsymbol{\xi}, t) \times \dot{\mathbf{x}}(\boldsymbol{\xi}, t)}{[d(\mathbf{x}, \boldsymbol{\xi})]^3}\cdot(\boldsymbol{\xi} - \mathbf{x})\, dv(\boldsymbol{\xi}) \\ &\quad + \frac{1}{4\pi}\oint_{\mathfrak{s}} \frac{\dfrac{\partial \dot{\mathbf{x}}(\boldsymbol{\xi}, t)}{\partial t} + \mathbf{w}(\boldsymbol{\xi}) \times \dot{\mathbf{x}}(\boldsymbol{\xi}, t)}{d(\mathbf{x}, \boldsymbol{\xi})}\cdot d\mathbf{s}(\boldsymbol{\xi}) + h_1. \end{aligned} \quad (71.3)$$

Equally perspicuous, perhaps, are the two more direct forms

$$\begin{aligned} \phi^* &= \frac{1}{4\pi}\int_{\mathfrak{v}} \frac{\operatorname{div} \ddot{\mathbf{x}}(\boldsymbol{\xi}, t)}{d(\mathbf{x}, \boldsymbol{\xi})}\, dv(\boldsymbol{\xi}) + h_2, \\ &= \frac{1}{4\pi}\int_{\mathfrak{v}} \frac{\ddot{\mathbf{x}}(\boldsymbol{\xi}, t)}{[d(\mathbf{x}, \boldsymbol{\xi})]^3}\, dv(\boldsymbol{\xi}) + \frac{1}{4\pi}\oint_{\mathfrak{s}} \frac{\ddot{\mathbf{x}}(\boldsymbol{\xi}, t)}{d(\mathbf{x}, \boldsymbol{\xi})}\, dv(\boldsymbol{\xi}) + h_2. \end{aligned} \quad (71.4)$$

---

[1] [1873, **1**, §4] [1879, **3**, p. 139].

[2] [1879, **3**, p. 139]. The result is given also in [1880, **5**, pp. 223–225] [1880, **7**, p. 276].

72. Lamb planes and Lamb surfaces. In a motion which is neither an irrotational nor a Beltrami motion the vorticity $\mathbf{w}$ and velocity $\dot{\mathbf{x}}$ are independent in direction, except possibly at certain singular points, lines, or surfaces, and hence at each regular point determine a plane, which we may call the *Lamb plane*, and whose normal is parallel to the Lamb vector $\mathbf{w} \times \dot{\mathbf{x}}$.

We pause in passing to consider the relation between the Lamb planes and the Mongian surfaces $h = \text{const.}$, $g = \text{const.}$ Multiplying (16.12) by (40.2) yields

$$\mathbf{w} \times \dot{\mathbf{x}} = \mathbf{w} \cdot \text{grad}\, h, \tag{72.1}$$

and hence the projections of $\dot{\mathbf{x}}$ and of grad h upon the tangent to the vortex-line are equal:[1]

$$\dot{\mathbf{x}}_{\mathbf{w}} = (\text{grad}\, h)_{\mathbf{w}} = \frac{dh}{dw} \tag{72.2}$$

This fact is the starting point of the following geometric construction for the Mongian surfaces, which is illustrated in Fig. 75.1. First, let any set of vortex-surfaces be taken as the surfaces $g = \text{const.}$ At a point P, extend $\mathbf{w}$ and project $\dot{\mathbf{x}}$ upon it, obtaining $\dot{\mathbf{x}}_{\mathbf{w}}$; then at the point A such

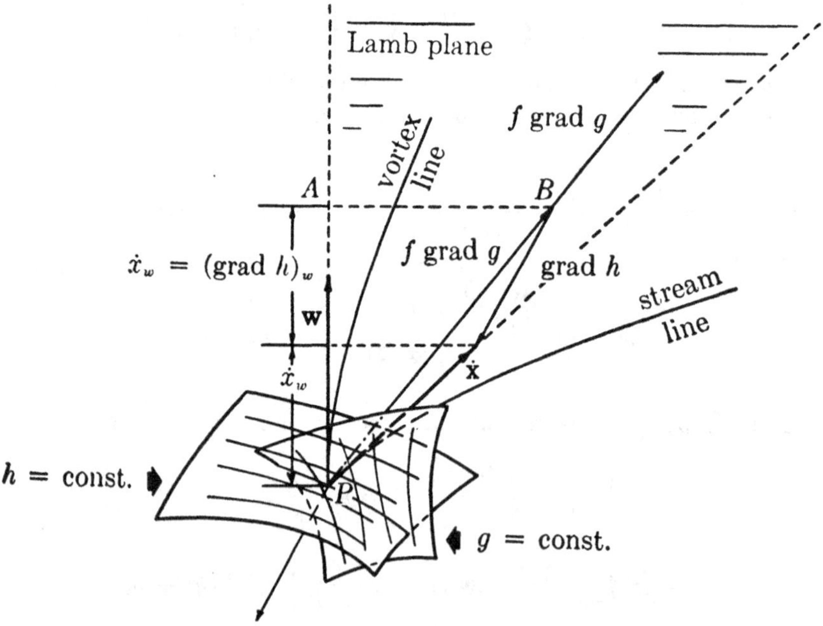

Fig. 75.1

[1] [1881, 2, p. 11].

that $PA = 2\dot{\mathbf{x}}_{\mathbf{w}}$, construct a plane perpendicular to $\mathbf{w}$. Construct the normal $\mathbf{n}$ to $g =$ const. at P. Then f is given by the ratio $f = PB/|\text{grad}\, g|$, and the vector whose origin is the point B at which the normal $\mathbf{n}$ intersects the plane and whose terminus is the terminus of $\dot{\mathbf{x}}$ is equal to grad h, so that the tangent plane to $h =$ const. at P is the plane perpendicular to this vector. This construction yields the value of f at each point and the tangent plane to $h =$ const. at each point, whence the surfaces $h =$ const. may be obtained.

Now the Lamb planes may or may not envelope a one-parameter family of surfaces. To grasp this fact one need only visualize at a given point the stream-surface containing the vortex-line and the vortex-surface containing the stream-line: these have a common tangent plane, but generally these two surfaces do not coincide. A necessary and sufficient condition for the existence of *Lamb surfaces*,[2] which are simultaneously vortex-surfaces and stream-surfaces, is that the Lamb vector $\mathbf{w} \times \dot{\mathbf{x}}$ be complex-lamellar and non-vanishing. By the Euler-Kelvin criterion (12.5) it is then necessary and sufficient that

$$\mathbf{w} \times \dot{\mathbf{x}} \cdot \text{curl}\,(\mathbf{w} \times \dot{\mathbf{x}}) = 0, \qquad \mathbf{w} \times \dot{\mathbf{x}} \neq 0, \tag{72.3}$$

a condition which may be put into the form

$$\mathbf{w} \times \dot{\mathbf{x}} \cdot (\dot{\mathbf{x}} \cdot \text{grad}\, \mathbf{w} - \mathbf{w} \cdot \text{grad}\, \dot{\mathbf{x}}) = 0. \tag{72.4}$$

By eliminating $\mathbf{w} \times \dot{\mathbf{x}}$ between (72.3) and Lagrange's acceleration formula (38.2) we may obtain a form of the condition for the existence for Lamb surfaces which though lacking the symmetry of (72.4) is nevertheless easier to apply, *viz.*

$$\mathbf{w} \times \dot{\mathbf{x}} \cdot \left(\text{curl}\, \ddot{\mathbf{x}} - \frac{\partial \mathbf{w}}{\partial t}\right) = 0. \tag{72.5}$$

By the d'Alembert-Euler condition (45.4) it follows then that *Lamb surfaces exist in any circulation-preserving motion with steady vorticity.*

Equivalently, for the existence of Lamb surfaces it is necessary and sufficient that there exist a non-constant scalar b and a non-vanishing scalar c such that

$$\mathbf{w} \times \dot{\mathbf{x}} = c\, \text{grad}\, b. \tag{72.6}$$

The surfaces $b =$ const. are the Lamb surfaces, and b must satisfy the differential system

$$(\mathbf{w} \times \dot{\mathbf{x}}) \times \text{grad}\, b = 0. \tag{72.7}$$

[2] These surfaces were introduced by Lamb [1878, 2] [1879, 1, §145] [1932, 1, §165]. *Cf.* [1893, 1, §§22–24] [1921, 2, §762]. The name *Bernoulli surfaces* was proposed by Caldonazzo [1924, 2] [1925, 3, §2] in a somewhat different sense.

Since both stream-lines and vortex-lines lie upon the Lamb surfaces (if these exist), these surfaces together with the stream-surfaces and vortex-surfaces normal to them form three one-parameter families of surfaces, and hence serve to define a natural curvilinear co-ordinate system.[3]

The following statements are but immediate applications of a classical theorem of Darboux.[4]

a. *In a complex-lamellar motion with Lamb surfaces, the vorticity is complex-lamellar if and only if the vortex-lines be lines of curvature both on the Lamb surfaces and on the surfaces normal to the velocity.*

b. *In a motion in which both the velocity and the vorticity are complex-lamellar, Lamb surfaces exist if and only if the stream-lines be lines of curvature on the surfaces normal to the vorticity, while the vortex-lines are lines of curvature on the surfaces normal to the velocity.*

c. *In a motion with Lamb surfaces and with complex-lamellar vorticity, the motion itself is complex-lamellar if and only if the stream-lines be lines of curvature both on the Lamb surfaces and on the surfaces normal to the vorticity.*

In a steady motion the Lamb surfaces, being stream-surfaces, are material surfaces (§19), and since they are also vortex-surfaces, it follows that *in any steady motion such that* curl $\ddot{\mathbf{x}}$ *is zero or normal to the Lamb vector there exist stationary material surfaces which are both stream-surfaces and vortex-surfaces.*

73. The line integral Bernoulli theorem. A curve which is everywhere tangent to the Lamb planes we shall call a *Lamb curve* $\mathfrak{c}_L$. Both stream-lines and vortex-lines are Lamb curves, and in a motion where Lamb surfaces exist any curve lying wholly upon some one of them is a Lamb curve. If we integrate Masotti's acceleration formula (51.10) along a Lamb curve we obtain the ***line integral Bernoulli theorem***:

$$\frac{1}{2}\dot{x}^2 + \chi \Big|_{\mathfrak{c}_L} = \int_{\mathfrak{c}_L} d\mathbf{x}\cdot\ddot{\mathbf{x}}: \tag{73.1}$$

in a motion with steady vorticity, the flow of acceleration along a Lamb curve at any instant equals the difference of the values of $\frac{1}{2}\dot{x}^2 + \chi$ *at the two ends of the curve.* In particular, *in a motion with steady vorticity the circulation of the acceleration around a closed Lamb curve is zero.* In a steady motion since $\chi = 0$ the line integral Bernoulli theorem gives a direct connection between speed and acceleration. For any particular

[3] Craig [1881, 2, pp. 5–6] used as co-ordinate surfaces the Lamb surfaces (which he incorrectly assumed always to exist), any independent family of stream-surfaces, and any independent family of vortex-surfaces.

[4] [1866, 1, ¶15].

dynamical model the acceleration is expressed in terms of other quantities (*cf. e.g.* §17), and (73.1) then shows directly the effect of these quantities upon the speed of flow.[1]

74. The curvilinear Bernoulli theorem. Now in general it is possible to express any vector, and in particular the acceleration $\ddot{\mathbf{x}}$, as the sum of a gradient plus a second field in an infinite number of ways:

$$\ddot{\mathbf{x}} = -\text{grad}\ \zeta + \ddot{\mathbf{x}}^*. \tag{74.1}$$

The function ζ may be the Stokes scalar potential ϕ^*, the Monge potential $-h^*$, or some quite different function. We shall assume that $\zeta \neq$ const., so that (74.1) really expresses a decomposition of the acceleration field, and we shall assume further that at least one possible choice of ζ be such that[1] $\ddot{\mathbf{x}}^* \neq \mathbf{w} \times \dot{\mathbf{x}}$. We have

$$\text{curl}\ \ddot{\mathbf{x}} = \text{curl}\ \ddot{\mathbf{x}}^*. \tag{74.2}$$

For reasons which will appear in the next chapter we shall call $\ddot{\mathbf{x}}^*$ the *diffusive acceleration,* taking care to recall ever that this field is determined only to within an arbitrary gradient. The numerous theorems to follow in whose statements reference to the diffusive acceleration occurs may be divided into two classes. Those which essentially employ only curl $\ddot{\mathbf{x}}^*$ are single statements, but those which employ $\ddot{\mathbf{x}}^*$ itself are really an infinity of statements, one for each admissible choice of $\ddot{\mathbf{x}}^*$.

In a motion with steady vorticity comparison of (74.1) with (51.10) yields

$$\text{grad}\ (\zeta + \chi + \tfrac{1}{2}\dot{x}^2) = \dot{\mathbf{x}} \times \mathbf{w} + \ddot{\mathbf{x}}^*. \tag{74.3}$$

Suppose $\dot{\mathbf{x}} \times \mathbf{w} \neq 0$, and at each point let $\mathbf{t}$ be a vector determined by the intersection of the Lamb plane with the plane normal to $\ddot{\mathbf{x}}^*$. Then by taking the dot product of (74.3) with $\mathbf{t}$ we obtain

$$\mathbf{t}\cdot\text{grad}\ (\zeta + \chi + \tfrac{1}{2}\dot{x}^2) = 0. \tag{74.4}$$

Now the field $\mathbf{t}$ is a tangent field for a certain congruence of Lamb curves, determined by the condition that they be normal to the field $\ddot{\mathbf{x}}^*$. From (74.4) follows then the ***curvilinear Bernoulli theorem:*** *in a motion with steady vorticity, let the curves* $\mathfrak{c}_L$ *be the Lamb curves normal to the diffusive acceleration field. Then*

$$\zeta + \chi + \tfrac{1}{2}\dot{x}^2 = f(\mathfrak{c}_L, t); \tag{74.5}$$

[1] An example is given by Carstoiu [1947, **8**, Ch. VI, §3].

[1] From a kinematical point of view the foregoing statements are trivial. Dynamically, however, a medium is defined by specifying the acceleration, and thus some one decomposition may have particular physical significance. For examples, see §§17, 77.

that is, along any one of these curves at any one instant the expression on the left has a constant value. It is possible that some admissible diffusive acceleration field $\ddot{\mathbf{x}}^*$ be parallel to the Lamb vector $\mathbf{w} \times \dot{\mathbf{x}}$ but unequal to it. In this case the result (74.5) holds for any Lamb curve.

In the special case of steady motion the curvilinear Bernoulli theorem (74.5) assumes a simpler form:

$$\mathfrak{z} + \tfrac{1}{2}\dot{x}^2 = f(\mathfrak{c}_\mathsf{L}). \tag{74.6}$$

Thus *upon each of the curves* $\mathfrak{c}_\mathsf{L}$ *there is a finite least upper bound* $\bar{\mathfrak{z}}(\mathfrak{c}_\mathsf{L})$ *for* $\mathfrak{z}$, *attained (if at all) at and only at a stagnation point*:

$$\bar{\mathfrak{z}}(\mathfrak{c}_\mathsf{L}) = f(\mathfrak{c}_\mathsf{L}), \tag{74.7}$$

so that (74.6) becomes

$$\mathfrak{z} + \tfrac{1}{2}\dot{x}^2 = \bar{\mathfrak{z}}. \tag{74.8}$$

If further there be a finite greatest lower bound $\underline{\mathfrak{z}}(\mathfrak{c}_\mathsf{L})$ for $\mathfrak{z}$ on some particular Lamb curve $\mathfrak{c}_\mathsf{L}$, then on that same curve *there must be a finite least upper bound* $\bar{\dot{x}}$ *for the speed*:

$$\tfrac{1}{2}\bar{\dot{x}}^2 = \bar{\mathfrak{z}} - \underline{\mathfrak{z}}. \tag{74.9}$$

An equivalent form for (74.6) then is

$$\tfrac{1}{2}(\bar{\dot{x}}^2 - \dot{x}^2) = \mathfrak{z} - \underline{\mathfrak{z}}. \tag{74.10}$$

From these last results follow the principal application of Bernoulli's theorem in hydrodynamics. One of these, for example, consists in the observation that if $\mathfrak{z}$ = const. upon one of the curves then the speed also must be constant upon that curve.

75. The superficial Bernoulli theorem. We consider now a motion in which Lamb surfaces exist and in which also the vorticity is steady. By inserting (72.6) and (74.1) into (51.10) we then obtain

$$\operatorname{grad}\,(\mathfrak{z} + \chi + \tfrac{1}{2}\dot{x}^2) = \ddot{\mathbf{x}}^* - c \operatorname{grad} b. \tag{75.1}$$

If further

$$\ddot{\mathbf{x}}^* = e \operatorname{grad} b, \tag{75.2}$$

then

$$\operatorname{grad}\,(\mathfrak{z} + \chi + \tfrac{1}{2}\dot{x}^2) = (e - c) \operatorname{grad} b, \tag{75.3}$$

whence it follows that $\mathfrak{z} + \chi + \frac{1}{2}\dot{x}^2$ is constant upon each of the Lamb surfaces b = const. We may state this result as the ***superficial Bernoulli theorem***: *in a motion where Lamb surfaces* $\mathfrak{s}_\mathsf{L}$ *exist and where the vorticity* *is steady, if it be possible to find a diffusive acceleration field* $\ddot{\mathbf{x}}^*$ *which is zero or normal to the Lamb surfaces, then*

$$\zeta + \chi + \tfrac{1}{2}\dot{x}^2 = f(\mathcal{S}_{\mathsf{L}}, t): \tag{75.4}$$

that is, the expression on the left is constant upon each of the Lamb surfaces at each instant.

Consider now the case of a circulation-preserving motion with steady vorticity. As was shown in §72, Lamb surfaces do indeed exist. By (48.1) we may put $\zeta = \phi^*$, $\ddot{\mathbf{x}}^* = 0$ in (74.1), and hence (51.10) becomes

$$\dot{\mathbf{x}} \times \mathbf{w} = \text{grad}\,(\chi + \phi^* + \tfrac{1}{2}\dot{x}^2), \tag{75.5}$$

The foregoing theorem applies and (75.4) follows. Conversely, suppose (75.5) holds; from Lagrange's acceleration formula (38.2) we then obtain

$$\ddot{\mathbf{x}} = \frac{\partial \dot{\mathbf{x}}}{\partial t} - \text{grad}\,(\chi + \phi^*), \tag{75.6}$$

and hence

$$\text{curl}\,\ddot{\mathbf{x}} = \frac{\partial \mathbf{w}}{\partial t}. \tag{75.7}$$

Thus if curl $\ddot{\mathbf{x}} = 0$ it follows that $\partial\mathbf{w}/\partial t = 0$, and conversely if $\partial\mathbf{w}/\partial t = 0$ then curl $\ddot{\mathbf{x}} = 0$. In summary of the foregoing analysis we may state then that *in a circulation-preserving motion with steady vorticity, Lamb surfaces* $\mathcal{S}_{\mathsf{L}}$ *exist, and*

$$\chi + \phi^* + \tfrac{1}{2}\dot{x}^2 = f(\mathcal{S}_{\mathsf{L}}, t). \tag{75.8}$$

Conversely, in a motion such that Lamb surfaces exist and (75.5) *holds, if the motion be circulation-preserving then also the vorticity must be steady, and if the vorticity be steady then also the motion must be circulation-preserving.* As a corollary follows the ***Lamb characterization of steady circulation-preserving motion*** (1878):[1] *for a circulation-preserving motion to be steady it is both necessary and sufficient that Lamb surfaces exist and that*

$$\left|\frac{d}{dn}(\phi^* + \tfrac{1}{2}\dot{x}^2)\right| = |\,\mathbf{w} \times \dot{\mathbf{x}}\,|, \tag{75.9}$$

where d/dn denotes differentiation in a direction normal to the Lamb surface. The formula (75.9) is but an alternative expression for (75.5) in the special case of steady motion. The statement that (75.8) actually holds in a circulation-preserving motion with steady vorticity we may

---

[1] [1878, 2] [1879, 1, §145]. A special case had been discovered previously by Cotterill [1876, 2]. The derivation above is based upon that of Basset [1888, 1, §39]. *Cf.* [1880, 5, p. 220] [1880, 6, pp. 344–347] [1893, 1, §§23–24].

call the ***Lamb-Masotti*[2] *form of Bernoulli's theorem.*** In particular, $\chi + \phi^* + \frac{1}{2}\dot{x}^2$ is a function of time only along each stream-line,[3] along each vortex-line,[4] and along any Lamb curve.[5]

Under conditions sufficient for the validity of (75.8), the finite upper and lower bounds discussed in §74 exist more generally and refer to a whole Lamb surface, rather than to a single Lamb curve.

76. The spatial Bernoulli theorem. In any irrotational motion, steady or not, we have the formula at Euler (48.3), and in any steady Beltrami motion we have (52.2). Recalling that in an unsteady irrotational motion $\chi = -\partial\phi/\partial t$, we may write both these results in a single formula, obtaining the ***spatial Bernoulli theorem***: *in any irrotational motion and in any steady Beltrami motion the acceleration-potential ϕ^* satisfies*

$$\chi + \phi^* + \tfrac{1}{2}\dot{x}^2 = 0, \tag{76.1}$$

where in the case of steady motion χ reduces to a constant. This result is a slight generalization of the classical Bernoulli-Euler theorem discussed in §68.

Under conditions sufficient for the validity of (76.1), the finite upper and lower bounds discussed in §74 exist more generally and refer to the whole motion rather than to a particular Lamb curve.

Conversely, suppose that in a circulation-preserving motion with steady vorticity the spatial Bernoulli theorem (76.1) holds. Since by (51.10) and (48.1) we have

$$\text{grad}\,(\tfrac{1}{2}\dot{x}^2 + \chi + \phi^*) + \mathbf{w} \times \dot{\mathbf{x}} = 0, \tag{76.2}$$

from (76.1) it must follow that

$$\mathbf{w} \times \dot{\mathbf{x}} = 0. \tag{76.3}$$

---

[2] [1927, **3**, §3].

[3] In a sense, the hydraulic statements of D. and J. Bernoulli may be regarded as assertions pertaining to a single stream-line. As a hydrodynamical theorem for plane steady flow, the "Bernoulli equation for the stream-lines" was first derived by Euler in 1751 [1767, **1**, §§18–20] by a remarkable analysis in a partially material description! The result in three-dimensional steady flow is also Euler's [1757, **3**, §§LI–LII], and in [1757, **3**, §§LVIII–IX] he showed that specialization of his general analytical expressions yields

$$\phi^* + \tfrac{1}{2}\dot{x}^2 = \int r\omega^2\,dr + \tfrac{1}{2}r^2\omega^2$$

for the flow energy of a stream-line of a steady simple vortex (48.4), as is evident from first principles. Stokes regarded the general theorem as well known [1842, **2**, p. 1]. *Cf.* also [1857, **1**, §5] [1888, **1**, §39].

[4] It was stated incorrectly by Cisotti [1923, **1**, §2] that these are the only curves upon which $\phi^* + \frac{1}{2}\dot{x}^2$ is constant in a steady motion of this type; the error was corrected by Segre [1923, **2**, §4, footnote].

[5] Certain Lamb curves occurring in certain special motions are remarked by Popov [1951, **14**].

We have thus proved the ***characterization of Stokes*** (1842):[1] *in a circulation-preserving motion with steady vorticity, in order that the spatial Bernoulli theorem hold it is both necessary and sufficient that the motion be an irrotational motion or a (steady) Beltrami motion.*

77. Appendix: some dynamical interpretations for Bernoulli theorems in fluid motions. In this appendix we write out a few of the most interesting of the many dynamical applications of the foregoing results.

A. *Classical hydrodynamics.* In a *barotropic flow of an inviscid fluid subject to conservative extraneous force,* by the Hankel-Kelvin theorem (§49) the motion is circulation-preserving, and hence if the vorticity be steady Lamb surfaces exist and Lamb's superficial Bernoulli theorem (75.8) is valid. By using the formula (49.4) for the acceleration-potential we then obtain

$$\chi + \upsilon + \int \frac{dp}{\rho} + \tfrac{1}{2}\dot{x}^2 = f(\mathfrak{S}_L, t). \tag{77.1}$$

To be able to apply the results of §74 concerning bounds, we must investigate the boundedness of the function $\chi + \zeta = \chi + \phi^* = \chi + \upsilon + \int dp/\rho$. Nothing can be said about χ except in steady motion, when $\chi = \text{const.}$ The force potential υ may be assumed to be bounded. For barotropic relations of the type which occur in practice (*e.g.* $\rho = \text{const.}$ for a homogeneous incompressible fluid, $p = k\rho^\gamma$ for isothermal or isentropic motion of a perfect gas) the quantity $\int dp/\rho$ is bounded. Thus in general there is a finite least upper bound $\bar{\dot{x}}$ for the speed on each Lamb surface of a steady motion.

In an *irrotational flow* or in a *Beltrami flow with steady vorticity* the spatial Bernoulli theorem (76.1) assumes the form

$$\chi + \upsilon + \int \frac{dp}{\rho} + \tfrac{1}{2}\dot{x}^2 = 0, \tag{77.2}$$

a slight generalization of (68.1). Conversely, if (77.2) holds for a barotropic motion of an inviscid fluid subject to extraneous force, it follows that $\mathbf{w} \times \dot{\mathbf{x}} = 0$.[1]

---

[1] Stokes [1842, 2, pp. 2–3] thus demonstrated the formula (76.3) for steady motion, but erroneously concluded therefrom that $w = 0$. See §52[2].

---

[1] As a consequence of this result it follows that if in a steady motion of this type there exist a curve along which both density and speed are constant, and which touches every Lamb surface at least once, then the flow is an irrotational or a Beltrami flow, since the superficial theorem (77.1) must then reduce to the spatial theorem (77.2). Thus in particular, as was noted by Lecornu [1919, 1] in correcting an earlier statement of Stokes [1842, 2, p. 3], all steady flows which originate in a quiet reservoir at uniform pressure are necessarily irrotational or Beltrami flows. Now the Lagrange-Cauchy velocity-potential theorem states that under these same circumstances a

B. *Gas dynamics.* For baroclinic motions of *inviscid fluids possessed of an equation of state and subject to conservative extraneous force* we employ Euler's dynamical equation in the form (17.9), where ξ is a state variable. We then choose $\zeta = -h^*$, where h^* is given by (17.10), and apply the curvilinear Bernoulli theorem (74.5):

$$\chi + \upsilon + \int \frac{1}{\rho}\frac{\partial p}{\partial \rho}\,d\rho + \tfrac{1}{2}\dot{x}^2 = f(c_L, t) \qquad (77.3)$$

where the Lamb curves c_L are those everywhere normal to the surfaces $\xi = \text{const}$. For each different choice of the state variable ξ the curvilinear Bernoulli theorem (77.3) refers to different Lamb curves and $\partial p/\partial \rho$ has a different meaning. *In the special case when heat conduction may be neglected,* from the energy equation it follows that the specific entropy η is constant along each stream-line of a steady motion, so that if we choose $\xi = \eta$ the curves c_L become the stream-lines. This simplest and most useful example of our general result is well known in the literature of gas dynamics.[2]

C. *Theory of viscous liquids.*[3] *For viscous incompressible fluids of uniform viscosity and subject to conservative extraneous force* we employ Navier's equation in the form (17.4). We consider first the *circulation-preserving case,* when there is a flexion-potential σ. In a motion with steady vorticity Lamb surfaces exist and the superficial Bernoulli theorem (75.8) holds as in classical hydrodynamics, except that now the acceleration-potential ϕ^* is given by (49.6), so that[4]

$$\chi + \frac{p}{\rho} + \upsilon + \frac{\mu}{\rho}\sigma + \tfrac{1}{2}\dot{x}^2 = f(\mathcal{S}_L, t). \qquad (77.4)$$

---

finite material portion of fluid once in irrotational flow remains ever in irrotational flow. Consequently, a flow of this type starting from rest in a finite vessel remains always irrotational. Not so, however, with a flow such that $\lim \dot{x} = 0$, $\lim w = 0$ at ∞, for by the second Gromeka-Beltrami theorem (52.4) we have $w/\rho\dot{x} = \text{const.}$ on each stream-line, so that for these flows it is quite possible that the particles may start from rest in a flow irrotational at ∞ yet acquire rotation, the only condition being that $\lim w/\rho\dot{x}$ as $r \to \infty$ shall exist and have a value other than zero. Thus a flow from a quiet infinite reservoir at uniform pressure may well be a Beltrami flow rather than an irrotational flow. This point was noted by Lecornu, though his discussion was not altogether adequate.

[2] I have not been able to trace its source.

[3] A curvilinear Bernoulli theorem for viscous compressible fluids is given in [1950, 2]. There is a sequence of Italian papers [1925, **3**] [1931, **1**] [1948, **5**] which purport to give forms of Bernoulli's theorem valid for motions of viscous compressible fluids but are based on a form of the dynamical equation [1918, **1**, §III] valid only subject to the tacit requirement that there exist a functional relation connecting $p + (\lambda + 2\mu)\vartheta$ and ρ, a circumstance physically most improbable unless the fluid be incompressible or inviscid.

[4] This result is obtained by Masotti [1927, **3**, §4].

In the particular case when curl $\mathbf{w} = 0$, (77.4) reduces to the same form as for a motion of an inviscid fluid of uniform density.

Now by the first Gromeka-Beltrami theorem (§52) and the analysis of §49 it follows that in order for a viscous liquid of the type under discussion to experience *a steady Beltrami motion* there must exist a flexion-potential and (49.6) must hold. The spatial Bernoulli theorem (76.1) then holds simultaneously and assumes the form[5]

$$\frac{p}{\rho} + \upsilon + \frac{\mu}{\rho}\sigma + \tfrac{1}{2}\dot{x}^2 = \text{const.} \tag{77.5}$$

This result can reduce to the same form as for inviscid fluids, *viz.*

$$\frac{p}{\rho} + \upsilon + \tfrac{1}{2}\dot{x}^2 = \text{const.}, \tag{77.6}$$

if and only if[6] $\sigma = 0$, *i.e.* curl $\mathbf{w} = 0$, but by (12.14) it follows then that $w = 0$, as was noticed by Sbrana.[7] The unlikelihood of irrotational motions of viscous liquids was discussed in §37.

The foregoing results concern only special and exceptional flows; the general Bernoulli theorem for viscous liquids is of curvilinear type. In the resolution (74.1) we choose

$$\zeta = \phi^*, \qquad \ddot{\mathbf{x}}^* = -\frac{\mu}{\rho}\,\text{curl}\,\mathbf{w}, \tag{77.7}$$

where ϕ^* is given by $(17.5)_1$. The Lamb curves of the curvilinear theorem then become those normal to the flexion field curl $\mathbf{w}$, and from (74.5) we conclude that[8]

$$\chi + \upsilon + \frac{p}{\rho} + \tfrac{1}{2}\dot{x}^2 = f(\mathfrak{c}_L, t): \tag{77.8}$$

the classical Bernoullian expression is constant at each instant along each Lamb curve which is everywhere normal to the lines of flexion. Various degenerate cases of this result occur in the literature.[9]

---

[5] This result was given by Craig [1880, **7**, p. 276].

[6] [1925, **3**].

[7] [1931, **1**, §2].

[8] [1950, **2**].

[9] If the lines of flexion be normal to the stream-lines, the curves $\mathfrak{c}_L$ are the stream-lines [1931, **1**, §2]. If the lines of flexion be normal to the vortex-lines, the curves $\mathfrak{c}_L$ are the vortex-lines [1948, **5**, §3]. In order for the theorem to hold on any Lamb curve whatever, it is necessary and sufficient that the lines of flexion be normal to the Lamb planes [1948, **5**, §3].

Chapter VIII. CONVECTION AND DIFFUSION OF VORTICITY

78. Beltrami's diffusion equation. Having discussed at some length the properties of the vorticity field at any given instant, we turn to analysis of the manner in which it changes in time. All results related to this subject rest upon a basic identity first obtained in full generality by Beltrami, which we shall now derive. By expanding (45.2) we have

$$\operatorname{curl} \ddot{\mathbf{x}} = \frac{\partial \mathbf{w}}{\partial t} + \dot{\mathbf{x}} \cdot \operatorname{grad} \mathbf{w} - \mathbf{w} \cdot \operatorname{grad} \dot{\mathbf{x}} + \mathbf{w} \operatorname{div} \dot{\mathbf{x}}, \tag{78.1}$$

a result which by (18.3) can be put into the form

$$\dot{\mathbf{w}} = \operatorname{curl} \ddot{\mathbf{x}} + \mathbf{w} \cdot \operatorname{grad} \dot{\mathbf{x}} - \mathbf{w} \operatorname{div} \dot{\mathbf{x}}, \tag{78.2}$$

whence by (24.3) follows the ***diffusion equation of Beltrami*** (1871):[1]

$$\frac{1}{J} \frac{\delta J \mathbf{w}}{\delta t} = \operatorname{curl} \ddot{\mathbf{x}} + \mathbf{w} \cdot \operatorname{grad} \dot{\mathbf{x}}, \tag{78.3}$$

or

$$\frac{1}{J} \frac{\delta J w^i}{\delta t} = \frac{\epsilon^{ijk}}{\sqrt{g}} \ddot{x}_{k,j} + w^j \dot{x}^i{}_{,j}. \tag{78.4}$$

Before discussing the significance of this result, we shall obtain numerous easy consequences and related expressions.

To form an equation for the rate of change of the vorticity magnitude, we multiply Beltrami's equation by $J\mathbf{w}$, from (29.1) thus obtaining[2]

$$\begin{aligned} \frac{1}{2} \frac{\delta}{\delta t} (Jw)^2 &= J\mathbf{w} \cdot \frac{\delta J \mathbf{w}}{\delta t}, \\ &= J^2[\mathbf{w} \cdot \operatorname{curl} \ddot{\mathbf{x}} + \mathbf{w} \cdot \mathbf{\Delta} \cdot \mathbf{w}]. \end{aligned} \tag{78.5}$$

---

[1] [1871, 1, §6] [1903, 1, §5] [1911, 1, §4]. The special cases of (78.2) or (78.3) appropriate to viscous incompressible fluids and to baroclinic motions of inviscid fluids, easily obtained by substituting into it the special forms for $\ddot{\mathbf{x}}$ given in §17, are familiar in the hydrodynamical literature. The former was almost certainly known to Stokes (see [1845, 1, §10]) and was apparently first published by St. Venant [1869, 2, §4] and Bobylew [1873, 1]. For incompressible but not homogeneous fluids, the latter was given by Trowbridge [1877, 2] (the paper is marred by numerous misprints); for compressible fluids, by Silberstein [1896, 1]. For the origin of the still more special case when $\operatorname{curl} \ddot{\mathbf{x}} = 0$, see §94[1].

[2] For the special case of a viscous incompressible fluid, this equation was derived by v. Kármán [1937, 4]. *Cf.* §94.

79. The generalized Ertel equation.[1] The Beltrami diffusion equation may be put into an apparently more general but in fact equivalent form suggested by a special case discovered by Ertel.[2] The basic idea of this analysis is to introduce into (78.3) an arbitrary tensor $\emptyset$, which may be identified with various quantities at pleasure.

By (78.4) and (18.9) we have

$$\begin{aligned}\frac{\delta}{\delta t}(J\mathbf{w}\cdot\text{grad}\,\emptyset) &= J\left\{[\text{curl}\,\ddot{\mathbf{x}} + \mathbf{w}\cdot\text{grad}\,\dot{\mathbf{x}}]\cdot\text{grad}\,\emptyset + \mathbf{w}\cdot\frac{\delta}{\delta t}\,\text{grad}\,\emptyset\right\},\\ &= J\left\{[\text{curl}\,\ddot{\mathbf{x}} + \mathbf{w}\cdot\text{grad}\,\dot{\mathbf{x}}]\cdot\text{grad}\,\emptyset\right. \qquad (79.1)\\ &\qquad \left. + \mathbf{w}\cdot\left[\text{grad}\,\frac{\delta\emptyset}{\delta t} - \text{grad}\,\dot{\mathbf{x}}\cdot\text{grad}\,\emptyset\right]\right\}.\end{aligned}$$

Hence

$$\frac{\delta}{\delta t}(J\mathbf{w}\cdot\text{grad}\,\emptyset) - J\mathbf{w}\cdot\text{grad}\,\frac{\delta\emptyset}{\delta t} = J\,\text{curl}\,\ddot{\mathbf{x}}\cdot\text{grad}\,\emptyset, \qquad (79.2)$$

which is the desired result. By putting $\emptyset = \mathbf{x}$ we may recover the Beltrami equation, of which this not inelegant identity thus appears as an equivalent form.

Consider now a substantially constant quantity $\emptyset$:

$$\frac{\delta\emptyset}{\delta t} = 0. \qquad (79.3)$$

From (79.2) then follows the ***generalized Euler-Ertel conservation theorem:***[3] *if it be possible to find a substantially constant function $\emptyset$ such that*

$$\text{curl}\,\ddot{\mathbf{x}}\cdot\text{grad}\,\emptyset = 0, \qquad (79.4)$$

then

$$J\mathbf{w}\cdot\text{grad}\,\emptyset = \text{const. for each particle}; \qquad (79.5)$$

equivalently,

$$Jw\,\frac{d\emptyset}{dw} = \text{const. for each particle}, \qquad (79.6)$$

where d/dw denotes the directional derivative along the vortex-line.

---

[1] The analysis of this section is extracted from [1951, **6** & **7**].

[2] [1942, **1–4**]. Ertel derived the special case of (79.2) appropriate to barotropic flow of an inviscid fluid subject to conservative extraneous force, gave a geometrical interpretation for the right-hand side, and applied the result to various meteorological situations.

[3] The result proved by Euler [1757, **3**, §LV], of course not expressed in these terms, is equivalent to (79.6) in the case of steady isochoric circulation-preserving motion. *Cf.* §98 and my history [1954, **1**, Part XIII]. Ertel (see note 2 above) considers only the circulation-preserving case, so that (79.6) holds for all substantially constant $\emptyset$.

By the d'Alembert-Euler condition (45.4), in a circulation-preserving motion the requirement (79.4) is satisfied by all functions $\emptyset$, while in a motion which is not circulation-preserving, for a scalar function $\emptyset$ it states that curl $\ddot{\mathbf{x}}$ shall be tangent to the surfaces $\emptyset$ = const.

More generally, for a circulation-preserving motion we have from (79.2) simply

$$\frac{1}{J}\frac{\delta}{\delta t}(J\mathbf{w}\cdot\text{grad }\emptyset) - \mathbf{w}\cdot\text{grad }\frac{\delta\emptyset}{\delta t} = 0, \qquad (79.7)$$

for any tensor $\emptyset$. We may call this result the ***Ertel commutation formula,*** since it shows that $J^{-1}\delta/\delta t$ commutes with $J\mathbf{w}\cdot$grad for circulation-preserving motions.

While (79.7) is restricted to the circulation-preserving case, a related result is valid in far greater generality. Putting the identity

$$\text{curl }\ddot{\mathbf{x}}\cdot\text{grad }\emptyset = \text{div }(\ddot{\mathbf{x}} \times \text{grad }\emptyset) \qquad (79.8)$$

into (79.2) and integrating over a volume $\mathfrak{v}$ yields

$$\begin{aligned}\int_{\mathfrak{v}}\left[\frac{1}{J}\frac{\delta}{\delta t}(J\mathbf{w}\cdot\text{grad }\emptyset) - \mathbf{w}\cdot\text{grad }\frac{\delta\emptyset}{\delta t}\right]dv &= \oint_{\mathfrak{s}} d\mathbf{s}\cdot(\ddot{\mathbf{x}} \times \text{grad }\emptyset), \\ &= \oint_{\mathfrak{s}} (d\mathbf{s} \times \ddot{\mathbf{x}})\cdot\text{grad }\emptyset.\end{aligned} \qquad (79.9)$$

By formulating conditions sufficient for the vanishing of the surface integral we conclude that *in a region such that on all finite boundaries* $\ddot{\mathbf{x}}_t = 0$, *while in any part extending to infinity the condition*

$$(\ddot{\mathbf{x}} \times \text{grad }\emptyset)_n = \bar{o}(r^{-2}), \qquad (79.10)$$

then the Ertel commutation formula holds in average over the whole motion:

$$\int_{\mathfrak{v}}\left[\frac{1}{J}\frac{\delta}{\delta t}(J\mathbf{w}\cdot\text{grad }\emptyset) - \mathbf{w}\cdot\text{grad }\frac{\delta\emptyset}{\delta t}\right]dv = 0. \qquad (79.11)$$

In particular, the result holds for any motion of a material confined within finite walls to which it adheres.

80. Rate of change of circulation. To compute the rate of change of the flow along a material curve $\mathfrak{C}$, we put $\mathbf{c} = \dot{\mathbf{x}}$ in (23.2), obtaining[1]

$$\begin{aligned}\frac{\delta}{\delta t}\int_{\mathfrak{C}} d\mathbf{x}\cdot\dot{\mathbf{x}} &= \int_{\mathfrak{C}} d\mathbf{x}\cdot[\ddot{\mathbf{x}} + \text{grad }\tfrac{1}{2}\dot{x}_{\mathbf{i}}^2], \\ &= \int_{\mathfrak{C}} d\mathbf{x}\cdot\ddot{\mathbf{x}} + \tfrac{1}{2}\dot{x}^2\Big|_{\mathfrak{C}}.\end{aligned} \qquad (80.1)$$

[1] [1869, 1, §59(c)] [1871, 1, §12].

When $\mathfrak{C}$ is a closed circuit, we have simply

$$\frac{\delta}{\delta t}\oint_{\mathfrak{C}} d\dot{\mathbf{x}}\cdot\dot{\mathbf{x}} = \oint_{\mathfrak{C}} d\mathbf{x}\cdot\ddot{\mathbf{x}}: \tag{80.2}$$

the material rate of change of circulation equals the circulation of the acceleration.[2] By Kelvin's transformation, an equivalent formula is[3]

$$\frac{\delta}{\delta t}\oint_{\mathfrak{C}} d\mathbf{x}\cdot\dot{\mathbf{x}} = \frac{\delta}{\delta t}\oint_{\mathfrak{S}} d\mathbf{s}\cdot\mathbf{w} = \int_{\mathfrak{S}} d\mathbf{s}\cdot\operatorname{curl}\ddot{\mathbf{x}}, \tag{80.3}$$

where $\mathfrak{S}$ is a surface bounded by $\mathfrak{C}$.

When the curve $\mathfrak{C}$ is a Lamb curve in a motion with steady vorticity, we may eliminate either the line integral or the speed from (80.1) and (73.1), thus obtaining

$$\begin{aligned}\frac{\delta}{\delta t}\int_{\mathfrak{C}_L} d\mathbf{x}\cdot\dot{\mathbf{x}} &= 2\int_{\mathfrak{C}_L} d\mathbf{x}\cdot\ddot{\mathbf{x}} - \chi\Big|_{\mathfrak{C}_L},\\ &= (\dot{x}^2 + \chi)\Big|_{\mathfrak{C}_L}.\end{aligned} \tag{80.4}$$

Hence in a steady motion the rate of change of the flow along a material Lamb curve is the difference between the squared speed at the two ends, and in particular the circulation about a closed material Lamb curve is constant. These last results become self-evident when the Lamb curve is a stream-line. Another interesting special case may be obtained by choosing as the Lamb curve a vortex-line: *the circulation about a closed material vortex-line in a steady motion is constant in time.* In interpreting this result one must take care to recollect first that closed material vortex-lines generally do not exist, and second that in any case they are not generally steady. A trivial example of the result stated by the theorem is furnished by a rotationally-symmetric motion: the vortex-lines are always circles and thus are closed material circuits, and since the motion is complex-lamellar the circulation about every vortex-line is always zero.

81. Circulation relative to a rotating co-ordinate system. Sometimes it is desirable to consider the circulation and the rate of change of circulation as apparent to an observer stationed in a frame of reference

[2] The special cases of (80.2) or (80.3) appropriate to viscous incompressible fluids were given by Poincaré [1893, 1, §150], V. Bjerknes [1902, 1, footnote], and Jaffé [1920, 1], while that appropriate to inviscid compressible fluids yields the celebrated theorem of V. Bjerknes [1898, 1 & 2] [1900, 2]. The general circulation theorem for viscous compressible fluids is discussed in [1928, 1] [1949, 1] [1949, 2, pp. 22-28].

[3] [1897, 2, §31].

which is itself in motion relative to that (the "stationary" system) in which the velocity $\dot{\mathbf{x}}$ and the acceleration $\ddot{\mathbf{x}}$ are measured. Let quantities in this second ("moving") co-ordinate system be distinguished by primes. Then by (16.9) we have

$$\oint_{\mathfrak{c}} d\mathbf{x}'\cdot\dot{\mathbf{x}}' = \oint_{\mathfrak{c}} d\mathbf{x}'\cdot\left[\dot{\mathbf{x}} - \frac{d\mathbf{c}}{dt} - \boldsymbol{\omega}\times\mathbf{r}'\right]. \tag{81.1}$$

Since $\mathfrak{c}$ is a given circuit, $d\mathbf{x}' = d\mathbf{x}$. Hence

$$\begin{aligned}\oint_{\mathfrak{c}} d\mathbf{x}'\cdot\dot{\mathbf{x}}' &= \oint_{\mathfrak{c}} d\mathbf{x}\cdot\dot{\mathbf{x}} - \oint_{\mathfrak{c}} d\mathbf{x}'\cdot\boldsymbol{\omega}\times\mathbf{r}',\\ &= \oint_{\mathfrak{c}} d\mathbf{x}\cdot\dot{\mathbf{x}} + \boldsymbol{\omega}\cdot\oint_{\mathfrak{c}} d\mathbf{x}'\times\mathbf{r}'.\end{aligned} \tag{81.2}$$

By putting $\mathbf{c} = \mathbf{r}'$ in (8.2) we obtain

$$\oint_{\mathfrak{c}} d\mathbf{x}'\times\mathbf{r}' = -2\int_{\mathfrak{s}} d\mathbf{s}'. \tag{81.3}$$

Hence

$$\boldsymbol{\omega}\cdot\oint_{\mathfrak{c}} d\mathbf{x}'\times\mathbf{r}' = -2\int_{\mathfrak{s}} d\mathbf{s}'\cdot\boldsymbol{\omega} = -2\omega S_{\perp} \tag{81.4}$$

where $S_{\perp}$ is the area inclosed by the projection of $\mathfrak{s}$ upon a plane perpendicular to $\boldsymbol{\omega}$. The formula (81.2) now becomes

$$\oint_{\mathfrak{c}} d\mathbf{x}'\cdot\dot{\mathbf{x}}' = \oint_{\mathfrak{c}} d\mathbf{x}\cdot\dot{\mathbf{x}} - 2\omega S_{\perp}. \tag{81.5}$$

Thus if there be no circulation about $\mathfrak{c}$ apparent to an observer in the rotating system, there must be a circulation of magnitude $2\omega S_{\perp}$ apparent to an observer in the stationary system.

We now compute the rate of change of $\oint d\mathbf{x}'\cdot\dot{\mathbf{x}}'$ as apparent to an observer in the moving system. Since the basic formula (80.2) holds in any system whatever, we have

$$\frac{\delta'}{\delta t'}\oint_{\mathfrak{C}} d\mathbf{x}'\cdot\dot{\mathbf{x}}' = \oint_{\mathfrak{C}} d\mathbf{x}'\cdot\ddot{\mathbf{x}}'. \tag{81.6}$$

Then from (16.10) and (47.2) follows

$$\frac{\delta'}{\delta t'}\oint_{\mathfrak{C}} d\mathbf{x}'\cdot\dot{\mathbf{x}}' = \oint_{\mathfrak{C}} d\mathbf{x}\cdot\ddot{\mathbf{x}} - \oint_{\mathfrak{C}} d\mathbf{x}'\cdot\left[\frac{d\boldsymbol{\omega}}{dt}\times\mathbf{r}' + 2\boldsymbol{\omega}\times\dot{\mathbf{x}}'\right]. \tag{81.7}$$

Now by (8.1), easy vectorial identities, and (27.2) we have

$$\oint_{\mathfrak{C}} d\mathbf{x}'\cdot\left[\frac{d\omega}{dt}\times\mathbf{r}' + 2\omega\times\dot{\mathbf{x}}'\right]$$

$$\begin{aligned} &= \int_{\mathfrak{S}} d\mathbf{s}'\cdot\operatorname{curl}'\left[\frac{d\omega}{dt}\times\mathbf{r}' + 2\omega\times\dot{\mathbf{x}}'\right], \\ &= 2\int_{\mathfrak{S}} d\mathbf{s}'\cdot\left[\frac{d\omega}{dt} - \omega\cdot\operatorname{grad}'\dot{\mathbf{x}}' + \omega\operatorname{div}'\dot{\mathbf{x}}'\right], \\ &= 2\frac{\delta'}{\delta t'}\int_{\mathfrak{S}} d\mathbf{s}\cdot\omega = 2\frac{\delta'}{\delta t'}(\omega\mathcal{S}_{\perp}). \end{aligned} \tag{81.8}$$

Hence (81.7) becomes

$$\frac{\delta'}{\delta t'}\oint_{\mathfrak{C}} d\mathbf{x}'\cdot\dot{\mathbf{x}}' = \frac{\delta}{\delta t}\oint_{\mathfrak{C}} d\mathbf{x}\cdot\dot{\mathbf{x}} - 2\frac{\delta'}{\delta t'}(\omega\mathcal{S}_{\perp}). \tag{81.9}$$

That is, *the rate of change of circulation about a material curve as apparent to an observer in the moving system equals the rate of change apparent to an observer in the stationary system diminished by twice the rate of change, as apparent to an observer in the moving system, of the product of the angular speed* ω *by the area* $\mathcal{S}_{\perp}$ *inclosed by the projection of* $\mathfrak{C}$ *upon a plane normal to the axis of rotation.* This elegant result, in the special case when ω = const. apparently regarded as already well known by Poincaré (1893),[1] is often applied to a study of the atmospheric circulation.

82. Rate of change of the moments of vorticity. We now calculate the rate of change of the moments of vorticity $\mathcal{W}_n$. For a material volume $\mathfrak{B}$ we have from (65.1) and (27.2)

$$\dot{\mathcal{W}}_n = \oint_{\mathfrak{S}} (d\mathbf{s}\cdot[\mathbf{w}\{\mathbf{r}^{(n)}\dot{\mathbf{x}}\} + \dot{\mathbf{w}}\mathbf{r}^{(n+1)} - \mathbf{w}\mathbf{r}^{(n+1)}\operatorname{div}\dot{\mathbf{x}}] - \operatorname{grad}\dot{\mathbf{x}}\cdot d\mathbf{s}\cdot\mathbf{w}\mathbf{r}^{(n+1)}). \tag{82.1}$$

Upon substituting for $\dot{\mathbf{w}}$ from Beltrami's formula (78.2), we find that by a happy circumstance two of the terms so introduced cancel the last two terms in (82.1), which becomes simply

$$\dot{\mathcal{W}}_n = \oint_{\mathfrak{S}} d\mathbf{s}\cdot\mathbf{w}\{\mathbf{r}^{(n)}\dot{\mathbf{x}}\} + \oint_{\mathfrak{S}} d\mathbf{s}\cdot\operatorname{curl}\ddot{\mathbf{x}}\,\mathbf{r}^{(n+1)}. \tag{82.2}$$

[1] [1893, 1, §158]. It was apparently rediscovered by V. Bjerknes [1902, 1].

By Green's transformation follows

$$\dot{\mathfrak{W}}_n = \oint_{\mathfrak{S}} d\mathbf{s}\cdot\mathbf{w}\{\mathbf{r}^{(n)}\dot{\mathbf{x}}\} + \int_{\mathfrak{V}} \{\mathbf{r}^{(n)} \text{ curl } \ddot{\mathbf{x}}\}\, dv. \tag{82.3}$$

Now by the transport theorem (25.4) we have

$$\dot{\mathfrak{W}}_n = \frac{\partial \mathfrak{W}_n}{\partial t} + \oint_{\mathfrak{S}} d\mathbf{s}\cdot\dot{\mathbf{x}}\{\mathbf{r}^{(n)}\mathbf{w}\}. \tag{82.4}$$

Application of $(4.4)_1$ and Green's transformation yields

$$\begin{aligned} 0 &= \int_{\mathfrak{V}} \text{div curl } (\mathbf{r}^{(n)}\ddot{\mathbf{x}})\, dv, \\ &= \oint_{\mathfrak{S}} d\mathbf{s}\cdot \text{curl } (\mathbf{r}^{(n)}\ddot{\mathbf{x}}), \\ &= -\oint_{\mathfrak{S}} \{\mathbf{r}^{(n)}(d\mathbf{s} \times \ddot{\mathbf{x}})\} + \oint_{\mathfrak{S}} d\mathbf{s}\cdot \text{curl } \ddot{\mathbf{x}}\, \mathbf{r}^{(n)}. \end{aligned} \tag{82.5}$$

By combining this result and (82.2) with (82.4) we finally obtain the not inelegant relation[1]

$$\frac{\partial \mathfrak{W}_n}{\partial t} = \oint_{\mathfrak{S}} d\mathbf{s}\cdot[\mathbf{w}\{\mathbf{r}^{(n)}\dot{\mathbf{x}}\} - \dot{\mathbf{x}}\{\mathbf{r}^{(n)}\mathbf{w}\}] + \oint_{\mathfrak{S}} \{\mathbf{r}^{(n)}(d\mathbf{s} \times \ddot{\mathbf{x}})\}. \tag{82.6}$$

The vorticity moment theorem of §65 states conditions sufficient that $\mathfrak{W}_0, \mathfrak{W}_1, \ldots, \mathfrak{W}_n$ shall vanish. From (82.6) we may now state weaker conditions sufficient that these moments remain constant in time: *if all finite boundaries of a continuously differentiable motion be stationary, and if upon them either of the conditions*

$$\mathbf{w}_\mathbf{n} = 0, \qquad \ddot{\mathbf{x}}_\mathbf{t} = 0, \tag{82.7}$$

or

$$\mathbf{w}_\mathbf{n} = 0, \qquad \ddot{\mathbf{x}}_\mathbf{t}^* = 0, \tag{82.8}$$

be satisfied, while in any portion of the material extending to infinity both of the conditions

$$\dot{\mathbf{x}}\mathbf{w}_\mathbf{n} = \bar{o}(r^{-n-2}), \qquad \mathbf{w}\dot{\mathbf{x}}_\mathbf{n} = \bar{o}(r^{-n-2}), \tag{82.9}$$

and either or both of the conditions

---

[1] The special case $n = 0$ is derived in [1948, 1], generalizing an earlier analysis of Jaffé [1921, 1] for viscous liquids. The general formula (82.6) is obtained in [1951, 1, §12].

$$\ddot{\mathbf{x}}_t = \bar{o}(r^{-n-2}), \qquad \ddot{\mathbf{x}}_t^* = \bar{o}(r^{-n-2}) \tag{82.10}$$

be satisfied, then the first $n + 1$ moments of vorticity are constant in time:

$$\mathcal{W}_0 = \text{const.}, \mathcal{W}_1 = \text{const.}, \ldots, \mathcal{W}_n = \text{const.} \tag{82.11}$$

In the case when all finite boundaries are stationary and the material adheres without slipping, we have $\ddot{\mathbf{x}}_t = 0$ and $\mathbf{w}_n = 0$, so that the only conditions of the theorem which remain to be considered are the order conditions (82.9) and (82.10) at infinity.

It may be expected that in a motion such that the theorem of §65 is valid for a certain value of n, the foregoing theorem may hold for $n + 1$ or $n + 2$.

83. Convection and diffusion of vorticity. The preceding sections have presented formulae for the rate of change of vorticity and associated quantities, and have deduced certain particular consequences of these formulae in special cases. We now take up the main function of the present chapter, the analysis of the mechanism of vorticity change in full generality. For convenience we list here the three basic formulae already derived:

$$\dot{\mathbf{w}} = \operatorname{curl} \ddot{\mathbf{x}} + \mathbf{w} \cdot \operatorname{grad} \dot{\mathbf{x}} - \mathbf{w} \operatorname{div} \dot{\mathbf{x}}, \tag{78.2}$$

$$\frac{\delta}{\delta t} \int_{\mathfrak{S}} d\mathbf{s} \cdot \mathbf{w} = \int_{\mathfrak{S}} d\mathbf{s} \cdot \operatorname{curl} \ddot{\mathbf{x}}, \tag*{(80.3)$_2$}$$

$$\frac{\delta}{\delta t} \int_{\mathfrak{B}} \{\mathbf{r}^{(n)} \mathbf{w}\}\, dv = \oint_{\mathfrak{S}} d\mathbf{s} \cdot \mathbf{w} \{\mathbf{r}^{(n)} \dot{\mathbf{x}}\} + \int_{\mathfrak{B}} \{\mathbf{r}^{(n)} \operatorname{curl} \ddot{\mathbf{x}}\}\, dv. \tag{82.3}$$

In all three right-hand sides there are terms depending only upon the present distribution of velocity and a term depending upon the temporal change of velocity field through curl $\ddot{\mathbf{x}}$. The mechanism of change of vorticity must then be of two types, the first independent of the acceleration and dependent only upon the present velocity and its spatial derivatives, and the second associated with the acceleration of the particles. These two separate rates of change we shall distinguish by calling them the *convective* and *diffusive* rates, respectively, and we shall often speak of ***convection*** and ***diffusion*** as the processes associated with these rates.[1] Thus the quantities

$$\operatorname{curl} \ddot{\mathbf{x}}, \qquad \int_{\mathfrak{S}} d\mathbf{s} \cdot \operatorname{curl} \ddot{\mathbf{x}}, \qquad \text{and} \qquad \int_{\mathfrak{B}} \{\mathbf{r}^{(n)} \operatorname{curl} \ddot{\mathbf{x}}\}\, dv \tag{83.1}$$

---

[1] This terminology derives ultimately from Jaffé [1921, **1**] and was introduced in the present sense in [1948, **3**].

are the *diffusive rates of change* of vorticity, flux of vorticity, and moments of vorticity, respectively, while the quantities

$$\mathbf{w}\cdot\text{grad}\,\dot{\mathbf{x}} - \mathbf{w}\,\text{div}\,\dot{\mathbf{x}}, \qquad 0, \qquad \text{and} \qquad \oint_{\mathfrak{S}} d\mathbf{s}\cdot\mathbf{w}\{\mathbf{r}^{(n)}\dot{\mathbf{x}}\} \tag{83.2}$$

are the corresponding *convective rates of change*,[2] associated respectively with material points, surfaces, and volumes. It is trivial to remark that $\ddot{\mathbf{x}}$ may be replaced by $\ddot{\mathbf{x}}^*$ in (83.1).

The basic characteristic of the convective mechanism is that it does not change the circulation of any material circuit. The circulation-preserving motions, defined in §46, are *motions in which convection alone occurs:* to these motions, and thus to a full elaboration of the nature of convection, is devoted the next chapter. The remainder of the present chapter analyses the nature of diffusion and the manner in which convection and diffusion co-operate to produce the most general type of continuous motion. *The mechanism of convection, then, carries vortex-lines as material lines and preserves the strength of the vortex-tubes, while the mechanism of diffusion tends to draw the vortex-lines away from the particles instantly situate upon them and alters the strengths of the vortex-tubes.*

Comparison of (78.2) and (47.6) shows that the convective rate of change of vorticity and the apparent rate of diffusion of vorticity induced by a rotating frame are of the same form. The connection between these two rates may be established as follows. In the primed or "moving" frame, as in any other, Beltrami's equation (78.2) is valid:

$$\frac{\delta'\mathbf{w}'}{\delta t'} = \mathbf{w}'\cdot\text{grad}'\,\dot{\mathbf{x}}' - \mathbf{w}'\,\text{div}\,\dot{\mathbf{x}}' + \text{curl}'\,\ddot{\mathbf{x}}'. \tag{83.3}$$

By (47.6) we thus obtain

$$\frac{\delta'}{\delta t'}(\mathbf{w}' + 2\boldsymbol{\omega}) = (\mathbf{w}' + 2\boldsymbol{\omega})\cdot\text{grad}'\,\dot{\mathbf{x}}' - (\mathbf{w}' + 2\boldsymbol{\omega})\,\text{div}'\,\dot{\mathbf{x}}' + \text{curl}\,\ddot{\mathbf{x}}, \tag{83.4}$$

where $\boldsymbol{\omega}$ is the angular velocity of the moving frame with respect to the stationary one. By (32.3), then,

$$\frac{\delta'\mathbf{w}}{\delta t'} = \mathbf{w}\cdot\text{grad}'\,\dot{\mathbf{x}}' - \mathbf{w}\,\text{div}'\,\dot{\mathbf{x}}' + \text{curl}\,\ddot{\mathbf{x}}. \tag{83.5}$$

---

[2] We must beware of associating the convective acceleration $\dot{\mathbf{x}}\cdot\text{grad}\,\dot{\mathbf{x}}$ with the mechanism of convection of vorticity. Since $\dot{\mathbf{x}}\cdot\text{grad}\,\dot{\mathbf{x}} \times \dot{\mathbf{x}} + \text{grad}\,\frac{1}{2}\dot{x}^2$, the convective acceleration necessarily coincides in part with the diffusive acceleration $\ddot{\mathbf{x}}^*$ unless the Lamb vector $\mathbf{w} \times \dot{\mathbf{x}}$ be lamellar. In the terminology §16, $\dot{\mathbf{x}}\cdot\text{grad}\,\mathbf{w}$ is the *convection of the vorticity*, while by $(83.2)_1$ $\mathbf{w}\cdot\text{grad}\,\dot{\mathbf{x}} - \mathbf{w}\,\text{div}\,\dot{\mathbf{x}}$ is the *convective rate of change of vorticity.*

We may express this equation in words as follows: *If an observer in a moving system would resolve the rate of change of vorticity with respect to a fixed system into convective and diffusive portions, he need only calculate the convective rate of change of the vorticity* $\mathbf{w}'$ *as apparent to himself and then replace therein* $\mathbf{w}'$ *by* $\mathbf{w}' + 2\boldsymbol{\omega}$, *where* $\boldsymbol{\omega}$ *is the angular velocity of his frame of reference with respect to the stationary one; the remainder is the diffusive portion.*

The vector curl $\ddot{\mathbf{x}}$ is the *spatial diffusion vector*, and its vector-lines we may call the *lines of diffusion*, for it is toward them that the vorticity is instantaneously drawn by the diffusive mechanism. In a circulation-preserving motion, these lines do not exist. The spatial diffusion vector curl $\ddot{\mathbf{x}}$ = curl $\ddot{\mathbf{x}}^*$ expresses the influence of the diffusive acceleration $\ddot{\mathbf{x}}^*$ in the change of vorticity.

In order that the spatial diffusion vector of a given motion have the same value for two different observers, it follows immediately from (47.6) that it is both necessary and sufficient that (47.7) be satisfied. The general solution of (47.7) was obtained in §47 on the supposition that $\boldsymbol{\omega}(t)$, the angular velocity of the frame of the moving observer, be given.

In the identity (79.9) put $\emptyset = \mathbf{x}$; there results

$$\int_{\mathfrak{B}} \left[\frac{1}{J}\frac{\delta J\mathbf{w}}{\delta t} - \mathbf{w}\cdot\text{grad}\,\dot{\mathbf{x}}\right] dv = \oint_{\mathfrak{S}} d\mathbf{s} \times \ddot{\mathbf{x}}. \tag{83.6}$$

By comparing the left hand side of this result with $(83.2)_1$ and then formulating conditions sufficient to ensure the vanishing of the surface integral on the right, we obtain the following theorem: *in a region such that upon all finite boundaries*

$$\ddot{\mathbf{x}}_t = 0 \qquad or \qquad \ddot{\mathbf{x}}_t^* = 0, \tag{83.7}$$

while in any portion of the material extending to infinity

$$\ddot{\mathbf{x}}_t = \bar{o}(r^{-2}) \qquad or \qquad \ddot{\mathbf{x}}_t^* = \bar{o}(r^{-2}), \tag{83.8}$$

the average rate change of vorticity is by convection only.[3]

In §43 we have seen that the circulation around any circuit lying upon a vortex-surface and reducible upon it is zero. If there exist material vortex-surfaces, circuits once lying upon them continue ever to do so, so that in such a motion there exist material circuits whose circulation is always zero. Such surfaces behave then as material membranes along which convection operates unaccompanied by diffusion, while the effect of diffusion as it were flows across them, for the circuits lying upon their flanks are protected from the action of diffusion, and

[3] [1948, 1] [1951, 7].

only by cutting them can a circuit experience a change of circulation. A special but important example is furnished by a motion endowed with steady Lamb surfaces, for these are then material vortex-surfaces. From $(83.1)_2$ it follows that in order for material vortex-surfaces to exist it is necessary that

$$\int_{\mathfrak{S}} d\mathbf{s} \cdot \operatorname{curl} \ddot{\mathbf{x}} = 0 \tag{83.9}$$

for any area $\mathfrak{S}$ upon them, and hence that

$$(\operatorname{curl} \ddot{\mathbf{x}})_n = 0. \tag{83.10}$$

In particular, then, for steady Lamb surfaces it is necessary that

$$\mathbf{w} \times \dot{\mathbf{x}} \cdot \operatorname{curl} \ddot{\mathbf{x}} = 0, \tag{83.11}$$

to which our earlier criterion (72.5) reduces in the case of steady motion.

The process of diffusion under certain circumstances may grant immunity to particular quantities whose fate it generally governs. For example, from (45.3) we see that in a motion where the lines of diffusion coincide with the vortex-lines, the motion of these latter is regulated by convection alone, and from (78.5) that if the lines of diffusion be normal to the vortex-lines, the vorticity magnitude is free from the effects of diffusion. Other quantities enjoying a privileged position with respect to diffusion in certain classes of motions will be discussed in §91.

84. Apparatus for the material resolution. Thus far in the study of the transport of vorticity we have employed the spatial description alone, but a clearer understanding of the mechanism of convection and diffusion rewards him whose patience is sufficient to work out the appropriate material expressions.

Our first step is to compute a material expression for the spatial diffusion vector curl $\ddot{\mathbf{x}}$. We have

$$\begin{aligned} x^i{}_{,J}\epsilon^{JKL}(\ddot{x}_i x^i{}_{,L})_{,K} &= x^i{}_{,J}\epsilon^{JKL}\ddot{x}_{i,K}x^i{}_{,L}, \\ &= \epsilon^{JKL}x^i{}_{,J}x^i{}_{,L}x^k{}_{,K}\ddot{x}_{i,k}, \\ &= (\det x^m{}_{,M})\epsilon^{iki}\ddot{x}_{i,k}. \end{aligned} \tag{84.1}$$

Put

$$D^I \equiv \frac{\epsilon^{IJK}}{\sqrt{G}}(\ddot{x}_i x^i{}_{,K})_{,J}. \tag{84.2}$$

Then by (14.7) we obtain from (84.1)

$$x^i{}_{,J}D^J = J\frac{\epsilon^{jki}}{\sqrt{g}}\ddot{x}_{i,k}, \tag{84.3}$$

or, equivalently,

$$D^I = JX^I{}_{,i} \frac{\epsilon^{jki}}{\sqrt{g}} \ddot{x}_{i,k}. \tag{84.4}$$

The vector whose components (84.2) are D^I may be written CURL (GRAD $\mathbf{x}\cdot\ddot{\mathbf{x}}$).

The formulae (84.3) and (84.4) show that CURL (GRAD $\mathbf{x}\cdot\ddot{\mathbf{x}}$) is proportional to the spatial diffusion vector curl $\ddot{\mathbf{x}}$, so that these two quantities vanish together. We may therefore expect that in the material description the vector CURL (GRAD $\mathbf{x}\cdot\ddot{\mathbf{x}}$) will replace curl $\ddot{\mathbf{x}}$ as the agent of diffusion, and accordingly we name it the *material diffusion vector*.

In particular, equivalent to the d'Alembert-Euler condition (45.4) as necessary and sufficient for circulation-preserving motion we have now the ***Hankel-Appell condition*** (1861, 1897)[1].

$$\text{CURL (GRAD } \mathbf{x}\cdot\ddot{\mathbf{x}}) = 0. \tag{84.5}$$

We shall require also a material expression for the vorticity itself. We have

$$\begin{aligned} w^i X^I{}_{,i} &= \frac{\epsilon^{ijk}}{\sqrt{g}} \dot{x}_{k,j} X^I{}_{,i}, \\ &= \frac{\epsilon^{ijk}}{\sqrt{g}} \dot{x}_{k,J} X^J{}_{,j} X^I{}_{,i}, \\ &= -\frac{\epsilon^{kji}}{\sqrt{g}} \dot{x}_{k,J} X^J{}_{,j} X^I{}_{,i}. \end{aligned} \tag{84.6}$$

Substituting on the right the result of exchanging the large with the small letters in (14.14) we obtain

$$w^i X^I{}_{,i} = -\frac{\epsilon^{IJK}}{J\sqrt{G}} x^i{}_{,J} \dot{x}_{i,K}. \tag{84.7}$$

Multiplying this result by $x^k{}_{,I}$ yields ***Beltrami's vorticity formula*** (1871):[2]

$$w^k = \frac{\epsilon^{IJK}}{J\sqrt{G}} x^i{}_{,J} x^k{}_{,K} \dot{x}_{i,I}, \tag{84.8}$$

which is the desired material expression for the vorticity.

---

[1] An equivalent condition of more complex form was derived by Lagrange [1762, **3**, §XLIV]. The equivalent condition δ [CURL (GRAD) $\mathbf{x}\cdot\ddot{\mathbf{x}}$)]$/\delta t = 0$ was deduced by Hankel [1861, **1**, §6] from the dynamical equations for barotropic motion of perfect fluids, subject to conservative extraneous force, but without statement of its significance. That the condition (84.5) holds in the same circumstances was noted by Appell [1897, **1**, §2].

[2] [1871, **1**, §6].

85. The basic vorticity formula. It now remains only to derive the connection between the vorticity and the material diffusion vector CURL (GRAD $\mathbf{x}\cdot\ddot{\mathbf{x}}$). From (84.2) follows

$$\begin{aligned} D^I &= \frac{\epsilon^{IJK}}{\sqrt{G}}\,\ddot{x}_{i,J}x^i{}_{,K}, \\ &= \frac{\epsilon^{IJK}}{\sqrt{G}}\,(\ddot{x}_{i,J}x^i{}_{,K} + \dot{x}_{i,J}\dot{x}^i{}_{,K}), \\ &= \frac{\delta}{\delta t}\left(\frac{\epsilon^{IJK}}{\sqrt{G}}\,\dot{x}_{i,J}x^i{}_{,K}\right). \end{aligned} \tag{85.1}$$

Integrating this formula from 0 to t along the path of a particle $\mathbf{X}$ yields

$$\frac{\epsilon^{IJK}}{\sqrt{G}}\,\dot{x}_{i,J}x^i{}_{,K} = W^I + \int_0^t D^I\,dt, \tag{85.2}$$

where $\mathbf{W}(\mathbf{X})$ is a vector constant of integration. If we multiply this equation by $x^i{}_{,I}/J$, by (84.8) we obtain

$$w^k = \left(W^I + \int_0^t D^I\,dt\right)\frac{x^i{}_{,I}}{J}, \tag{85.3}$$

or

$$\mathbf{w} = \left[\mathbf{W} + \int_0^t \text{CURL}\,(\text{GRAD}\,\mathbf{x}\cdot\ddot{\mathbf{x}})\,dt\right]\cdot\frac{\text{GRAD}\,\mathbf{x}}{J}. \tag{85.4}$$

Since at $t = 0$ we have $x^i{}_{,I} = \delta^i{}_I$, $J = 1$, the constant $\mathbf{W}$ is simply the value of the vorticity at the initial instant $t = 0$. The not inelegant relation (85.4) may be called the ***basic vorticity formula***, since from it all properties of the vorticity are directly and easily discovered. In the derivation we have followed the method of Cauchy (1815),[1] who thus obtained the special case (94.3), to be discussed later. The basic vorticity equation may be derived alternatively by direct integration of Beltrami's diffusion equation (78.3).[2]

---

[1] [1827, **1**, 1st part, Sect. I, §4]. The special form which (85.3) assumes for baroclinic motions of an inviscid fluid subject to non-conservative extraneous force was derived by Appell [1917, **1**, §1] [1921, **2**, §814]; an elaborate treatment is given by Lichtenstein [1925, **1**, Ch. I, §5] [1927, **2**, Ch. 1, §2] [1929, **1**, Ch. 10, §§1–2]; the special case for an incompressible fluid is discussed by Villat [1930, **1**, pp. 10–12]. That which holds for viscous incompressible fluids was derived by Carstoiu [1946, **1**] [1947, **8**, Ch. V, §6]. For a discussion of the form assumed in a viscous compressible fluid, see [1948, **4**] [1947, **8**, Ch. VI, §3].

[2] [1948, **3**].

The first term on the right hand side of (85.3) represents the change in vorticity by convection, while the second term represents the effect of diffusion. *The convected vorticity depends only upon the initial vorticity and upon the relative displacement of an arbitrarily small neighborhood containing the particle* **X**; that is, upon the value of GRAD **x**. *Thus the mechanism of convection is independent of all temporal rates of change and of the paths of the particles, being completely determined at any given time by the displacements of the particles from their initial positions.* In two different motions having the same initial and final configurations and the same initial vorticity, the convected vorticity is identical, although the intermediate history of the two motions may be quite different. With diffusion, however, it is entirely otherwise. The diffused vorticity is the resultant of the cumulative effect of the material diffusion vector CURL (GRAD $\mathbf{x}\cdot\ddot{\mathbf{x}}$) encountered by the particle at each point along its path, and thus is *essentially dependent upon the history of the motion.* The quantity $\int_0^t$ CURL (GRAD $\mathbf{x}\cdot\ddot{\mathbf{x}}$) dt, or $\int_0^t \mathbf{D}\, dt$, may be called the *resultant diffusion vector* for the particle **X** in the time interval $(0, t)$.

The diffusion produced by a given motion $\mathbf{x} = \mathbf{x}_1(\mathbf{X}, t)$ in the time interval $(0, t)$ may be reduced to an equivalent convection process in the following way. Let a second motion $\mathbf{x} = \mathbf{x}_2(\mathbf{X}, t)$ in which there is no diffusion at all (*i.e.*, a circulation-preserving motion) start from an initial configuration at time 0 identical with that of the given motion, $\mathbf{x}_1(\mathbf{X}, 0) = \mathbf{x}_2(\mathbf{X}, 0) = \mathbf{X}$, and end at time t in a configuration again identical with that of the given motion, $\mathbf{x}_1(\mathbf{X}, t) = \mathbf{x}_2(\mathbf{X}, t)$. Then by (85.3) *the diffused vorticity of the first motion equals the convected vorticity of the second motion, provided the initial vorticity of each particle of the second motion be set equal to the value of the resultant diffusion vector for the particle* **X** *as computed from the first motion.*

Particular models of continua, such as elastic bodies or viscous fluids, are defined by giving a particular functional form to the dependence of the acceleration $\ddot{\mathbf{x}}$ upon other variables (*cf.* §17). Thus *the process of diffusion is different in each special type of continuum, while that of convection is the same for all.* We may go even further. The foregoing analysis justifies the statement that *the dynamical specification of a medium consists in the statement of what conditions give rise to diffusion of vorticity and how great the quantity of diffusion then results.*[3]

[3] Any of the models of §17 may serve as an example. For an inviscid fluid subject to conservative extraneous force (17.7) yields the spatial diffusion vector

$$\text{curl}\ \ddot{\mathbf{x}} = \text{grad}\,\frac{1}{\rho} \times \text{grad}\, p$$

showing that diffusion results if and only if there be baroclinic stratification, the quantitative measure thereof being most conveniently expressed in terms of the solenoids of V. Bjerknes (see §80[2] for references). A rigorous discussion of these solenoids is given in [1949, 2, pp. 24-26].

86. Effect of diffusion upon flux of vorticity and upon circulation. We have seen that the mechanism of convection does not alter the circulation around any material circuit, or the flux of vorticity through any material surface. We may now study the effect of diffusion upon these same quantities. By the basic formula (85.4) we have

$$\mathbf{w}\cdot d\mathbf{s} = \left[\mathbf{W} + \int_0^t \text{CURL}\,(\text{GRAD}\,\mathbf{x}\cdot\ddot{\mathbf{x}})\,dt\right]\cdot\frac{\text{GRAD}\,\mathbf{x}}{J}\cdot d\mathbf{s}. \tag{86.1}$$

Hence from $(15.6)_1$ follows

$$\mathbf{w}\cdot d\mathbf{s} = \left[\mathbf{W} + \int_0^t \text{CURL}\,(\text{GRAD}\,\mathbf{x}\cdot\ddot{\mathbf{x}})\,dt\right]\cdot d\mathbf{S}. \tag{86.2}$$

Kirchhoff's proof[1] of the third Helmholtz theorem (§46) is equivalent to annulling the material diffusion vector CURL (GRAD $\mathbf{x}\cdot\ddot{\mathbf{x}}$) in (86.2), whence follows

$$\mathbf{w}\cdot d\mathbf{s} = \mathbf{W}\cdot d\mathbf{S}, \tag{86.3}$$

an elegant formula which might be put into words as follows: the specific flux of vorticity for a particle does not change. Comparison of this special case with the general formula (86.2) yields the significance of the resultant diffusion vector $\int_0^t$ CURL (GRAD $\mathbf{x}\cdot\ddot{\mathbf{x}}$) dt: *at a material point situate upon a material surface, the length of the projection of the resultant diffusion vector upon the direction at time* 0 *normal to the material surface equals the increase in the time interval* $(0, t)$ *of the flux of vorticity through*

[1] More precisely, both Kirchhoff [1876, 1, Vorlesung 15, §3] and Stokes (note added in 1883 reprint of [1848, 2]) gave indications of proofs of the second and third Helmholtz theorems following from Cauchy's formula (94.3) below. The details were worked out by Appell [1897, 1, §8]. (It is perhaps simpler to replace (86.2) by a counterpart relating normal vectors $F_{,i}$ and $F_{,I}$ to a material surface $F = 0$.) It was Dirichlet's [1860, 1, Introd.] remarking the inherent limitations in the spatial method which revived interest in the material description. *Cf.* §14[2]. Apparently at the instigation of Riemann, in 1860 the University of Göttingen announced a competition [1861, 1, Introd.]: "Lagrangiani modi utilitates adhuc fere penitus neglecti clarissimus Dirichlet indicavit in commentatione postuma 'de problemate quodam hydrodynamica' inscripta; sed ab explicatione earum uberiore morbo supremo impeditus esse videtur. Itaque postulat ordo theoriam motus fluidorum aequationibus Lagrangianis superstructum eamque eo saltem perductam, ut leges motus rotatorii a clarissimo Helmholtz alio modo erutae inde redundent." In awarding the prize to Hankel, the philosophical faculty remarked that his essay [1861, 1] did not altogether meet the requirements, since "Bei der Entwickelung der allgemeinen Gesetze der Wirbelbewegung wird freilich Lagrange's Betrachtungsweise ohne Noth verlaßen . . . ." The announcement's specific demand that the material ("Lagrangian") method be developed far enough to yield Helmholtz's theorems, and the phrase "ohne Noth" in the statement of the award, suggest that the proofs later sketched by Kirchhoff and Stokes were known to Riemann in 1860.

the material surface per unit initial area. In particular, it follows that the *flux of vorticity through a material surface assumes at time t its value at time* 0, *without necessarily having maintained this value throughout the interval* (0, *t*), *if and only if*[2]

$$\int_0^t \text{CURL}\,(\text{GRAD}\;\mathbf{x}\cdot\ddot{\mathbf{x}})\,dt = 0 \tag{86.4}$$

for all particles upon it: that is, if and only if the resultant diffusion vector vanish.

To ascertain the effect of diffusion upon circulation, we integrate (86.2) over a surface and then apply Kelvin's transformation. If we write $\dot{\mathbf{X}}$ for the value of $\dot{\mathbf{x}}$ at the time $t = 0$, we obtain

$$\begin{aligned}\oint_{\mathfrak{C}} d\mathbf{x}\cdot\dot{\mathbf{x}} - \oint_{\mathfrak{C}} d\mathbf{X}\cdot\dot{\mathbf{X}} &= \iint_{\mathfrak{S}\,0}^{t} d\mathbf{S}\cdot\,\text{CURL}\,(\text{GRAD}\;\mathbf{x}\cdot\ddot{\mathbf{x}})\,dt,\\ &= \int_0^t\oint_{\mathfrak{C}} d\mathbf{X}\cdot\text{GRAD}\;\mathbf{x}\cdot\ddot{\mathbf{x}}\,dt.\end{aligned} \tag{86.5}$$

Thus the circulation around a material curve is increased[3] by the amount $\oint d\mathbf{X}\cdot\text{grad}\;\mathbf{x}\cdot\ddot{\mathbf{x}}$ in each interval of time dt. As a criterion for the circulation to resume its initial value after the lapse of the interval, without necessarily having maintained it in the interim, we have again (86.4). By $(15.1)_1$ we may obtain from (86.5) the equation

$$\oint_{\mathfrak{C}} d\mathbf{x}\cdot\dot{\mathbf{x}} - \oint_{\mathfrak{C}} d\mathbf{X}\cdot\dot{\mathbf{X}} = \int_0^t\oint_{\mathfrak{C}} d\mathbf{x}\cdot\ddot{\mathbf{x}}\,dt, \tag{86.6}$$

which is an integrated form of our earlier spatial expression (80.2). Thus we may see a third and very easy method of deducing the basic vorticity equation (85.3): *viz.* by integrating (80.2), thus obtaining (86.6), whence follows (86.5), from which in turn by Kelvin's transformation and the continuity of motion follows (85.3).

87. Effect of diffusion upon the vortex-lines. Write

$$\mathbf{Y} \equiv \mathbf{W} + \int_0^t \mathbf{D}\,dt. \tag{87.1}$$

---

[2] [1948, **3**].

[3] Special cases of this result are given in [1897, **1**, §3] [1921, **2**, §816] [1927, **2**, Ch. 2, §5] [1929, **1**, Ch. 10, §4] [1947, **8**, Ch. V, §9].

Forming the cross product $d\mathbf{X} \times \mathbf{Y}$, by (85.2) and $(15.1)_4$ we obtain

$$\begin{aligned}\sqrt{G}\,\epsilon_{IJK}\,dX^J Y^K &= \epsilon_{IJK}\epsilon^{KLM} X^J{}_{,i}\dot{x}_{j,L} x^j{}_{,M}\,dx^i,\\ &= \epsilon_{IJK}\epsilon^{KLM} X^J{}_{,i}\dot{x}_{j,k} x^k{}_{,L} x^j{}_{,M}\,dx^i.\end{aligned} \tag{87.2}$$

By use of $(14.14)_1$ and $(14.14)_2$ we reduce this result to the form

$$\begin{aligned}\sqrt{G}\,\epsilon_{IJK}\,dX^J Y^K &= \frac{J\sqrt{G}}{\sqrt{g}}\,\epsilon_{IJK}\epsilon^{lkj}\dot{x}_{j,k} X^K{}_{,l} X^J{}_{,i}\,dx^i,\\ &= J\sqrt{G}\,\epsilon_{IJK} w^l X^K{}_{,l} X^J{}_{,i}\,dx^i, \qquad (87.3)\\ &= \sqrt{g}\,\epsilon_{jil}\,dx^i w^l x^j{}_{,I},\end{aligned}$$

or

$$d\mathbf{X} \times \left(\mathbf{W} + \int_0^t \text{CURL}\,(\text{GRAD}\,\mathbf{x}\cdot\ddot{\mathbf{x}})\,dt\right) = \text{GRAD}\,\mathbf{x}\cdot(d\mathbf{x} \times \mathbf{w}), \tag{87.4}$$

whence[1]

$$d\mathbf{x} \times \mathbf{w} = \text{grad}\,\mathbf{X}\cdot\left[d\mathbf{X} \times \left(\mathbf{W} + \int_0^t \text{CURL}\,(\text{GRAD}\,\mathbf{x}\cdot\ddot{\mathbf{x}})\,dt\right)\right]. \tag{87.5}$$

Now if a material curve $\mathfrak{C}$ be initially a vortex-line, for each tangent $d\mathbf{X}$ we shall have $d\mathbf{X} \times \mathbf{W} = 0$, so that (87.5) becomes

$$d\mathbf{x} \times \mathbf{w} = \text{grad}\,\mathbf{X}\cdot\left[d\mathbf{X} \times \int_0^t \text{CURL}\,(\text{GRAD}\,\mathbf{x}\cdot\ddot{\mathbf{x}})\,dt\right]. \tag{87.6}$$

Kirchhoff's proof[2] of the second Helmholtz theorem is equivalent to the observation that in a circulation-preserving motion (87.6) reduces to

$$d\mathbf{x} \times \mathbf{w} = 0: \tag{87.7}$$

a material line initially a vortex-line remains ever a vortex-line. For the general case (87.5) gives a quantitative measure of how far a material line which at time 0 was a vortex-line has been turned away from the vortex-line at time t by diffusion. We may conclude also that *a material line which at time* 0 *is a vortex-line will be again a vortex-line at time t, without necessarily having been a vortex-line throughout the interval* (0, t), *if and only if for all particles upon it*

$$d\mathbf{X} \times \int_0^t \text{CURL}\,(\text{GRAD}\,\mathbf{x}\cdot\ddot{\mathbf{x}})\,dt = 0: \tag{87.8}$$

---

[1] A special case is given by Carstoiu [1947, 8, Ch. IV, §5].

[2] For references see §86[1].

that is, if and only if the resultant diffusion vector for each particle be parallel to the initial direction of the vortex-line through that particle.

In particular, from (87.8) it follows that *a necessary and sufficient condition for a material line which is initially a vortex-line to remain a vortex-line throughout the interval is*

$$d\mathbf{X} \times \text{CURL}\,(\text{GRAD}\,\mathbf{x}\cdot\ddot{\mathbf{x}}) = 0 \tag{87.9}$$

for each element $d\mathbf{X}$ *tangent to the initial curve.* This condition is a material equivalent of the spatial condition (45.3), and may be deduced from it by direct application of the formulae of transformation of §14.

88. Properties of $d\mathbf{x} \times \mathbf{w}$. Carstoiu[1] has observed that the vector $d\mathbf{x} \times \mathbf{w}$, where $d\mathbf{x}$ is an arbitrary element of arc satisfying (15.1), has some interesting formal properties.

The first of these is the fact that $d\mathbf{x} \times \mathbf{w}$ satisfies (87.5), an equation formally very similar to the basic vorticity formula (85.4). An equivalent differential equation follows from (78.2):

$$\begin{aligned}\frac{\delta}{\delta t}(d\mathbf{x} \times \mathbf{w}) &= (d\mathbf{x}\cdot\text{grad}\,\dot{\mathbf{x}}) \times \mathbf{w} \\ &\quad + d\mathbf{x} \times [\text{curl}\,\ddot{\mathbf{x}} + \mathbf{w}\cdot\text{grad}\,\dot{\mathbf{x}} - \mathbf{w}\,\text{div}\,\dot{\mathbf{x}}];\end{aligned} \tag{88.1}$$

by the identity (1.15) we thus obtain[2]

$$\frac{\delta}{\delta t}(d\mathbf{x} \times \mathbf{w}) = d\mathbf{x} \times \text{curl}\,\ddot{\mathbf{x}} - \text{grad}\,\dot{\mathbf{x}}\cdot(d\mathbf{x} \times \mathbf{w}), \tag{88.2}$$

an equation formally very similar to Beltrami's diffusion equation (78.3).

Second, if we multiply (87.4) by another element of arc $d\bar{\mathbf{X}}$, by (15.1) we obtain

$$d\bar{\mathbf{X}}\cdot d\mathbf{X} \times \left[\mathbf{W} + \int_0^t \text{CURL}\,(\text{GRAD}\,\mathbf{x}\cdot\ddot{\mathbf{x}})\,dt\right] = d\bar{\mathbf{x}}\cdot d\mathbf{x} \times \mathbf{w}. \tag{88.3}$$

From this result we see that a direction normal to $d\mathbf{X} \times [\mathbf{W} + \int_0^t \text{CURL}\,(\text{GRAD}\,\mathbf{x}\cdot\ddot{\mathbf{x}})\,dt]$ at time 0 is carried at time t into a direction normal to $d\mathbf{x} \times \mathbf{w}$; in the special case of a circulation-preserving motion,[3] a direction normal to $d\mathbf{X} \times \mathbf{W}$ remains ever normal to $d\mathbf{x} \times \mathbf{w}$. An alternative form of (88.3) is

$$d\bar{\mathbf{X}} \times d\mathbf{X}\cdot\left[\mathbf{W} + \int_0^t \text{CURL}\,(\text{GRAD}\,\mathbf{x}\cdot\ddot{\mathbf{x}})\,dt\right] = d\bar{\mathbf{x}} \times d\mathbf{x}\cdot\mathbf{w}. \tag{88.4}$$

---

[1] [1947, 8, Ch. IV].

[2] A special case is given by Carstoiu [1947, 8, Ch. IV, §3].

[3] [1947, 8, Ch. IV, §4].

Now it follows from the hypothesis of continuity of motion that two material curves initially tangent to one another remain ever tangent, *i.e.*, from $d\bar{\mathbf{X}} \times d\mathbf{X} = 0$ it follows that $d\bar{\mathbf{x}} \times d\mathbf{x} = 0$. The result (88.4) shows more generally how the angle between two material lines is affected by convection and diffusion of vorticity. It is really a reformulation of the flux equation (86.1).

Third, write

$$\mathbf{H} \equiv d\mathbf{S} \times \left(d\mathbf{X} \times \left[\mathbf{W} + \int_0^t \text{CURL}\,(\text{GRAD}\,\mathbf{x}\cdot\ddot{\mathbf{x}})\,dt\right]\right). \tag{88.5}$$

By (15.5) and (87.3) we readily calculate

$$x^i{}_{,I}H^I = x^l{}_{,I}\frac{\epsilon^{IJK}}{J\sqrt{G}}x^j{}_{,J}\,ds_j\sqrt{g}\,\epsilon_{kml}\,dx^m w^l x^k{}_{,K}, \tag{88.5a}$$

whence by $(14.14)_1$ we obtain

$$\begin{aligned} x^i{}_{,I}H^I &= \frac{\epsilon^{ijl}}{\sqrt{g}}X^K{}_{,l}x^k{}_{,K}\,ds_j\sqrt{g}\,\epsilon_{kml}\,dx^m w^l, \\ &= \frac{\epsilon^{ijl}}{\sqrt{g}}\delta^k{}_l\,ds_j\sqrt{g}\,\epsilon_{kml}\,dx^m w^l, \\ &= \frac{\epsilon^{ijk}}{\sqrt{g}}ds_j\sqrt{g}\,\epsilon_{kml}\,dx^m w^l, \end{aligned} \tag{88.6}$$

a result which we can write vectorially as

$$\left[d\mathbf{S} \times \left(d\mathbf{X} \times \left[\mathbf{W} + \int_0^t \text{CURL}\,(\text{GRAD}\,\mathbf{x}\cdot\ddot{\mathbf{x}})\,dt\right]\right)\right]\cdot\text{GRAD}\,\mathbf{x} = d\mathbf{s} \times (d\mathbf{x} \times \mathbf{w}). \tag{88.7}$$

From this formula it follows that a material surface now normal to $d\mathbf{x} \times \mathbf{w}$ was initially normal to $d\mathbf{X} \times [\mathbf{W} + \int_0^t \text{CURL}\,(\text{GRAD}\,\mathbf{x}\cdot\ddot{\mathbf{x}})\,dt]$, and conversely. If we now multiply (88.7) by a second element of surface $d\bar{\mathbf{s}}$, by $(15.6)_1$ follows

$$J\,d\mathbf{S} \times \left[d\mathbf{X} \times \left(\mathbf{W} + \int_0^t \ldots\right)\right]\cdot d\bar{\mathbf{S}} = d\mathbf{s} \times (d\mathbf{x} \times \mathbf{w})\cdot d\bar{\mathbf{s}}. \tag{88.8}$$

Further properties of $d\mathbf{x} \times \mathbf{w}$ and of $d\mathbf{x} \times \text{curl}\,\ddot{\mathbf{x}}$ have been investigated by Carstoiu.

89. Appell's generalization of convection. The notion of convection of vorticity has been generalized by Appell[1] in an ingenious fashion. By an analysis parallel to Cauchy's in material variables he proved that *given any family of lines, furnished with continuously turning tangents, which in a given motion* $\dot{\mathbf{x}}$ *are material lines, there exists a continuously differentiable field* $\mathbf{v}$ *whose circulation about any material circuit is constant and whose vortex-lines are the given material lines.* We shall now establish Appell's result by a shorter spatial derivation. Given any such family of material lines, it is easy to find at any instant a continuously differentiable solenoidal field $\mathbf{w}'$ of which they are vector-lines (§11); this field will then satisfy the Helmholtz-Zorawski criterion (28.7), so that there exists a scalar function b such that

$$\dot{\mathbf{w}}' - \mathbf{w}'\cdot\text{grad}\,\dot{\mathbf{x}} = b\mathbf{w}'. \tag{89.1}$$

Now if c be any non-zero scalar function whatever, the field $\tilde{\mathbf{w}} \equiv c\mathbf{w}'$ has the same vector-lines as $\mathbf{w}'$, and it too will be solenoidal if c be chosen as a solution of

$$\text{grad}\,c\cdot\mathbf{w}' = 0. \tag{89.2}$$

But from (89.1) we have

$$\dot{\tilde{\mathbf{w}}} - \tilde{\mathbf{w}}\cdot\text{grad}\,\dot{\mathbf{x}} = (\dot{c} + cb)\mathbf{w}'. \tag{89.3}$$

If c' be any particular solution of the first order linear ordinary differential equation

$$\dot{c} + cb = -c\,\text{div}\,\dot{\mathbf{x}}, \tag{89.4}$$

then $c \equiv f(\mathbf{X})c'$ gives a family of solutions. Let f be chosen as a solution of the first order linear partial differential equation

$$\text{grad}\,c\cdot\mathbf{w}' = \text{grad}\,f\cdot c'\mathbf{w}' + f\,\text{grad}\,c'\cdot\mathbf{w}' = 0. \tag{89.5}$$

Now both (89.2) and (89.4) are simultaneously satisfied and (89.3) reduces to the form of the Zorawski criterion (28.1). Thus the flux of $\tilde{\mathbf{w}}$ through any material surface is constant in time, and since $\tilde{\mathbf{w}}$ is solenoidal there exists a field $\mathbf{v}$ such that $\tilde{\mathbf{w}} = \text{curl}\,\mathbf{v}$. By Kelvin's transformation it follows that the circulation of $\mathbf{v}$ about each material circuit is constant.

90. The generalized convection vector. Convection and the notions associated with it may be partially generalized in a different and more fruitful way. In the present section we introduce the general apparatus necessary for this generalization; in the following section we exhibit two cases in which the diffusion mechanism transports certain quantities

[1] [1899, 1, §§1–10].

associated with the vorticity in a manner closely analogous to that in which the vorticity itself is transported in a corresponding class of motions of pure convection; and thereafter we append an explanation of how these results materalize in gas dynamics.

The underlying idea of this generalization consists in introducing a class of vector fields $\mathbf{v}_C$ proportional to the velocity:

$$\mathbf{v}_C \equiv \frac{\dot{\mathbf{x}}}{v_0}, \tag{90.1}$$

where v_0 is any non-vanishing substantially constant scalar:

$$\dot{v}_0 = 0, \qquad v_0 \neq 0. \tag{90.2}$$

Any such field $\mathbf{v}_C$ we shall call a *generalized convection vector*[1] and v_0 we shall call the *defining parameter.* We introduce also the curl of the convection vector:

$$\mathbf{w}_C \equiv \text{curl } \mathbf{v}_C. \tag{90.3}$$

Then we have identically

$$\mathbf{w} = \text{curl } v_0\mathbf{v}_C = v_0\mathbf{w}_C + \text{grad } v_0 \times \mathbf{v}_C, \tag{90.4}$$

and hence quite independently of the condition (90.2) we obtain

$$\mathbf{v}_C \cdot \mathbf{w}_C = v_0^{2}\dot{\mathbf{x}} \cdot \mathbf{w}. \tag{90.5}$$

By (12.5) it follows that *the generalized convection vector is complex-lamellar if and only if the motion be complex-lamellar.*

The researches of Neményi & Prim (1947-1949)[2] have drawn attention to motions in which

$$\mathbf{v}_C \times \mathbf{w}_C = 0, \tag{90.6}$$

the possibility $\mathbf{w}_C = 0$ not being excluded. The special case $v_0 = 1$ is an irrotational or Beltrami motion, and the class of motions satisfying the generalization (90.6) they call *generalized Beltrami motions.* The remainder of this section presents results equivalent to those of Neményi & Prim concerning this interesting type of motion.

---

[1] In previous works [1951, 5] [1952, 2] I have called $\mathbf{v}_C$ the "generalized Crocco vector," but in fact in the work of Crocco [1936, 1] v_0 is an absolute constant used only as a normalizing factor. The basic idea in the analysis above is that v_0 shall be only substantially constant; it need not be of the same dimensions as $\mathbf{v}$ and is not used for normalization. The idea of using such a vector is due to Hicks, Guenther, & Wasserman [1947, 6, Introd.], who take v_0 as the ultimate speed $\hat{\dot{x}}$ for the streamlines in a steady flow of a perfect gas devoid of heat flux and subject to no extraneous force (*cf.* §§74, 92). A more general class of convection vectors is considered by Hicks [1949, 8].

[2] [1948, 9, §5] [1949, 4 & 9].

We first establish the connection between generalized Beltrami motions and irrotational or ordinary Beltrami motions. By (90.4) and $(90.2)_1$ we obtain

$$\dot{\mathbf{x}} \times \mathbf{w} = v_0{}^2 \mathbf{v}_C \times \mathbf{w}_C + \dot{x}^2 \operatorname{grad} \log v_0 + \dot{\mathbf{x}} \frac{\partial \log v_0}{\partial t}. \tag{90.7}$$

From this identity it is obvious that *a generalized Beltrami motion in which the defining parameter is either uniform or steady is an irrotational or Beltrami motion if and only if the defining parameter be both uniform and steady.* This result implies broadly that the class of generalized Beltrami motions is more extensive than that of irrotational and Beltrami motions. We may calculate the angle ψ between the vortex-line and the stream-line in the following way. If $\mathbf{w}_C \times \mathbf{v}_C = 0$, the two summands on the right-hand side of (90.4) are perpendicular, so that

$$w^2 = v_0{}^2 w_C{}^2 + (\operatorname{grad} \log v_0 \times \dot{\mathbf{x}})^2. \tag{90.8}$$

Simultaneously (90.7) becomes

$$\dot{\mathbf{x}} \times \mathbf{w} = \dot{x}^2 \operatorname{grad} \log v_0 + \dot{\mathbf{x}} \frac{\partial \log v_0}{\partial t}. \tag{90.9}$$

By eliminating grad log v_0 between these two relations we obtain

$$w^2 = v_0{}^2 w_C{}^2 + |\operatorname{grad} \log v_0|^2 \dot{x}^2, \tag{90.10}$$

while squaring (90.9) yields

$$(\dot{\mathbf{x}} \times \mathbf{w})^2 = \dot{x}^2 \left[\dot{x}^2 (\operatorname{grad} \log v_0)^2 - \left(\frac{\partial \log v_0}{\partial t} \right)^2 \right]. \tag{90.11}$$

The angle ψ is then obtained by combining (90.10) and (90.11):

$$\csc \psi = \frac{\dot{x} w}{|\dot{\mathbf{x}} \times \mathbf{w}|} = \left[\frac{v_0{}^2 w_C{}^2 + |\operatorname{grad} \log v_0|^2 \dot{x}^2}{\dot{x}^2 |\operatorname{grad} \log v_0|^2 - \left(\dfrac{\partial \log v_0}{\partial t} \right)^2} \right]^{\frac{1}{2}}, \tag{90.12}$$

whence follows

$$\tan^2 \psi = \frac{v_C{}^2 |\operatorname{grad} \log v_0|^2 - \left[\dfrac{\partial}{\partial t} \left(\dfrac{1}{v_0} \right) \right]^2}{w_C{}^2 + \left[\dfrac{\partial}{\partial t} \left(\dfrac{1}{v_0} \right) \right]^2}, \tag{90.13}$$

a formula which in the case when the defining parameter is steady reduces to the elegant form

$$\tan \psi = \pm \frac{v_C}{w_C} |\operatorname{grad} \log v_0|. \tag{90.14}$$

The result (90.11) implies a *lower bound for the speed of a generalized Beltrami motion, viz.*

$$\dot{x} \geqq \frac{\left|\dfrac{\partial \log v_0}{\partial t}\right|}{|\operatorname{grad} \log v_0|}, \tag{90.15}$$

for whose validity it is assumed that the point in question is not a stagnation point or a point where grad $v_0 = 0$.

The result (90.9) also contains some further information. By comparing (90.9) with (72.6) we conclude that *in a generalized Beltrami motion whose defining parameter is steady but not uniform, the surfaces upon which that parameter is constant are Lamb surfaces.*

By putting (90.9) into Lagrange's acceleration formula (38.2) we obtain

$$\begin{aligned} \ddot{\mathbf{x}} &= \frac{\partial \dot{\mathbf{x}}}{\partial t} - \dot{x}^2 \operatorname{grad} \log v_0 - \dot{\mathbf{x}} \frac{\partial \log v_0}{\partial t} + \operatorname{grad} \tfrac{1}{2}\dot{x}^2, \\ &= v_0 \frac{\partial \mathbf{v}_C}{\partial t} + v_0^2 \operatorname{grad} \tfrac{1}{2} v_C^2. \end{aligned} \tag{90.16}$$

Hence *in a generalized Beltrami motion whose convection vector is steady the acceleration is complex-lamellar, its normal surfaces being the surfaces of constant magnitude of the convection vector.*

Taking the curl of (90.16) yields

$$\operatorname{curl} \ddot{\mathbf{x}} = \operatorname{grad} v_0 \times \frac{\partial \mathbf{v}_C}{\partial t} + v_0 \frac{\partial \mathbf{w}_C}{\partial t} + \tfrac{1}{2} \operatorname{grad} v_0^2 \times \operatorname{grad} v_C^2. \tag{90.17}$$

On the assumption that the time derivatives be zero, we now consider in turn the situations which annul the remaining term on the right. First, if v_0 be uniform then from $\mathbf{w}_C \times \mathbf{v}_C = 0$ it follows that $\mathbf{w} \times \dot{\mathbf{x}} = 0$. Second, we may have $v_C^2 = \text{const.}$ Third, if the surfaces $v_0 = \text{const.}$ coincide with the surfaces $v_C = \text{const.}$, these in turn are surfaces of constant speed. By our earlier results, then, Lamb surfaces exist and are surfaces of constant speed, and the acceleration is complex-lamellar and normal to these surfaces. In summary of these results we state that *a generalized Beltrami motion whose convection vector is steady is a circulation-preserving motion if and only if*

a) *it be an irrotational or a Beltrami motion, or*
b) *its convection vector be of uniform magnitude, or*
c) *the surfaces of constant speed be Lamb surfaces and the acceleration be normal to them.*

Although as indicated by this result the generalized Beltrami motions usually fail to be circulation-preserving, yet in some types of such motions the mechanism of diffusion operates in a fashion closely analogous to convection, as we shall now discover.

91. Generalized convection theorems. Consider first a steady generalized Beltrami motion, and let υ be any solution of

$$\operatorname{div}(\upsilon \mathbf{v}_C) = 0. \tag{91.1}$$

By applying the theorem of Gromeka and Beltrami derived in §12 and by noting that vector sheets of $\mathbf{v}_C$ are also stream-surfaces we then obtain the ***convection theorem for steady generalized Beltrami motions:*** *the surfaces*

$$\frac{w_C}{\upsilon v_C} = \text{const.} \tag{91.2}$$

are stream-surfaces; in particular, (91.2) *holds on each stream-line.* The analogy to convection is immediate, and the result is an interesting generalization of the second Gromeka-Beltrami theorem (52.3).

The corresponding generalized convection theorem for complex-lamellar motions lies more deeply.[1] For any function f we readily obtain the identity

$$\operatorname{curl}\left[f\left(\frac{1}{v_0}\frac{\partial \mathbf{v}_C}{\partial t} + \mathbf{w}_C \times \mathbf{v}_C\right)\right] = \operatorname{grad}\frac{f}{v_0} \times \frac{\partial \mathbf{v}_C}{\partial t} - \frac{\mathbf{w}_C}{v_0}\frac{\partial f}{\partial t} + \frac{1}{v_0}\frac{\partial f \mathbf{w}_C}{\partial t}$$

$$+ \mathbf{v}_C \cdot \operatorname{grad} f\mathbf{w}_C - f\mathbf{w}_C \cdot \operatorname{grad} \mathbf{v}_C + f\mathbf{w}_C \operatorname{div} \mathbf{v}_C - \mathbf{v}_C \operatorname{div}(f\mathbf{w}_C). \tag{91.3}$$

Now let f be chosen as a solution of

$$\operatorname{grad}\frac{f}{v_0} \times \frac{\partial \mathbf{v}_C}{\partial t} = \frac{\mathbf{w}_C}{v_0}\frac{\partial f}{\partial t}, \tag{91.4}$$

and let υ^{1/v_0} be any permissible density for $\mathbf{v}_C$, *i.e.* let υ be any solution of

$$\frac{1}{v_0}\frac{\delta \log \upsilon}{\delta t} = -\operatorname{div} \mathbf{v}_C, \tag{91.5}$$

or equivalently

$$\frac{1}{v_0}\frac{\partial \upsilon}{\partial t} + \operatorname{div}(\upsilon \mathbf{v}_C) = 0. \tag{91.6}$$

Our identity (91.3) then reduces to

$$\frac{1}{\upsilon}\operatorname{curl}\left[f\left(\frac{1}{v_0}\frac{\partial \mathbf{v}_C}{\partial t} + \mathbf{w}_C \times \mathbf{v}_C\right)\right]$$

$$= \frac{1}{v_0}\frac{\delta}{\delta t}\left(\frac{f\mathbf{w}_C}{\upsilon}\right) - \frac{f\mathbf{w}_C}{\upsilon}\cdot \operatorname{grad} \mathbf{v}_C - \frac{\mathbf{v}_C}{\upsilon}\operatorname{div}(f\mathbf{w}_C), \tag{91.7}$$

[1] The following analysis is reproduced from [1951, 5].

a generalization of Beltrami's diffusion equation (78.3), to which it reduces when $v_0 = 1$, $f = 1$, $v = j^{-1}$.

We now form the scalar product of (91.7) with $f\mathbf{w}_C/v$, thus obtaining a generalization of (78.5):

$$\frac{1}{2v_0}\frac{\delta}{\delta t}\left(\frac{fw_C}{v}\right)^2 = \frac{f\mathbf{w}_C}{v}\cdot\text{grad}\ \mathbf{v}_C\cdot\frac{f\mathbf{w}_C}{v} + \frac{f}{v^2}\mathbf{v}_C\cdot\mathbf{w}_C\mathbf{w}_C\cdot\text{grad}\ f$$
$$+ \frac{f\mathbf{w}_C}{v^2}\cdot\text{curl}\left[f\left(\frac{1}{v_0}\frac{\partial\mathbf{v}_C}{\partial t} + \mathbf{w}_C \times \mathbf{v}_C\right)\right]. \qquad (91.8)$$

By (90.5) the assumption that the motion is complex-lamellar, now employed for the first time, annuls the second term on the right in (91.8). Let us assume further that the third term is zero:

$$\mathbf{w}_C\cdot\text{curl}\left[f\left(\frac{1}{v_0}\frac{\partial\mathbf{v}_C}{\partial t} + \mathbf{w}_C \times \mathbf{v}_C\right)\right] = 0. \qquad (91.9)$$

We then obtain the equation

$$\frac{1}{2v_0}\frac{\delta}{\delta t}\left(\frac{f\mathbf{w}_C}{v}\right)^2 = \frac{f\mathbf{w}_C}{v}\cdot\text{grad}\ \mathbf{v}_C\cdot\frac{f\mathbf{w}_C}{v}. \qquad (91.10)$$

We now impose the further requirement that the vortex-lines be steady. Then it is possible to choose co-ordinates at a single point[2] in such a way that

$$ds^2 = h^2(dx^1)^2 + g_{22}(dx^2)^2 + g_{33}(dx^3)^2,$$
$$(w)^2 = w^1w_1, \qquad \frac{\delta x^1}{\delta t} = \dot{x}^1 = 0, \qquad \frac{\partial h}{\partial t} = 0. \qquad (91.11)$$

That is, the x^1 co-ordinate curve is tangent to the vortex-line at the point in question. From (91.10) we have then

$$\frac{1}{v_0}\frac{\delta}{\delta t}\log\frac{fw_C}{v} = v_C{}^1{}_{,1},$$
$$= \frac{\partial v_C{}^1}{\partial x^1} + \Gamma^1_{1k}v_C{}^k, \qquad (91.12)$$

whence follows, since $v_C{}^1 = 0$,

---

[2] The neighboring x^1 co-ordinate curves need not be vortex-lines. Thus (91.11) does not impose any restriction on the class of motions considered.

$$\begin{aligned}\frac{\delta}{\delta t}\log\frac{f\mathbf{w}_C}{\upsilon} &= \Gamma^1_{12}\frac{\delta x^2}{\delta t} + \Gamma^1_{13}\frac{\delta x^3}{\delta t},\\ &= \frac{\partial \log h}{\partial x^2}\frac{\delta x^2}{\delta t} + \frac{\partial \log h}{\partial x^3}\frac{\delta x^3}{\delta t},\\ &= \frac{\partial \log h}{\partial x^1}\frac{\delta x^1}{\delta t} + \frac{\partial \log h}{\partial x^2}\frac{\delta x^2}{\delta t} + \frac{\partial \log h}{\partial x^3}\frac{\delta x^3}{\delta t} + \frac{\partial \log h}{\partial t},\\ &= \frac{\delta \log h}{\delta t}.\end{aligned} \tag{91.13}$$

Hence

$$\frac{\delta}{\delta t}\left(\frac{f w_C}{h\upsilon}\right) = 0. \tag{91.14}$$

The equation just derived expresses a simple conservation law. Going back and collecting the assumptions we have made to derive it, we obtain the ***generalized vorticity convection theorem for complex-lamellar motions***: *given a complex-lamellar motion with steady vortex-lines, let the x^1 co-ordinate curve at the point in question be tangent to the vortex-line and let $dx^1/ds = h^{-1}$; let v_0 be any substantially constant function, let $\mathbf{v}_C$ and $\mathbf{w}_C$ be defined by*

$$\mathbf{v}_C \equiv \frac{\dot{\mathbf{x}}}{v_0}, \qquad \mathbf{w}_C \equiv \operatorname{curl} \mathbf{v}_C; \tag{90.1,3}$$

let f be any solution of

$$\operatorname{grad}\frac{f}{v_0} \times \frac{\partial \mathbf{v}_C}{\partial t} = \frac{\mathbf{w}_C}{v_0}\frac{\partial f}{\partial t}; \tag{91.4}$$

and let υ be any solution of

$$\frac{1}{v_0}\frac{\partial \upsilon}{\partial t} + \operatorname{div}(\upsilon \mathbf{v}_C) = 0; \tag{91.6}$$

if further it be possible to find among the functions v_0 and f satisfying these conditions a pair such that also

$$\mathbf{w}_C \cdot \operatorname{curl}\left[f\left(\frac{1}{v_0}\frac{\partial \mathbf{v}_C}{\partial t} + \mathbf{w}_C \times \mathbf{v}_C\right)\right] = 0, \tag{91.9}$$

then

$$\frac{f w_C}{h\upsilon} = \text{const.} \tag{91.15}$$

for each particle. Of the conditions of this theorem only the requirement of complex-lamellar motion with steady vortex-lines and the one equation (91.9) are truly restrictive, the others being rather in the nature of definitions of the class of admissible functions v_0, f, and v.

Both the results (91.2) and (91.15) are theorems of convective type, since they state that certain quantities are carried along with the particles as if they were native properties. They are generalized rather than true vorticity convection theorems, however, because for their validity the diffusion mechanism need not be absent, curl $\ddot{\mathbf{x}}$ being in general unequal to zero. They are rather to be regarded as statements that subject to certain conditions the diffusion of the actual vorticity $\mathbf{w}$ behaves somewhat like convection of the generalized vorticity $\mathbf{w}_C$.

92. Appendix: generalized convection theorems in gas dynamics.[1] We now apply the results of the previous section to gas dynamics. A gas for which there is an equation of state of the type

$$\rho = P(p)H(\eta), \qquad H'(\eta) \neq 0, \tag{92.1}$$

is a *Prim gas.*[2] A *perfect gas* is the special case $P(p) = p^{1/\gamma}$, $H(\eta) = C \exp [(\eta - \eta_0)/c_p]$, where c_p and c_v are the specific heats, assumed constant, and $\gamma = c_p/c_v$. For steady flow of an inviscid gas devoid of heat flux and subject to no extraneous force a special case of the curvilinear Bernoulli theorem (77.3) holds, *viz.*

$$\int_{\rho_0}^{\rho} \left(\frac{\partial p}{\partial \rho}\right)_\eta \frac{d\rho}{\rho} + \tfrac{1}{2}\dot{x}^2 = \tfrac{1}{2}\bar{x}^2, \tag{92.2}$$

where the ultimate speed $\bar{x}$ is constant along each stream-line. In the definition (90.1) of the convection vector we may then take $v_0 = \bar{x}$. For a Prim gas the Bernoulli equation (92.2) then becomes

$$1 - v_C^{\,2} = \frac{\Pi(p)}{\Pi(p_0)}, \tag{92.3}$$

where $\Pi(p)$ is defined by

$$P(p) = \frac{1}{\Pi'(p)} \tag{92.4}$$

and p_0 is the *stagnation pressure,* like $\bar{x}$ a function only of the stream-lines. The Eulerian continuity equation (24.5) becomes[3]

$$\text{div}\,[P(p)\mathbf{v}_C] = 0. \tag{92.5}$$

[1] The analysis of this section is reproduced from [1952, **2**, §§9–10].

[2] [1949, **9**].

[3] Special cases were given by Crocco [1936, **1**, eq. 6] and by Hicks, Guenther, & Wasserman [1947, **6**, eq. 4.4].

It is possible to derive from Euler's dynamical equation (17.7) the basic condition[4]

$$\mathbf{v}_C \times \mathbf{w}_C = \tfrac{1}{2}(1 - v_C^2) \text{ grad } \log \Pi(p_0), \tag{92.6}$$

whence it follows that *a steady flow of a Prim gas, devoid of heat flux and subject to no extraneous force, is a generalized Beltrami flow if and only if the stagnation pressure be uniform.* In all cases we have from (92.6)

$$\text{curl}\left[\frac{\mathbf{v}_C \times \mathbf{w}_C}{1 - v_C^2}\right] = 0. \tag{92.7}$$

Consider first a generalized Beltrami flow. By comparing (92.5) with (91.1) we see that the choice $v = P(p)$ is permissible, and hence by (91.2) we obtain the ***vorticity theorem of Neményi & Prim*** (1947):[5] *in a steady flow of a Prim gas devoid of heat flux and subject to no extraneous force, if the stagnation pressure be uniform then*

$$\frac{w_C}{P(p)v_C} = \text{const.} \tag{92.8}$$

along each stream-line; equivalently,

$$\frac{w_C \Pi'(p)}{\sqrt{\Pi(p_0) - \Pi(p)}} = \text{const.} \tag{92.9}$$

The second form follows from the first by (92.3).

Second, consider a steady complex-lamellar flow. We may again choose $v = P(p)$, and (92.7) shows that the condition (91.9) may be satisfied by the choice $f = (1 - v_C^2)^{-1} = \Pi(p_0)/\Pi(p)$. By (91.15) then follows the ***generalized Crocco-Prim vorticity theorem:***[6] *in a steady complex-lamellar flow of a Prim gas devoid of heat flux and subject to no extraneous force, we have*

$$\frac{w_C}{h}\frac{d \log \Pi(p)}{dp} = \text{const.} \tag{92.10}$$

along each stream-line, the function h being defined by $(91.11)_1$. In plane flow $h = 1$; in rotationally-symmetric flow $h = r$, where r is the distance from the axis of symmetry; and for a perfect gas $d \log \Pi(p)/dp = p^{-1}$. If the ultimate speed $\bar{\bar{x}}$ be uniform, the quantity w_C may be replaced by the vorticity magnitude w.

---

[4] A special case is given by Hicks, Guenther, & Wasserman [1947, **6**, eq. 4.2].

[5] The special case for perfect gases is given in [1949, **4**, Th. 6].

[6] [1951, **5**]. The special cases of this result valid in plane and in rotationally-symmetric flows of a perfect gas when $\bar{\bar{x}}$ is uniform were discovered by Crocco [1936, **1**]. Prim removed the restrictions to uniform $\bar{\bar{x}}$ [1948, **8**] and to perfect gases [1952, **4**, eqq. (65), (194)].

Chapter IX. Circulation-Preserving Motions

Par cette découverte, toute la Méchanique des fluides fut réduite à un seul point d'analyse; & si les équations qui la renferment étoient intégrables, on pourroit dans tous les cas déterminer complettement les circonstances du mouvement & de l'action d'un fluide mu par des forces quelconques; malheureusement elles sont si rebelles, qu'on n'a pu jusqu'à présent en venir à bout que dans des cas très-limités.

—Lagrange [1788, **1**, 2e partie, sect. 10].

93. Kelvin's proofs of the Helmholtz theorems. This final chapter concerns circulation-preserving motions. As we have shown in §46, this class of motions is characterized by the second and third Helmholtz theorems: (2) *in a circulation-preserving motion, the vortex-lines are material lines*, and (3) *in a motion such that the vortex-lines are material, in order that the strengths of all vortex-tubes remain constant in time it is necessary and sufficient that the motion be circulation-preserving.* In §§45-46 these theorems were demonstrated as special cases of more general results of Zorawski, and the results of the analysis of §§86-87 in material variables yielded as special cases a second pair of proofs, essentially those of Kirchhoff. It is fitting that we open our detailed analysis of circulation-preserving motions with the elegant proofs of these fundamental and characterizing theorems which were conceived by Kelvin (1869),[1] proofs which rest directly on the circulation-preserving property.

To prove the second Helmholtz theorem, we compute the circulation around a material circuit $\mathfrak{C}$ which at time 0 lies entirely upon a given vortex-surface $\mathfrak{W}$ and is reducible upon it. By (43.2), the circulation about $\mathfrak{C}$ is zero at time 0. At time t the particles initially comprising the vortex-surface $\mathfrak{W}$ constitute a new surface $\mathfrak{w}$, which we do not know to be a vortex-surface. Upon it lies $\mathfrak{c}$, the present locus of the particles comprising $\mathfrak{C}$. Since the motion is circulation-preserving, the circulation about $\mathfrak{c}$ is zero, and hence by Kelvin's transformation (33.1) we have

$$0 = \oint_{\mathfrak{c}} d\mathbf{x}\cdot\dot{\mathbf{x}} = \int_{\mathfrak{s}} d\mathbf{s}\cdot\mathbf{w}, \qquad (93.1)$$

[1] [1869, 1, §§60(f)–60(i); §§59(d), 60(q)]. The third theorem had been proved previously in essentially the same way by Hankel [1861, 1, §9], who had used a different method for the more difficult second theorem [*ibid.*, §11].

where $\mathfrak{s}$ is the portion of $\mathfrak{w}$ inclosed by $\mathfrak{c}$. But $\mathfrak{C}$ is an arbitrary circuit upon $\mathfrak{W}$; hence by the continuity of motion $\mathfrak{c}$ is an arbitrary circuit upon $\mathfrak{w}$; hence $\mathfrak{s}$ is an arbitrary portion of $\mathfrak{w}$, and hence finally (93.1) implies that $d\mathbf{s}\cdot\mathbf{w} = 0$ upon $\mathfrak{w}$. It follows then that $\mathfrak{w}$ is a vortex-surface. Since vortex-lines are the curves of intersection of vortex-surfaces, and since all vortex-surfaces are material surfaces, the vortex-lines are material lines.

To prove the third Helmholtz theorem, we recall that by Kelvin's transformation the strength of a vortex-tube equals the circulation about any curve once embracing it, and from Helmholtz's second theorem, in a circulation-preserving motion a material curve once lying upon a vortex-tube always lies upon the same vortex-tube. Since the circulation about this curve is constant during the motion, it follows that the strength of the vortex-tube is constant during the motions. Conversely, if the vortex-lines be material and the strengths of the vortex-tubes constant in time, it follows by Kelvin's transformation that the circulation about any circuit once embracing a vortex-tube is constant in time, but since any circuit defines a vortex-tube, the motion is circulation-preserving.

The Helmholtz theorems are of the nature of first integrals. While they vividly present some aspects of the nature of circulation-preserving motions, there are many questions concerning these most classical of motions which remain unanswered to this day. One of these concerns the nature of the vortex-lines. In general, the configurations assumed by the vortex-lines are very elaborate (*cf.* §10[2]). Does the condition of circulation-preserving motion in any way restrict their possible shapes, as the condition of complex-lamellar motion so markedly regulates the behavior of the stream-lines (§50)?

94. Résumé of further properties of circulation-preserving motions already demonstrated. At various points in the preceding chapters we have already deduced certain properties of circulation-preserving motions, and various of our general equations reduce to a simpler form for motions of this type. These results we now resume.

First, as *necessary and sufficient conditions for circulation-preserving motion* we have both the d'Alembert-Euler condition

$$\operatorname{curl} \ddot{\mathbf{x}} = 0 \tag{45.4}$$

and the Hankel-Appell condition

$$\text{CURL}\,(\text{GRAD}\,\mathbf{x}\cdot\ddot{\mathbf{x}}) = 0. \tag{84.5}$$

By combining (45.4) and the two forms (78.2) and (78.3) of Beltrami's diffusion equation, as alternative necessary and sufficient conditions we obtain the ***d'Alembert-Euler vorticity equation*** (1750-1755):[1]

$$\dot{\mathbf{w}} = \mathbf{w}\cdot\text{grad}\,\dot{\mathbf{x}} - \mathbf{w}\,\text{div}\,\dot{\mathbf{x}}, \tag{94.1}$$

[1] For rotationally-symmetric motion the equation was obtained by d'Alembert [1752, 1, §48] in a rather obscure form as a consequence of curl $\ddot{\mathbf{x}} = 0$; it was similarly obtained in full generality by Euler in his first paper on fluid dynamics [1761, 1, §59]; two of the three components were derived by Lagrange [1762, **3**, Ch. XLII]. This result was rediscovered by Stokes [1845, 1, §13] [1848, 2]. The special case valid in isochoric motions is usually called *Helmholtz's equation*, since Helmholtz [1858, 1, §2] based his proofs of his second and third vorticity theorems upon it. The form (94.2) appears in [1874, 1].

It is obvious that a motion is circulation-preserving if and only if, at every instant and for every volume $\mathfrak{v}$, the integral

$$\int_{\mathfrak{v}} (\text{curl}\,\ddot{\mathbf{x}})^2\, dv$$

has the value 0, and hence is a minimum with respect to its values for all other acceleration fields compatible with any constraints we may choose to impose. Equivalently, for a given instantaneous velocity field $\dot{\mathbf{x}}$ with vorticity $\mathbf{w}$, a motion is circulation-preserving if and only if $\partial\mathbf{w}/\partial t$ be such as to render the Gaussian "constraint"

$$\mathcal{C} \equiv \int_{\mathfrak{v}} \left[\frac{\partial\mathbf{w}}{\partial t} - \text{curl}\,(\dot{\mathbf{x}}\times\mathbf{w})\right]^2 dv$$

a minimum (zero) for arbitrary $\mathfrak{v}$. Suppose, however, we let $\mathfrak{v}$ be a single volume, fixed once and for all. Then the sufficiency of the minimizing of the Gaussian constraint is no longer trivial. For the isochoric case, it has been established by Delval [1950, **7**, §5], but only subject to boundary conditions in general incompatible. By simply omitting the surface term which he proposes, we shall ameliorate his result, as follows now.

Since the vorticity field must be solenoidal, we have the side condition div $(\partial\mathbf{w}/\partial t) = 0$ to be satisfied by all competing fields $\partial\mathbf{w}/\partial t$. Writing λ for the corresponding Lagrangian multiplier, we get

$$\delta\mathcal{C} + \delta\int_{\mathfrak{v}} \lambda\,\text{div}\,\frac{\partial\mathbf{w}}{\partial t}\, dv = 0$$

as our asserted variational equation. Equivalently,

$$\delta\int_{\mathfrak{v}} \left\{\left[\frac{\partial\mathbf{w}}{\partial t} - \text{curl}\,(\dot{\mathbf{x}}\times\mathbf{w})\right]^2 - \text{grad}\,\lambda\cdot\frac{\partial\mathbf{w}}{\partial t}\right\} dv - \delta\oint_{\mathfrak{s}} d\mathbf{s}\cdot\frac{\partial\mathbf{w}}{\partial t}\lambda = 0.$$

Since it is only $\partial\mathbf{w}/\partial t$ which is varied, the Euler equation for the volume integral is

$$\frac{\partial\mathbf{w}}{\partial t} - \text{curl}\,(\dot{\mathbf{x}}\times\mathbf{w}) - \tfrac{1}{2}\,\text{grad}\,\lambda = 0, \tag{*}$$

while that for the surface integral is

$$\lambda\, d\mathbf{s} = 0. \tag{**}$$

Taking the divergence of (*) yields $\nabla^2\lambda = 0$. Hence, by (**), λ is a harmonic function

or[2]

$$\frac{\delta J\mathbf{w}}{\delta t} = J\mathbf{w}\cdot\text{grad } \dot{\mathbf{x}}. \tag{94.2}$$

We have also Ertel's alternative form:

$$\frac{\delta}{\delta t}(J\mathbf{w}\cdot\text{grad } \emptyset) = J\mathbf{w}\cdot\text{grad } \frac{\delta \emptyset}{\delta t}, \tag{79.7}$$

some of whose consequences are included in the discussion in §79, and more will appear in §98.

From (84.5) and the basic vorticity formula (85.4) follows another necessary and sufficient condition for circulation-preserving motion, the celebrated ***Cauchy vorticity formula*** (1815):[3]

$$\mathbf{w} = \mathbf{W}\cdot\frac{\text{GRAD } \mathbf{x}}{J}. \tag{94.3}$$

This elegant expression, which is easily shown to be the general solution of the d'Alembert-Euler equation (94.2), embodies a complete and explicit description of convection. It lay unnoticed for thirty years after its discovery, and although both Stokes[4] and Kirchhoff[5] appreciated its central importance in classical hydrodynamics, even in recent

---

which vanishes upon $\mathfrak{s}$. For the case when $\mathfrak{v}$ is simply connected, it follows that $\lambda = 0$, and (*) reduces to (94.1) for the isochoric case.

Moreau [1952, **3**, §7c] claims that $\delta \int_{\mathfrak{v}} \ddot{\mathbf{x}}^2 \, dv = 0$, where $\dot{\mathbf{x}}$ is varied subject to the side condition div $\dot{\mathbf{x}} = 0$ and where $\dot{\mathbf{x}}_n$ is prescribed upon the boundary $\mathfrak{s}$, leads to the isochoric case of (94.1). He does not provide a proof, nor have I been able to construct one.

[2] A quantity analogous to the vorticity has been sought by Hill [1885, **1**] in his study of circulation-preserving motions in n-dimensional spaces. He found that there is a vector quantity satisfying an equation similar to (94.2) if n be odd, a scalar quantity if n be even. To specify the local rate of rotation completely, of course, the $n(n - 1)/2$ components of $\dot{x}_{i,j} - \dot{x}_{j,i}$ are required; thus the quantities studied by Hill can fulfill in the n-dimensional case the function of the vorticity in the three-dimensional case in part at best. It is interesting to notice in this connection that the configuration-space of dynamics may be of either an odd or an even number of dimensions, while the phase-space of statistical mechanics is always of an even number of dimensions.

[3] [1827, **1**, 1st part, Sect. 1, §4]. A proof of sufficiency is given by Crudeli [1918, **1**, §I]. Cauchy's formula asserts that $\mathbf{w}$ is carried as a vector density by the motion, but no use has ever been made of this fact.

[4] [1845, **1**, §11] [1846, **1**, §1] [1848, **2**, pp. 37–41].

[5] *Cf.* §86[1].

expositions of the subject it is rarely given the prominence it deserves.[6] The interpretation of Cauchy's formula has been discussed already in the analysis of convection in §85, and the result itself will be employed several times in the remainder of this chapter.

An easy consequence of (94.2) is ***d'Alembert's theorem on rotationally-symmetric motions:***[7] *in a circulation-preserving rotationally-symmetric motion,* Jw = const. *for each particle if and only if* $w = 0$. For in rotationally-symmetric motion, the direction of $\mathbf{w}$ does not change for a given particle, so that it suffices to find all motions in which $J\mathbf{w}$ = const. for each particle. We must then solve $\mathbf{w}\cdot\text{grad}\,\dot{\mathbf{x}} = 0$. But $\mathbf{w}\cdot\text{grad}\,\dot{\mathbf{x}} = 0$ if and only if $w = 0$. In plane motion, however, $\mathbf{w}\cdot\text{grad}\,\dot{\mathbf{x}} = 0$ always, and no such result as d'Alembert's theorem might suggest is true.

A second easy consequence is the ***d'Alembert-Lagrange theorem:***[8] *a d'Alembert motion* (51.6) *in which*

$$T(t) \neq \frac{1}{A + Bt}, \tag{94.4}$$

where A and B are constants, is circulation-preserving if and only if it be irrotational. To prove this result, we note that putting (51.6) into (94.1) yields

$$\frac{T'}{T^2}\,\text{curl}\,\mathbf{u} = \mathbf{z}, \tag{94.5}$$

where $\mathbf{z}$ is a steady field and so is curl $\mathbf{u}$. Hence either curl $\mathbf{u} = 0$ or $T'/T^2 = B$. The latter alternative is excluded by (94.4).

In a circulation-preserving motion (78.5) reduces to ***Warren's equation*** (1870)[9]

$$\tfrac{1}{2}\frac{\delta}{\delta t}(Jw)^2 = J^2\mathbf{w}\cdot\mathbf{\Delta}\cdot\mathbf{w}. \tag{94.6}$$

---

[6] Lagrange in the *Méchanique Analitique* [1788, 1] [1815, 1] omitted (94.1), while (94.3) had not yet been discovered. Poisson in his *Traité de Mécanique* [1833, 1] omitted both (94.1) and (94.3), although he discussed at length (see §104 below) a question which can be answered by one glance at the latter, or by more careful study of the former. These two treatises contained the principal expositions of hydrodynamics available in the first half of the nineteenth century. Thus neither Stokes nor Helmholtz knew d'Alembert and Euler's result, which both rediscovered; Helmholtz apparently never learned of Cauchy's analysis, while Stokes noticed it only after his own work was complete. Helmholtz also in 1858 was apparently ignorant of the prior work of Stokes.

[7] [1752, 1, §49]. d'Alembert considered only the special case $J = 1$, w = overall const.

[8] d'Alembert's attempt [1761, 2, §X] is wrong both in statement and in proof; it was corrected by Lagrange [1762, 3, §XLII]. Neither, of course, put the result in terms of circulation.

[9] [1870, 1].

In order for the vorticity magnitude to experience a stationary value for a given particle, it is necessary and sufficient that $\mathbf{w}\cdot\mathbf{\Delta}\cdot\mathbf{w} = 0$. When $w \neq 0$, this condition may be fulfilled in two ways: either, as remarked by Carstoiu,[10] the deformation quadric at the point and time may be an hyperboloid, upon whose asymptotic cone the vorticity lies, or else the deformation quadric may be a cylinder whose generators are parallel to the vorticity, as for example in any plane motion. It would be interesting to delimit the class of motions in which Jw = const.; the more general case in which Jw = const. for each particle is characterized in §97.

The problem of determining all circulation-preserving motions in which $J\mathbf{w}$ = const. for each particle is easier. By (94.2), we have only to solve

$$\mathbf{w}\cdot\operatorname{grad}\dot{\mathbf{x}} = 0, \tag{94.7}$$

a condition which states that the velocity vector is constant along each vortex-line. Whether or not any circulation-preserving motions which satisfy this condition and are not plane or irrotational exist, I do not know. A further remark in this connection is made in §101.

In *steady* circulation-preserving motion (94.2) assumes the form of a simple reciprocal theorem:

$$\dot{\mathbf{x}}\cdot\operatorname{grad} J\mathbf{w} = J\mathbf{w}\cdot\operatorname{grad}\dot{\mathbf{x}}, \tag{94.8}$$

or (provided we observe due caution in defining the notion of directional derivative of a vector field)

$$\dot{x}\,\frac{dJ\mathbf{w}}{d\dot{x}} = Jw\,\frac{d\dot{\mathbf{x}}}{dw}. \tag{94.9}$$

95. Superposability of circulation-preserving motions.[1] Supposing given two circulation-preserving motions $\dot{\mathbf{x}}_1$ and $\dot{\mathbf{x}}_2$, we may seek a condition that their vector sum $\dot{\mathbf{x}} \equiv \dot{\mathbf{x}}_1 + \dot{\mathbf{x}}_2$ also be circulation-preserving. Such a pair of motions will be called *superposable.* If $\dot{\mathbf{x}}_1$ be superposable on itself, it will be called *self-superposable.* Fundamental for the analysis is the identity

$$\operatorname{curl}\ddot{\mathbf{x}} = \operatorname{curl}\ddot{\mathbf{x}}_1 + \operatorname{curl}\ddot{\mathbf{x}}_2 + \operatorname{curl}(\mathbf{w}_1 \times \dot{\mathbf{x}}_2 + \mathbf{w}_2 \times \dot{\mathbf{x}}_1), \tag{95.1}$$

[10] [1946, **2**, §4] [1946, **5**] [1947, **8**, Ch. III]. The tempting statement concerning the applicability of vortex-surfaces in [1946, **6**] is unfortunately false.

[1] The problem of superposability was first treated by Ballabh [**1940**, **2**] at the suggestion of Strang. His results are confined to the case of a viscous incompressible fluid, for which superposability is a relatively much more severe restriction than in the circulation-preserving case, and are summarized in the second paragraph of Note 2 below.

which follows at once from Lagrange's acceleration formula (38.2). By inspection of this identity we obtain the following ***superposibility condition***: *the circulation-preserving motions* $\dot{\mathbf{x}}_1$ *and* $\dot{\mathbf{x}}_2$ *are superposable if and only if*[2]

$$\text{curl}\,(\mathbf{w}_1 \times \dot{\mathbf{x}}_2 + \mathbf{w}_2 \times \dot{\mathbf{x}}_1) = 0. \tag{95.2}$$

---

[2] The theorem in the text may be put in terms of a condition that the acceleration field corresponding to $\dot{\mathbf{x}}_1 + \dot{\mathbf{x}}_2$ be lamellar. It is natural to extend the inquiry to the case of complex-lamellar acceleration. The result has physical interest since, as is plain from a glance at (17.7), *an acceleration field is associated with a possible motion of an inviscid fluid subject to conservative extraneous force if and only if it be complex-lamellar* [1757, **3**, §XVII]. From (12.5) and (95.1) it is immediate that the required necessary and sufficient condition is

$$\ddot{\mathbf{x}}_1 \cdot \text{curl}\,\ddot{\mathbf{x}}_2 + \ddot{\mathbf{x}}_2 \cdot \text{curl}\,\ddot{\mathbf{x}}_1 + (\ddot{\mathbf{x}}_1 + \ddot{\mathbf{x}}_2) \cdot \text{curl}\,(\mathbf{w}_1 \times \dot{\mathbf{x}}_2 + \mathbf{w}_2 \times \dot{\mathbf{x}}_1) = 0.$$

For self-superposability there is the much simpler necessary and sufficient condition

$$\ddot{\mathbf{x}} \cdot \text{curl}\,(\mathbf{w} \times \dot{\mathbf{x}}) = 0,$$

or, equivalently (since $\ddot{\mathbf{x}} \cdot \text{curl}\,\ddot{\mathbf{x}} = 0$),

$$\ddot{\mathbf{x}} \cdot \frac{\partial \mathbf{w}}{\partial t} = 0.$$

Thus *a flow of an inviscid fluid subject to conservative extraneous force is self-superposable if and only if the local rate of change of vorticity be normal to the acceleration.*

From Navier's dynamical equation (17.4), supposing the extraneous force conservative, we obtain the integrability condition

$$\text{curl}\,\ddot{\mathbf{x}} = -\nu\,\text{curl}\,\text{curl}\,\mathbf{w},$$

where $\nu \equiv \mu/\rho$ is the *kinematic viscosity*. From inspection of (95.1) we then obtain the following theorem: *let* $\dot{\mathbf{x}}_1$ *and* $\dot{\mathbf{x}}_2$ *be flows of homogeneous viscous incompressible fluids of kinematic viscosities* ν_1 *and* ν_2, *subject to conservative extraneous force; then necessary and sufficient that* $\dot{\mathbf{x}} \equiv \dot{\mathbf{x}}_1 + \dot{\mathbf{x}}_2$ *be such a flow of a similar fluid of kinematic viscosity* ν *is the condition*

$$(\nu - \nu_1)\,\text{curl}\,\text{curl}\,\mathbf{w}_1 + (\nu - \nu_2)\,\text{curl}\,\text{curl}\,\mathbf{w}_2 + \text{curl}\,(\mathbf{w}_1 \times \dot{\mathbf{x}}_2 + \mathbf{w}_2 \times \dot{\mathbf{x}}_1) = 0.$$

In particular, if $\nu_1 = \nu_2 = \nu$ this condition reduces to (95.2), a result derived by Ballabh [1940, 2, §1] and Strang [1948, **13**, §2]. Combining this result with that given in the text above yields the following theorem: *let two motions satisfy* (95.2); *then if each be circulation-preserving, so is their vector sum, while if each be possible flows of a single homogeneous incompressible viscous fluid, subject to conservative extraneous force, so also is their vector sum.* A number of the results stated in the text for the circulation-preserving case carry over to viscous fluids and were first derived in this connection by Ballabh [1940, 2, §1]. Putting $\dot{\mathbf{x}}_1 = \dot{\mathbf{x}}_2$ again yields (46.2), and hence the following result: *given a motion of a homogeneous viscous incompressible fluid subject to conservative extraneous force, then, the vorticity being assumed steady, the motion is self-superposable if and only if it be circulation-preserving,* i.e. *if and only if it admit a flexion-potential* (§49).

The reader will easily construct examples of superposable motions. It is trivial to remark that any two or irrotational motions or Beltrami motions with the same abnormality are superposable, and hence that any steady Beltrami or irrotational motion is self-superposable.[3] Also, if $\mathbf{w}_1 = \lambda\dot{\mathbf{x}}_2$ while $\mathbf{w}_2 = \lambda\dot{\mathbf{x}}_1$, then $\dot{\mathbf{x}}_1$ and $\dot{\mathbf{x}}_2$, supposed circulation-preserving, are superposable. If $\dot{\mathbf{x}}_1$ and $\dot{\mathbf{x}}_2$ be superposable, and if both $\dot{\mathbf{x}}_1$ and $\dot{\mathbf{x}}_2$ be also self-superposable, then any linear combination of $\dot{\mathbf{x}}_1$ and $\dot{\mathbf{x}}_2$ is self-superposable. If $\dot{\mathbf{x}}_1$, $\dot{\mathbf{x}}_2$, and $\dot{\mathbf{x}}_1 + \dot{\mathbf{x}}_2$ be each self-superposable, then $\dot{\mathbf{x}}_1$ and $\dot{\mathbf{x}}_2$ are superposable. If $\dot{\mathbf{x}}_3$ and $\dot{\mathbf{x}}_1$ be superposable, and also $\dot{\mathbf{x}}_3$ and $\dot{\mathbf{x}}_2$ be superposable, then $\dot{\mathbf{x}}_3$ and any linear combination of $\dot{\mathbf{x}}_1$ and $\dot{\mathbf{x}}_2$ are superposable. It is not true, however, that any two steady Beltrami motions are superposable, or that an irrotational motion is superposable upon an arbitrary circulation-preserving motion.

The problem of finding fields $\dot{\mathbf{x}}_1$ and $\dot{\mathbf{x}}_2$ satisfying (95.2) is considered in some detail by Strang[4] and Ergun.[5]

Putting $\dot{\mathbf{x}}_1 = \dot{\mathbf{x}}_2$ in (95.2) yields (46.2), and hence, taking into account a theorem of §46, the result that *any circulation-preserving motion with steady vorticity is self-superposable.*

96. Lamb's description of convection. In a circulation-preserving motion the formula (83.5), which expresses the time rate of change of the vorticity in one frame as apparent to an observer in another, reduces to

$$\frac{\delta' \mathbf{w}}{\delta t'} = \mathbf{w} \cdot \text{grad}' \, \dot{\mathbf{x}}' - \mathbf{w} \, \text{div}' \, \dot{\mathbf{x}}', \tag{96.1}$$

or

$$\frac{\delta' J\mathbf{w}}{\delta t'} = J\mathbf{w} \cdot \text{grad}' \, \dot{\mathbf{x}}'. \tag{96.2}$$

This result expresses a remarkable type of invariance: to calculate the rate of change of $\mathbf{w}$ we may simply replace $\mathbf{w}'$ by $\mathbf{w}$, or, in other words, since $\mathbf{w}' = \mathbf{w} - 2\boldsymbol{\omega}$, the formula for rate of change of vorticity is invariant under a substitution of $\mathbf{w} + \mathbf{f}(t)$ for $\mathbf{w}$. Thus *a motion is circulation-preserving if and only if in calculating the material rate of change of vorticity the effect of any rigid rotation may be systematically neglected.*

The foregoing invariance theorem is equivalent to an intrinsic characterization given by Lamb. For the primed system let us select a Cartesian frame whose axes are the principal axes of extension and whose angular velocity relative to the stationary frame is $\boldsymbol{\omega} = \frac{1}{2}\mathbf{w}$. Then in this frame

[3] Any pair of Beltrami motions with equal values of Ω satisfies (95.2) and any Beltrami motion satisfies curl $(\mathbf{w} \times \dot{\mathbf{x}}) = 0$, but the restriction to steady motion follows from the theorem of Beltrami given in §52. *Cf.* [1940, **2**, §4].

[4] [1948, **13**, §§4–10].

[5] [1949, **12**].

the vorticity is zero: $\mathbf{w}' = \mathbf{w} - 2\boldsymbol{\omega} = 0$, and furthermore the rates of shearing are zero (§21), so that the velocity gradient reduces to the simple form

$$\| \text{grad}' \, \dot{\mathbf{x}}' \| = \begin{Vmatrix} \Delta_1' & 0 & 0 \\ 0 & \Delta_2' & 0 \\ 0 & 0 & \Delta_3' \end{Vmatrix}. \tag{96.3}$$

The formula (96.2) becomes then

$$\frac{\delta' J w_i}{\partial t'} = J w_i \Delta_i', \tag{96.4}$$

or

$$\frac{\delta'}{\delta t'} (\log J w_i) = \Delta_i'. \tag{96.5}$$

The vector $J\mathbf{w}$ is calculated in the frame with respect to which the motion is circulation-preserving, while the rates of change, both of vorticity and of length, are calculated with respect to the moving frame. Since the principal axes themselves, which are here taken as co-ordinate axes in the moving frame, are rotating at angular velocity $\frac{1}{2}\mathbf{w}$ with respect to the stationary frame (§32), the moving frame may be regarded as rigidly attached to the principal axes of extension at all times. The formula (96.5) then embodies ***Lamb's description of convection*** (1885):[1] *a motion is circulation-preserving in a given frame of reference if and only if the logarithmic material rate of change of the projection of* $J\mathbf{w}$ *upon any one of the principal axes of extension appears to an observer whose frame of reference is rigidly attached to these axes to equal the corresponding principal rate of extension.* Lamb's result (96.5) may be put into another form by expressing the rates of extension Δ_i' in terms of the lengths l_i of line segments instantaneously parallel to the axes of extension:

$$\frac{\delta'}{\delta t'} (\log J w_i) = \lim_{l_i \to 0} \frac{\delta'}{\delta t'} \log l_i, \tag{96.6}$$

or

$$\lim_{l_i \to 0} \frac{\delta'}{\delta t'} \left(\log \frac{J w_i}{l_i} \right) = 0. \tag{96.7}$$

Hence

$$\frac{\delta'}{\delta t'} \left(\log \frac{J w_i}{l_i} \right) = \epsilon(l_i), \tag{96.8}$$

[1] [1885, 2].

where $\epsilon(l_i) \to 0$ as $l_i \to 0$. Integration from 0 to t now yields

$$\frac{Jw_i}{l_i} = \frac{W_i}{L_i} e^{\epsilon(l_i)t}, \tag{96.9}$$

whence finally

$$\frac{Jw_i}{l_i} = \frac{W_i}{L_i} [1 + \eta(L_i)t], \tag{96.10}$$

where $\eta(L_i) \to 0$ as $L_i \to 0$. This last formula shows that the ratio Jw_i/l_i, where w_i is the projection of the vorticity upon a principal axis of extension and l_i is the length of a line segment tangent to that axis, can be held arbitrarily close to a constant value in any finite time interval by choosing l_i small enough.

97. Appell's description of convection.[1] Let $\mathfrak{C}$ be any finite material line. Then by (21.5) and (94.6) we have

$$\begin{aligned} \frac{\delta}{\delta t} \int_{\mathfrak{C}} \frac{dx}{Jw} &= \frac{1}{2} \int_{\mathfrak{C}} \left[\frac{1}{Jw\, dx} \frac{\delta}{\delta t} (dx^2) - \frac{dx}{(Jw)^3} \frac{\delta}{\delta t} (Jw)^2 \right], \\ &= \int_{\mathfrak{C}} \frac{1}{Jw} \left[\frac{d\mathbf{x} \cdot \mathbf{\Delta} \cdot d\mathbf{x}}{dx} - \frac{dx\, \mathbf{w} \cdot \mathbf{\Delta} \cdot \mathbf{w}}{w^2} \right]. \end{aligned} \tag{97.1}$$

By Helmholtz's second theorem we may let $\mathfrak{C}$ be a finite portion of a vortex-line. Then

$$dx\, \mathbf{w} = d\mathbf{x}\, w, \tag{97.2}$$

and (97.1) reduces to

$$\frac{\delta}{\delta t} \int_{\mathfrak{C}} \frac{dx}{Jw} = 0. \tag{97.3}$$

This result is to be compared with (96.7), which may be regarded as its differential form.

Now in §45 we have seen that a motion in which the vortex-lines are material is not necessarily circulation-preserving, but need only satisfy

$$\mathbf{w} \times \text{curl}\, \ddot{\mathbf{x}} = 0. \tag{45.3}$$

In such a motion all the steps leading to (97.3) may be carried out, except that when we use the more general equation (78.5) in place of (94.6) we obtain not (97.3) but

$$\frac{\delta}{\delta t} \int_{\mathfrak{C}} \frac{dx}{Jw} = \int_{\mathfrak{C}} dx \frac{J\mathbf{w} \cdot \text{curl}\, \ddot{\mathbf{x}}}{w^3}. \tag{97.4}$$

---

[1] The analysis of this section was suggested by the rather more limited result given in [1921, 2, §760].

If (97.3) holds for all finite portions of all vortex-lines, by (97.4) it must follow that

$$\mathbf{w}\cdot\operatorname{curl}\ddot{\mathbf{x}} = 0. \tag{97.5}$$

Comparison of (45.3) and (97.5) yields curl $\ddot{\mathbf{x}} = 0$, and hence by the d'Alembert-Euler condition the motion is circulation-preserving. We may summarize the preceding analysis as follows: *in a circulation-preserving motion*

$$\int_{\mathfrak{C}} \frac{dx}{Jw} = \text{const. in time} \tag{97.6}$$

for any finite portion $\mathfrak{C}$ *of any vortex-line. Conversely, if the vortex-lines be material lines and* (97.6) *hold for any arbitrary portion of each of them, then the motion is circulation-preserving.*

From (97.6) follows a generalization of a result of Carstoiu:[2] *in a circulation-preserving motion, a necessary and sufficient condition that every finite material portion of every vortex-line be of length constant in time is that*

$$Jw = \text{const.} \tag{97.7}$$

for each particle.

98. The Euler and Ertel-Rossby convection theorems. Permanence of complex-lamellar motion. We notice first that the Ertel formula (79.7) can be interpreted as characterizing the circulation-preserving case. For general $\emptyset$ this fact is obvious, since (79.7) is equivalent to (94.2). But, by (79.2), we may restrict $\emptyset$ considerably and still obtain the desired result, since for given $\emptyset$ (79.7) is equivalent to

$$\operatorname{curl}\ddot{\mathbf{x}}\cdot\operatorname{grad}\emptyset = 0. \tag{98.1}$$

Thence to conclude that curl $\ddot{\mathbf{x}} = 0$, all we need is to be able to select at each point scalars $\emptyset$ having gradients parallel to three different directions. Thus it follows that *if there exist a class C of scalars a such that for each of three different directions* $\mathbf{e}_1$, $\mathbf{e}_2$, $\mathbf{e}_3$ *at each point the gradient of some one a is parallel to* $\mathbf{e}_1$, *another to* $\mathbf{e}_2$, *and a third to* $\mathbf{e}_3$, *and such that for all* $a \in C$ *Ertel's equation holds:*

$$\frac{\delta}{\delta t}(J\mathbf{w}\cdot\operatorname{grad} a) = Jw\cdot\operatorname{grad}\frac{\delta a}{\delta t}, \tag{98.2}$$

then the motion is circulation-preserving, and (79.7) *holds for all* $\emptyset$.

In the foregoing statement we may further restrict our class C to substantially constant functions a, replacing (98.2) by

---

[2] [1946, **6**] [1947, **8**, Ch. III, §3].

$$Jw \frac{da}{dw} = \text{const.} \tag{98.3}$$

for each particle.

In any steady motion, there exists at each point a material stream-surface having arbitrary normal direction. Hence it follows that *a steady motion is circulation-preserving if and only if, for three independent families of steady stream-surfaces* a = const., *we have*

$$Jw \frac{da}{dw} = \text{const.} \tag{98.4}$$

along each stream-line.

A closely related result, involving only *two* families of stream-surfaces, say a = const. and b = const., was derived by Euler.[1] Starting with a representation of type (11.8), *viz.*

$$\dot{\mathbf{x}} = Jc(a, b) \operatorname{grad} a \times \operatorname{grad} b, \tag{98.5}$$

where a = const. and b = const. are any two independent families of steady stream-surfaces, he obtained the identity

$$\begin{aligned} \mathbf{w} \times \dot{\mathbf{x}} \cdot d\mathbf{x} &= \mathbf{w} \cdot (Jc \operatorname{grad} a \times \operatorname{grad} b) \times d\mathbf{x}, \\ &= Jc\mathbf{w} \cdot [(\operatorname{grad} a \cdot d\mathbf{x}) \operatorname{grad} b - (\operatorname{grad} b \cdot d\mathbf{x}) \operatorname{grad} a], \\ &= Jc\mathbf{w} \cdot [\operatorname{grad} b \, da - \operatorname{grad} a \, db]. \end{aligned} \tag{98.6}$$

In the circulation-preserving case, by (75.5) follows

$$\begin{aligned} -d(\phi^* + \tfrac{1}{2}\dot{x}^2) &= -\operatorname{grad} (\phi^* + \tfrac{1}{2}\dot{x}^2) \cdot d\mathbf{x}, \\ &= \mathbf{w} \times \dot{\mathbf{x}} \cdot d\mathbf{x}. \end{aligned} \tag{98.7}$$

Comparing (98.6) and (98.7) yields Euler's formulae

$$\begin{aligned} \frac{\partial(\phi^* + \tfrac{1}{2}\dot{x}^2)}{\partial a} &= -Jc\mathbf{w} \cdot \operatorname{grad} b = -Jcw \frac{db}{dw} \\ \frac{\partial(\phi^* + \tfrac{1}{2}\dot{x}^2)}{\partial b} &= Jc\mathbf{w} \cdot \operatorname{grad} a = Jcw \frac{da}{dw}, \end{aligned} \tag{98.8}$$

whence follows as condition of integrability

$$\frac{\partial}{\partial a} (Jc\mathbf{w} \cdot \operatorname{grad} a) + \frac{\partial}{\partial b} (Jc\mathbf{w} \cdot \operatorname{grad} b) = 0, \tag{98.9}$$

where w is to be thought of as expressed by $\mathbf{w} = \operatorname{curl} (Jc \operatorname{grad} a \times \operatorname{grad} b)$. That the right-hand sides of (98.8) shall be functions of a

[1] [1757, **3**, §§LIV–LVI]. Euler worked out in detail the application to simple vortices [§§LVII–LVIII], hoping (in vain) that this analysis might suggest a general process of integration.

and b only, and hence that an integrating factor c satisfying (98.9) shall exist, follows from our general results above.

But for the motion to be circulation-preserving it is not in general sufficient that (98.4) hold for only two families of stream-surfaces. To obtain a converse in this case, suppose c already determined by (98.5), and suppose this same c satisfies (98.9) also. Then there exists a function $\psi^*(a, b)$ such that (98.8) is satisfied with ψ^* replacing $\phi^* + \frac{1}{2}\dot{x}^2$, and hence by the identity (98.6) follows

$$-d\psi^* = \mathbf{w} \times \dot{\mathbf{x}} \cdot d\mathbf{x}, \tag{98.10}$$

or

$$-\text{grad } \psi^* = \mathbf{w} \times \dot{\mathbf{x}}. \tag{98.11}$$

Hence

$$-\text{grad } (\psi^* - \tfrac{1}{2}\dot{x}^2) = \mathbf{w} \times \dot{\mathbf{x}} + \text{grad } \tfrac{1}{2}\dot{x}^2 = \ddot{\mathbf{x}}. \tag{98.12}$$

Thus the motion is circulation-preserving, its acceleration-potential being $\phi^* = \psi^* - \frac{1}{2}\dot{x}^2$. Such being the case, by the reasoning of the foregoing paragraph (98.9) must hold for all choices of a, b, c in (98.5).

In summary of these results, we have ***Euler's characterization of steady convection***: *In a steady motion, let it be possible to find two independent families of stream-surfaces* a = const., b = const. *such that* (98.4) *holds; with c determined by* (98.5), *if* (98.9) *holds also then it follows that the motion is circulation-preserving. Conversely, in a circulation-preserving motion* (98.4) *and* (98.9) *hold for any functions a, b such that* a = const., b = const. *are steady stream-surfaces.*

A very interesting special case of the Ertel theorem (94.3) is obtained by choosing $\emptyset$ to be the particular function

$$\Psi \equiv \int_0^t (\tfrac{1}{2}\dot{x}^2 - \phi^*)\, dt, \tag{98.13}$$

where the integration is to be carried out for a fixed particle $\mathbf{X}$. Then the Ertel theorem yields

$$\frac{\delta}{\delta t}(J\mathbf{w}\cdot\text{grad } \Psi) = J\mathbf{w}\cdot\text{grad } (\tfrac{1}{2}\dot{x}^2 - \phi^*). \tag{98.14}$$

By this result and (94.2) we have

$$\frac{\delta}{\delta t}[J\mathbf{w}\cdot(\dot{\mathbf{x}} - \text{grad } \Psi)] = J\mathbf{w}\cdot[\text{grad } \dot{\mathbf{x}}\cdot\dot{\mathbf{x}} + \ddot{\mathbf{x}} - \text{grad } (\tfrac{1}{2}\dot{x}^2 - \phi^*)] = 0. \tag{98.15}$$

Integrating this equation from 0 to t yields the ***convection theorem of Ertel & Rossby*** (1949):[2]

$$J\mathbf{w}\cdot(\dot{\mathbf{x}} - \text{grad } \Psi) = \mathbf{W}\cdot\dot{\mathbf{X}} \tag{98.16}$$

[2] [1949, 10 & 11].

where $\dot{\mathbf{X}} \equiv \dot{\mathbf{x}}_{t=0}$. In contrast to the Ertel theorem (94.3) or its equivalent, (79.6), the formula (98.16) is not sufficient to characterize convection.

A rather vague statement made long ago by Earnshaw[3] suggests that *in a circulation-preserving motion, a particle once in complex-lamellar motion remains ever in complex-lamellar motion.* I am unable to prove this statement, and in fact I doubt its truth. It is, however, a natural generalization of the Lagrange-Cauchy velocity potential theorem, which we shall discuss at length in §§104-106. The formula (98.16) suggests a means of approaching the problem. If the motion be initially complex-lamellar at the place occupied by the particle $\mathbf{X}$, then $\mathbf{W} \cdot \dot{\mathbf{X}} = 0$. For permanence of complex-lamellar motion we must have then

$$\mathbf{w} \cdot \operatorname{grad} \Psi = 0. \tag{98.17}$$

The following result then emerges: *in a circulation-preserving motion such that the surfaces* $\Psi =$ const. *are vortex-surfaces, a particle once in complex-lamellar motion is ever in complex-lamellar motion. Conversely, in any complex-lamellar circulation-preserving motion the surfaces* $\Psi =$ const. *are vortex-surfaces.* Earnshaw's conjecture would follow if (98.17) could be shown to be a consequence of $\mathbf{W} \cdot \dot{\mathbf{X}} = 0$. For a plane or axially-symmetric motion the conclusion is trivial. However, in assuming a motion to be plane or axially-symmetric the hydrodynamical literature always begs the question: if a motion be plane or axially-symmetric at one instant, will it so remain? Proof of the somewhat more general conjecture of Earnshaw is one of the outstanding unsolved problems of classical hydrodynamics.

99. Ertel & Köhler's description of steady convection.[1] In the foregoing section we have seen how Euler applied his representation (11.8) to the case of steady convection. We shall now consider a fuller use of it by Ertel & Köhler. By the result of Lamb given in §75, in a steady circulation-preserving motion which is neither an irrotational nor a Beltrami motion there are Lamb surfaces $b =$ const.:

$$\mathbf{w} \times \dot{\mathbf{x}} = \operatorname{grad} b, \qquad b \neq \text{const.} \tag{99.1}$$

Let the surfaces $\alpha =$ const. be any stream-surfaces other than the Lamb surfaces, and let the surfaces $\beta =$ const. be any vortex-surfaces other than the Lamb surfaces. Since both $\dot{\mathbf{x}}/J$ and $\mathbf{w}$ are solenoidal, Euler's representation (11.8) is available for each:

$$\dot{\mathbf{x}} = Jf(\alpha, b) \operatorname{grad} \alpha \times \operatorname{grad} b, \tag{99.2}$$

$$\mathbf{w} = g(\beta, b) \operatorname{grad} \beta \times \operatorname{grad} b. \tag{99.3}$$

[3] [1837, 1, §4].

[1] The results of Ertel & Köhler [1949, **6**] are derived and stated in a quite different way.

By taking the cross-product of these two results we obtain

$$\mathbf{w} \times \dot{\mathbf{x}} = -Jfg \frac{\partial(\alpha, b, \beta)}{\partial(x, y, z)} \operatorname{grad} b, \tag{99.4}$$

whence by (99.1) follows the ***Ertel-Köhler equation*** (1949):

$$Jfg \frac{\partial(\alpha, \beta, b)}{\partial(x, y, z)} = 1. \tag{99.5}$$

Equivalently,

$$Jfg = \frac{\partial(x, y, z)}{\partial(\alpha, \beta, b)}, \tag{99.6}$$

or

$$fg = \frac{\partial(X, Y, Z)}{\partial(\alpha, \beta, b)}. \tag{99.7}$$

We now seek an interpretation for the equation (99.6). First, let the surfaces $\alpha = 0, \pm 1, \pm 2, \ldots$ and the surfaces $b = 0, \pm 1, \pm 2, \ldots$ be described; these surfaces decussate the portion of space under consideration into tubes, which we shall name *standard stream-tubes.*[2] Now for an element of area $d\mathbf{s}$ on any surface $\mathbf{x} = \mathbf{x}(l, m)$, we have by (15.3) and (99.2)

$$d\mathbf{s}\cdot\dot{\mathbf{x}} = Jf \frac{\partial \mathbf{x}}{\partial l} \times \frac{\partial \mathbf{x}}{\partial m}\cdot \operatorname{grad} \alpha \times \operatorname{grad} b \, dl \, dm. \tag{99.8}$$

But since α and b are independent we may choose $l = \alpha$, $m = b$, obtaining

$$d\mathbf{s}\cdot\dot{\mathbf{x}} = Jf \frac{\partial \mathbf{x}}{\partial \alpha} \times \frac{\partial \mathbf{x}}{\partial b}\cdot \operatorname{grad} \alpha \times \operatorname{grad} b \, d\alpha \, db. \tag{99.9}$$

It is easy to see that identically for any continuously differentiable functions $\alpha(\mathbf{x})$, $b(\mathbf{x})$ we have

$$\frac{\partial \mathbf{x}}{\partial \alpha} \times \frac{\partial \mathbf{x}}{\partial b}\cdot \operatorname{grad} \alpha \times \operatorname{grad} b = 1. \tag{99.10}$$

Hence (99.9) becomes

$$d\mathbf{s}\cdot\dot{\mathbf{x}} = Jf \, d\alpha \, db. \tag{99.11}$$

Now $d\alpha \, db$ when integrated over an area yields the number of standard stream-tubes crossing the area. Therefore (99.11) shows that Jf is the flux of $\dot{\mathbf{x}}$ per standard stream-tube.[3] Analogous treatment of (99.3)

---

[2] This concept and several of those subsequently introduced fail of invariance under changes of parametrization or of units of length and time. The theorem finally resulting, however, is of course invariant.

[3] More precisely, it is the ratio of the flux per unit area to the number of standard stream-tubes per unit area.

shows that g is the flux of $\mathbf{w}$ per analogously defined *standard vortex-tube*. We shall call the region common to a single standard stream-tube and a single standard vortex-tube a *standard cell*. Since the volume of a region may be obtained by integrating $\partial(x, y, z)/\partial(\alpha, \beta, b)\, d\alpha\, d\beta\, db$, the quantity $\partial(x, y, z)/\partial(\alpha, \beta, b)$ is itself the volume of a standard cell.[4] By applying these interpretations to the quantities appearing in (99.6) we thus obtain ***Ertel & Köhler's description of steady convection***: *in a steady circulation-preserving motion, the product of the flux of velocity through a standard stream-tube by the flux of vorticity through a standard vortex-tube equals the volume of a standard cell.* This statement, of course, is only approximate in the sense of infinitesimals.

100. The vorticity convection theorem for complex-lamellar motions.[1] For complex-lamellar motions with steady vortex-lines the convective mechanism assumes a particularly simple form. In the generalized convection theorem of §91 we put $v_0 = 1$, $f = 1$. Then $\mathbf{v}_C = \dot{\mathbf{x}}$ and we may satisfy the condition (91.6) by the choice $v = J^{-1}$. By (45.2) the condition (91.9) reduces to[2]

$$\mathbf{w} \cdot \operatorname{curl} \ddot{\mathbf{x}} = 0. \tag{100.1}$$

Hence it follows from the theorem of §91 that *in a complex-lamellar motion with steady vortex-lines, in order that*

$$\frac{Jw}{h} = \text{const.} \tag{100.2}$$

for each particle it is necessary and sufficient that either the motion be circulation-preserving or else the lines of diffusion be normal to the vortex-lines.

In a plane motion $h = 1$ and the vortex-lines are steady. The resulting special case then yields the ***d'Alembert vorticity theorem*** (1761):[3] *in a plane circulation-preserving motion*

$$Jw = \text{const.} \tag{100.3}$$

[4] More precisely, the reciprocal of the number of standard cells per unit volume.

[1] [1951, 5].

[2] By (49.5) it follows that for the convection theorem (100.2) to hold for a viscous fluid of uniform density and viscosity subject to conservative extraneous force, it is necessary and sufficient that

$$\mathbf{w} \cdot \operatorname{curl} \operatorname{curl} \mathbf{w} = 0.$$

In a plane or a rotationally-symmetric motion this condition can be satisfied only if $\operatorname{curl} \operatorname{curl} \mathbf{w} = 0$, *i.e.*, only if the motion be circulation-preserving.

[3] d'Alembert [1761, 2, §XIII] treated only the case of steady isochoric motion. The result is usually attributed to Helmholtz [1858, 1, §5], whose statement is limited to isochoric motion; it follows also from an equation of Stokes [1842, 2, eq. 10]. *Cf.* [1878, 2].

for each particle. In a rotationally-symmetric motion the vortex-lines are steady and we have $h = r$, where r is the distance from the axis of symmetry. The resulting special case then yields the ***Svanberg vorticity theorem*** (1841):[4] *in a circulation-preserving rotationally-symmetric motion*

$$\frac{Jw}{r} = \text{const.} \tag{100.4}$$

for each particle.

It was these theorems, stating that Jw or Jw/r is carried by the motion as if it were some native property of the particles, which originally motivated the name "convection" for the vorticity transfer process in a circulation-preserving motion.

101. The transformation of Clebsch. The condition of circulation-preserving motion permits a single integration of the kinematical equations. Cauchy's vorticity formula

$$\mathbf{w} = \mathbf{W}\cdot\frac{\text{GRAD } \mathbf{x}}{J} \tag{94.3}$$

is in fact a first integral, since it is free of second material time derivatives. It is possible to express this first integral in two other rather different forms, in which appears the velocity itself rather than the vorticity. The first of these, *the transformation of Clebsch,* is easiest derived as a direct consequence of the existence of an acceleration-potential ϕ^*:

$$\ddot{\mathbf{x}} = -\text{grad } \phi^*; \tag{48.1}$$

the second, *the transformation of Weber,* from the corresponding material expression

$$\text{GRAD } \mathbf{x}\cdot\ddot{\mathbf{x}} = -\text{GRAD } \phi^*, \tag{101.1}$$

which follows from (48.1) by the rule for differentiating functions of functions.

We now proceed to the analysis of (48.1), to this end employing Duhem's representation (40.4) for the acceleration in terms of the Monge potentials f, g, h of the velocity. Now in any motion we have

$$\mathbf{w} = \text{grad } f \times \text{grad } g, \tag{101.2}$$

and the surfaces $f = \text{const.}$, $g = \text{const.}$ are vortex-surfaces. Since in a circulation-preserving motion by the second Helmholtz theorem the

[4] The result of Svanberg [1841, 2, §4] states that (100.4) holds for any circulation-preserving motion in which $\ddot{\theta} = 0$ (*cf.* §18). It may be proved from the theorem stated above by superposing a rigid rotation upon a rotationally symmetric motion The result stated in the text is usually attributed to Helmholtz [1858, 1, §6]; it follows also from an equation of Stokes [1842, 2, eq. 21]. *Cf.* [1878, 2].

vortex-lines are material, it is natural to conjecture, and we shall now accordingly prove, that it is possible to choose the functions g and h in such a way that the surfaces f = const., g = const. are also material surfaces, that is, in such a way that

$$\dot{f} = 0, \qquad \dot{g} = 0. \tag{101.3}$$

Let $F(\mathbf{X})$ and $G(\mathbf{X})$ be any pair of continuously differentiable functions satisfying (101.2) at $t = 0$. First, as was stated in §11, for one of the functions, say g, in (101.2) at time t we may select any function such that the surfaces g = const. are vortex-surfaces, and thus a permissible choice is given by

$$g(\mathbf{x}, t) = G(\mathbf{X}(\mathbf{x}, t)), \tag{101.4}$$

whereby $(101.3)_2$ is satisfied. By (11.5) for the two functions F and f we have

$$F = -\int \frac{\mathbf{E} \times \mathbf{W} \cdot d\mathbf{X}}{\mathbf{E} \; \text{GRAD} \; G}, \qquad f = -\int \frac{\mathbf{e} \times \mathbf{w} \cdot d\mathbf{x}}{\mathbf{e} \cdot \text{grad}\, G}, \tag{101.5}$$

where $\mathbf{E}$ and $\mathbf{e}$ are arbitrary continuous fields, which we may select independently and at pleasure, subject only to the restriction that the denominators in (101.5) be non-vanishing in the regions considered, and in each integral the path of integration is the same, *viz.*, any curve lying wholly upon one of the surfaces $g = G$ = const. By employing in turn Cauchy's vorticity formula (94.3), $(15.1)_3$, $(14.13)_2$, the result of interchanging large and small letters in $(14.14)_2$, and $(14.14)_1$ we obtain the identities[1]

$$\begin{aligned}
\frac{\mathbf{e} \times \mathbf{w} \cdot d\mathbf{x}}{\mathbf{e} \cdot \text{grad}\, G} &= \frac{\sqrt{g}\, \epsilon_{ijk} e^j w^k \, dx^i}{e^j G_{,j}}, \\
&= \frac{2\sqrt{G}\, \epsilon_{ijk} e^j W^L x^k{}_{,L} \, dX^M x^i{}_{,M}}{e^j \epsilon^{IJK} \epsilon_{jmn} G_{,I} x^m{}_{,J} x^n{}_{,K}}, \\
&= -\frac{2e^j \sqrt{G}\, \epsilon_{MIL} X^I{}_{,j} W^L \, dX^M}{e^j \epsilon_{jmn} G_{,I} \epsilon^{mpn} X^I{}_{,p}}, \\
&= \frac{e^j X^I{}_{,j} \sqrt{G}\, \epsilon_{ILM} W^L \, dX^M}{e^j X^I{}_{,j} G_{,I}}, \\
&= \frac{(\mathbf{e} \cdot \text{grad}\, \mathbf{X}) \cdot (\mathbf{W} \times d\mathbf{X})}{(\mathbf{e} \cdot \text{grad}\, \mathbf{X}) \cdot \text{GRAD}\, G}.
\end{aligned} \tag{101.6}$$

Hence if we make the choice

$$\mathbf{E} \equiv \mathbf{e} \cdot \text{grad}\, \mathbf{X}, \tag{101.7}$$

[1] The symbol $\sqrt{G}$ stands as usual for $\sqrt{\det G_{IJ}}$ and is not to be confused with the scalar G occurring in $\mathbf{e} \cdot \text{grad}\, G$, *etc.*

then comparison of $(101.5)_2$ with $(101.5)_1$ yields

$$f(\mathbf{x}, t) = -\int \frac{\mathbf{E} \times \mathbf{W} \cdot d\mathbf{X}}{\mathbf{E} \cdot \text{GRAD}\, G} = F(\mathbf{X}), \tag{101.8}$$

so that $(101.3)_1$ also is satisfied. The possibility of such a choice of the functions f and g was first proved by Clebsch (1859)[2] in a different way.

Duhem's acceleration formula (40.4) now reduces to

$$\ddot{\mathbf{x}} = \text{grad}\left[\tfrac{1}{2}\dot{x}^2 + \frac{\partial h}{\partial t} + f\frac{\partial g}{\partial t}\right]. \tag{101.9}$$

If we absorb into h a function of time only whose value, by (16.12), cannot affect the velocity, comparison of (101.9) with (48.1) yields

$$\phi^* + \tfrac{1}{2}\dot{x}^2 + \frac{\partial h}{\partial t} + f\frac{\partial g}{\partial t} = 0. \tag{101.10}$$

This result may be called ***Clebsch's transformation***:[3] it expresses the condition for circulation-preserving motion in integrated form and wholly in terms of the spatial variables. According to the definition of §69, Clebsch's transformation is a Bernoullian theorem, but unlike other results of this type it is not limited in its validity to motions with steady vorticity. But more than this: when combined with (101.3) and with the three components of (48.1), Clebsch's transformation yields a system of six first order partial differential equations for the seven quantities $\dot{x}$, $\dot{y}$, $\dot{z}$, f, g, h, ϕ^*, so that if one more equation be added, the system is rendered determinate.[4]

Another form of Clebsch's transformation may be obtained by noticing that from $(101.3)_2$ and (16.12) we have

$$\frac{\partial g}{\partial t} = -\dot{x}\cdot\text{grad}\, g = -[\text{grad}\, h + f\,\text{grad}\, g]\cdot\text{grad}\, g, \tag{101.11}$$

while by squaring (16.12) follows

$$\tfrac{1}{2}\dot{x}^2 = \tfrac{1}{2}(\text{grad}\, h)^2 + f\,\text{grad}\, g\cdot\text{grad}\, h + \tfrac{1}{2}f^2(\text{grad}\, g)^2. \tag{101.12}$$

---

[2] [1859, 1, §3]. Other proofs are given in [1861, 1, §VI] [1881, 1, §III and pp. 168–210] [1888, 1, §§33–35] [1906, 1, §166] [1921, 2, §799].

[3] Clebsch [1857, 1, §5] derived an equivalent formula from the equations for perfect fluids. See also [1871, 1, §13] [1881, 1, §III].

[4] In classical hydrodynamics ϕ^* is specified by (49.4) as a function of $\mathbf{x}$, ρ, and t. The second of these, rather than ϕ^*, is to be regarded as the seventh variable, and the one further equation required is the Eulerian continuity equation, which becomes $\delta \log \rho/\delta t + \text{div}\,\dot{\mathbf{x}} = 0$. It would be interesting to use Clebsch's transformation to study circulation-preserving flows of viscous fluids: the whole class of such flows may be obtained by putting (49.6) into (101.10) and then replacing the flexion-potential σ by an arbitrary harmonic function.

Putting these two results into (101.10) yields[5]

$$\phi^* = -\frac{\partial h}{\partial t} - \tfrac{1}{2}(\text{grad } h)^2 + \tfrac{1}{2}f^2(\text{grad } g)^2. \tag{101.13}$$

A part of the apparent simplicity of Clebsch's transformation is illusory, since in steady motion the function h and g generally cannot be steady. In fact comparison of (101.9) with Lagrange's acceleration formula (38.2) in the case of steady motion yields

$$\mathbf{w} \times \dot{\mathbf{x}} = \text{grad}\left[\frac{\partial h}{\partial t} + f\frac{\partial g}{\partial t}\right]. \tag{101.14}$$

Hence h and g can be steady if and only if the motion is an irrotational or a Beltrami motion, in which case Clebsch's transformation (101.10) reduces to the same form as does the spatial Bernoulli theorem (48.3).[6]

Carstoiu has pointed out a simple and interesting consequence of (101.3). In §94 we have noted that $\mathbf{w}\cdot\text{grad } \dot{\mathbf{x}} = 0$ is necessary and sufficient that $J\mathbf{w} = \text{const.}$ for each particle. If it is impossible to find a region in which $\mathbf{w}$ is zero or parallel to a fixed plane, then no matter how Cartesian axes x, y, z be selected we shall have $w_x w_y w_z \neq 0$ in general. Then in a sub-region sufficiently small that (101.2) holds with single-valued f and g, (101.14) yields

$$\frac{\partial(\dot{x}, f, g)}{\partial(x, y, z)} = 0; \tag{101.15}$$

from (101.2) and the condition $w_x w_y w_z \neq 0$ it follows that $\dot{x} = \dot{x}(f, g)$, whence by (101.3) $\ddot{\mathbf{x}} = 0$. Thus *in order for the velocity vector to be constant on each vortex-line of a rotational circulation-preserving motion, it is necessary that the vortex-lines be plane curves or the acceleration be zero.* Further work would be required in order to characterize circulation-preserving motions in which $J\mathbf{w} = \text{const.}$ for each particle.

Finally we consider the more general case in which the function f and g are not required to satisfy (101.3). By taking the curl of Duhem's acceleration formula (40.4) and comparing the result with the d'Alembert-Euler condition (45.4), we then obtain

$$\text{grad } \dot{f} \times \text{grad } g = \text{grad } \dot{g} \times \text{grad } f \tag{101.16}$$

as a necessary and sufficient condition to be satisfied by the Monge potentials of the velocity in order that the motion be circulation-preserving. That is, the vectors grad $\dot{g}$, grad $\dot{f}$, grad g, and grad f must

[5] Similar but false equations are given in [1842, 3, §8] [1881, 2, p. 12].

[6] This same result follows also from the fact that the surfaces $g = \text{const.}$ are material vortex-surfaces, and hence in a steady motion in order to be themselves steady they must be Lamb surfaces.

all lie in the plane perpendicular to the vorticity **w**, and the area inclosed by the parallelogram whose edges are grad $\dot{g}$ and grad f must equal the area enclosed by that whose edges are grad $\dot{f}$ and grad g. If we introduce the abbreviation

$$\lambda \equiv \phi^* + \tfrac{1}{2}\dot{x}^2 + \frac{\partial h}{\partial t} + f\,\frac{\partial g}{\partial t}, \tag{101.17}$$

then by forming grad λ and simplifying the result with the aid of (40.4) and (48.1) we obtain[7]

$$\text{grad}\,\lambda = \dot{g}\;\text{grad}\,f - \dot{f}\;\text{grad}\,g. \tag{101.18}$$

Hence

$$\frac{\partial(\lambda, f, g)}{\partial(x, y, z)} = \text{grad}\,\lambda\cdot\text{grad}\,f \times \text{grad}\,g = 0. \tag{101.19}$$

Since f and g are necessarily independent, from this result follows the equation of Duhem (1901):[8]

$$\lambda = \lambda(f, g, t). \tag{101.20}$$

Hence

$$\text{grad}\,\lambda = \frac{\partial\lambda}{\partial f}\,\text{grad}\,f - \frac{\partial\lambda}{\partial g}\,\text{grad}\,g. \tag{101.21}$$

Comparison with (101.18) now yields the equations of T. Stuart (1900)[9] and Lamb (1906)[10]

$$\frac{\partial\lambda}{\partial g} = \dot{f}, \qquad \frac{\partial\lambda}{\partial f} = -\dot{g}, \tag{101.22}$$

which are formally suggestive of Hamilton's canonical equations for the motion of a system of mass-points.[11]

If we have a circulation-preserving motion with acceleration-potential ϕ^*, that f, g, λ, h satisfy (101.16), (101.17), (101.20), and (101.22) is plainly not sufficient that f, g, h be Monge potentials of any given velocity field, since h is indeterminate up to an arbitrary steady function. The contribution of any h in (16.12), however, is only an irrotational motion. Thus a solution f, g, λ, h always yields a circulation-preserving

---

[7] [1901, 2, §2].

[8] [1901, 2, §3].

[9] According to [1924, 1, §167] [1932, 1, §167].

[10] [1906, 1, §166] [1932, 1, §167].

[11] Duhem [1901, 2, §§4–7] discussed the general theory of the integration of the system (101.3) (101.10) by analogy to Jacobi's method in analytical dynamics. Masuda [1953, 8] has used the Hamilton-Jacobi theory to obtain some of the general theorems of this chapter. Such general methods, failing to take account of the first integral (94.3) or its various equivalents, lead to an elaborate and awkward treatment.

motion. Clebsch's transformation (101.10) results from use of the particular solution $\lambda = 0$.

102. The transformation of Weber. The foregoing analysis has constructed a first integral of the kinematical equations for the case of circulation-preserving motion, following the spatial plan. We now give a corresponding material treatment. For any motion we have

$$\begin{aligned} \text{GRAD } \mathbf{x} \cdot \ddot{\mathbf{x}} &= \frac{\delta}{\delta t}(\text{GRAD } \mathbf{x} \cdot \dot{\mathbf{x}}) - \text{GRAD } \dot{\mathbf{x}} \cdot \dot{\mathbf{x}}, \\ &= \frac{\delta}{\delta t}(\text{GRAD } \mathbf{x} \cdot \dot{\mathbf{x}}) - \text{GRAD } \tfrac{1}{2}\dot{x}^2. \end{aligned} \tag{102.1}$$

Comparison with (101.1) yields

$$\text{GRAD } (\tfrac{1}{2}\dot{x}^2 - \phi^*) = \frac{\delta}{\delta t}(\text{GRAD } \mathbf{x} \cdot \dot{\mathbf{x}}). \tag{102.2}$$

Integrating this expression from 0 to t with $\mathbf{X}$ held constant, we obtain

$$\int_0^t \text{GRAD } (\tfrac{1}{2}\dot{x}^2 - \phi^*)\, dt = \text{GRAD } \mathbf{x} \cdot \dot{\mathbf{x}} - \dot{\mathbf{X}}, \tag{102.3}$$

where $\dot{\mathbf{X}}$ is the value of $\dot{\mathbf{x}}$ when $t = 0$. If we introduce the function Ψ defined by (98.13), this last result may be written

$$\text{GRAD } \Psi = \text{GRAD } \mathbf{x} \cdot \dot{\mathbf{x}} - \dot{\mathbf{X}}. \tag{102.4}$$

This formula, which is the material counterpart of the spatial equation (101.10), may be called ***Weber's transformation***;[1] it expresses the condition for circulation-preserving motion in an integrated material form. The single vectorial equation (102.4) constitutes three partial differential equations of first order for the four scalar functions x, y, z, Ψ, so that if a fourth equation be added, the system is rendered determinate.[2]

For the case of steady motion, an ingenious modification of Weber's transformation has been constructed by Ertel.[3] Instead of (98.13), write

$$\Psi(\tau, t) \equiv \int_\tau^t (\tfrac{1}{2}\dot{x}^2 - \phi^*)\, dt, \tag{102.5}$$

---

[1] Weber [1868, 1, §2] derived an equivalent formula from the equations for perfect fluids. This simple proof is essentially that given by Appell [1897, 1, §7].

[2] In classical hydrodynamics by (49.4) we obtain ϕ^* as a known function of ρ, $\mathbf{x}$, and t, and hence Ψ is a known function of $\mathbf{x}$, $\dot{\mathbf{x}}$, t, and ρ, the last being regarded as an additional dependent variable. The one further equation then required is the material continuity equation $\rho J = \rho_0$.

[3] [1952, 5].

where τ may vary arbitrarily from one particle to another: $\tau = \tau(X)$. Denote by primes functions evaluated at time $t = \tau$. In particular, write

$$\mathbf{x}' \equiv \mathbf{x}(\mathbf{X}, \tau(\mathbf{X})) \tag{102.6}$$

for the place occupied by the particle $\mathbf{X}$ at time τ, and write $\dot{\mathbf{x}}'$ for its velocity. Differentiating (102.5) yields

$$\text{GRAD } \mathbf{x}' = (\text{GRAD } \mathbf{x})' + (\text{GRAD } \tau)\dot{\mathbf{x}}', \tag{102.7}$$

whence

$$\text{GRAD } \mathbf{x}' \cdot \dot{\mathbf{x}}' = (\text{GRAD } \mathbf{x} \cdot \dot{\mathbf{x}})' + \dot{x}'^2 \text{ GRAD } \tau. \tag{102.8}$$

But

$$\begin{aligned}\text{GRAD } \Psi &= \int_\tau^t \text{GRAD } (\tfrac{1}{2}\dot{x}^2 - \phi^*)\, dt - (\tfrac{1}{2}\dot{x}^2 - \phi^*) \text{ GRAD } \tau, \\ &= \int_\tau^t \text{GRAD } (\tfrac{1}{2}\dot{x}^2 - \phi^*)\, dt \\ &\qquad - (\tfrac{1}{2}\dot{x}^2 + \phi^*) \text{ GRAD } (t - \tau) - \dot{x}'^2 \text{ GRAD } \tau,\end{aligned} \tag{102.9}$$

where we have used the fact that $\frac{1}{2}\dot{x}^2 + \phi^*$ is constant on each streamline (§75). Hence by (102.8) follows

$$\begin{aligned}\text{GRAD } \Psi = \int_\tau^t \text{GRAD } (\tfrac{1}{2}\dot{x}^2 - \phi^*)\, dt \\ - (\tfrac{1}{2}\dot{x}^2 + \phi^*) \text{ GRAD } (t - \tau) - \text{GRAD } \mathbf{x}' \cdot \dot{\mathbf{x}}' + (\text{GRAD } \mathbf{x} \cdot \dot{\mathbf{x}})'.\end{aligned} \tag{102.10}$$

Integrating (102.2) from τ to t yields

$$\int_\tau^t \text{GRAD } (\tfrac{1}{2}\dot{x}^2 - \phi^*)\, dt = \text{GRAD } \mathbf{x} \cdot \dot{\mathbf{x}} - (\text{GRAD } \mathbf{x} \cdot \dot{\mathbf{x}})', \tag{102.11}$$

whence by (102.10) follows

$$\text{GRAD } \Psi = \text{GRAD } \mathbf{x} \cdot \dot{\mathbf{x}} - (\tfrac{1}{2}\dot{x}^2 + \phi^*) \text{ GRAD } (t - \tau) - \text{GRAD } \mathbf{x}' \cdot \dot{\mathbf{x}}', \tag{102.12}$$

generalizing (102.4). Multiplying this result by grad $\mathbf{X}$ yields finally

$$\dot{\mathbf{x}} = \text{grad } \Psi + (\tfrac{1}{2}\dot{x}^2 + \phi^*) \text{ grad } (t - \tau) + \dot{\mathbf{x}}' \cdot \text{grad } \mathbf{x}'. \tag{102.13}$$

The next step is to formulate conditions sufficient that the last term shall vanish. If we consider particles $\mathbf{X}$ comprising a certain surface $\mathbf{X} = \mathbf{X}(l, m)$, then the places $\mathbf{x}'$ will also constitute a surface. Then grad x', grad y', and grad z' will all be tangent to that surface. For $\dot{\mathbf{x}}'\cdot$grad $\mathbf{x}'$ to vanish, it is sufficient that the velocity $\dot{\mathbf{x}}'$ be normal to the surface. If such a surface exists, we are then to choose $\tau(\mathbf{X})$ as the time when the particle $\mathbf{X}$ crosses it. Summarizing these results, we obtain ***Ertel's potential theorem***: *in a steady circulation-preserving motion such that every stream-line is normal to a certain surface, let $t - \tau$ be the time which the particle now at* $\mathbf{x}$ *has travelled since crossing that surface. Then a set of Monge potentials for the velocity is*

$$h = \Psi \equiv \int_\tau^t (\tfrac{1}{2}\dot{x}^2 - \phi^*)\,dt,$$

$$f = \tfrac{1}{2}\dot{x}^2 + \phi^*, \tag{102.14}$$

$$g = t - \tau.$$

We note that the Lamb-Stuart equations (101.22) are satisfied since $\dot{f} = 0$, $\lambda = \lambda(f) = f$, $\dot{g} = -1$.

Put

$$S \equiv \Psi + (\tfrac{1}{2}\dot{x}^2 + \phi^*)(t - \tau). \tag{102.15}$$

Then evidently when (102.14) is valid we have the alternative set

$$h = S, \qquad f = \tau - t, \qquad g = \tfrac{1}{2}\dot{x}^2 + \phi^*. \tag{102.16}$$

Now by (102.5) and the fact that $\tfrac{1}{2}\dot{x}^2 + \phi^*$ is constant in time for each particle we see that

$$S = \int_\tau^t \dot{x}^2\,dt = \int_0^x dx\,\dot{x} = \int_0^x d\mathbf{x}\cdot\dot{\mathbf{x}}, \tag{102.17}$$

where x is arc-length along the stream-line, measured from its point of intersection with the normal surface. The right hand side of (102.17) is simply the flow along the stream-line (§36). Then the Monge resolution asserted by (102.16) is equivalent to

$$\dot{\mathbf{x}} = \text{grad}\left(\int_0^x d\mathbf{x}\cdot\dot{\mathbf{x}}\right) - (t - \tau)\,\text{grad}\,(\tfrac{1}{2}\dot{x}^2 + \phi^*). \tag{102.18}$$

These Monge potentials h, f, g have simple kinematical interpretations: flow along the stream-line, time for traversing the stream-line, and flow energy of the stream-line.

Now consider the special case when the stream-lines in a certain region are all closed curves, of length, say, l, where l varies in general from one to another. Since the motion is steady, the time required for a particle to traverse each stream-line once is the same for all particles upon the stream-line, and may be called its *period*, T. If we replace x by $x + l$ and t by $t + T$, $\dot{\mathbf{x}}$ must remain unchanged, so that by (102.18)

$$\dot{\mathbf{x}} = \operatorname{grad}\left(\int_0^{x+l} d\mathbf{x}\cdot\dot{\mathbf{x}}\right) - (t + T - \tau)\operatorname{grad}\left(\tfrac{1}{2}\dot{x}^2 + \phi^*\right). \qquad (102.19)$$

Subtracting (102.18) yields

$$0 = \operatorname{grad}\left(\int_x^{x+l} d\mathbf{x}\cdot\dot{\mathbf{x}}\right) - T\operatorname{grad}\left(\tfrac{1}{2}\dot{x}^2 + \phi^*\right). \qquad (102.20)$$

But $\int_x^{x+l} d\mathbf{x}\cdot\dot{\mathbf{x}} = \oint d\mathbf{x}\cdot\dot{\mathbf{x}}$, the circulation around the stream-line. By (102.20), this quantity if not constant is a function of $\frac{1}{2}\dot{x}^2 + \phi^*$, and hence is constant on each Lamb surface (§75). Moreover, from (102.20) follows

$$T = \frac{d(\oint d\mathbf{x}\cdot\dot{\mathbf{x}})}{d(\frac{1}{2}\dot{x}^2 + \phi^*)}, \qquad (102.21)$$

provided $\oint d\mathbf{x}\cdot\dot{\mathbf{x}} \neq$ const. We have thus obtained the elegant ***Ertel theorem on circulating motions:***[4] *in a steady rotational circulation-preserving motion whose stream-lines are closed curves intersecting a certain surface orthogonally, the circulation of the stream-lines is a function of their energy* $\frac{1}{2}\dot{x}^2 + \phi^*$ *only. Hence it is the same for all stream-lines lying upon the same Lamb surface. Moreover, their periods are given by* (102.21).

The reader will easily verify Ertel's theorem for the case of a steady simple vortex (48.4). For the normal surface, select any azimuthal plane. It is evident that the period of the stream-line of radius r is $2\pi/\omega$. But by (48.5) and the formula in §75[3] we get in the rotational case

$$\frac{d(\oint d\mathbf{x}\cdot\dot{\mathbf{x}})}{d(\phi^* + \frac{1}{2}\dot{x}^2)} = \frac{\dfrac{d(\oint dx\cdot\dot{x})}{dr}}{\dfrac{d(\phi^* + \frac{1}{2}\dot{x}^2)}{dr}} = \frac{2\pi(2r\omega + r^2\omega')}{2r\omega^2 + r^2\omega\omega'} = \frac{2\pi}{\omega}, \qquad (102.22)$$

in conformity with Ertel's theorem.

103. Remarks on the extension of the transformations of Clebsch and Weber to arbitrary continuous media. It is easy to extend the transformation of Clebsch and Weber to arbitrary motions of a continuous medium, but in their most general form they do not effect an integration of the kinematical equations, but merely constitute alternative forms. The general Clebsch transformation consists in the formula

$$\operatorname{curl}\ddot{\mathbf{x}} = \operatorname{grad}\dot{f}\times\operatorname{grad}g - \operatorname{grad}\dot{g}\times\operatorname{grad}f, \qquad (103.1)$$

[4] [1950, **10**]. Our proof is taken from [1952, **5**, §V].

taken in conjunction with (16.12). The general Weber transformation is obtain by integrating (102.1):

$$\text{GRAD}\ \mathbf{x}\cdot\dot{\mathbf{x}} = \dot{\mathbf{X}} + \int_0^t \text{GRAD}\ \mathbf{x}\cdot\ddot{\mathbf{x}}\,dt + \text{GRAD}\int_0^t \tfrac{1}{2}\dot{x}^2\,dt. \tag{103.2}$$

Special cases of this last result were derived by Appell (1917)[1] and Carstoiu.[2]

Multiplying (103.2) by grad $\mathbf{X}$ yields

$$\dot{\mathbf{x}} = \text{grad}\int_0^t \tfrac{1}{2}\dot{x}^2\,dt + \text{grad}\ \mathbf{X}\cdot\left[\dot{\mathbf{X}} + \int_0^t \text{GRAD}\ \mathbf{x}\cdot\ddot{\mathbf{x}}\,dt\right], \tag{103.3}$$

which in the case of a circulation-preserving motion reduces to

$$\dot{\mathbf{x}} = \text{grad}\,\Psi + \text{grad}\ \mathbf{X}\cdot\dot{\mathbf{X}}. \tag{103.4}$$

From (103.3) it is not difficult to derive again the basic vorticity formula (85.4).

104. The two controversies concerning the Lagrange-Cauchy velocity-potential theorem. In 1781 Lagrange[1] presented his celebrated ***velocity-potential theorem***: *if a velocity potential exist at one time in a motion of an inviscid incompressible fluid, subject to conservative extraneous force, it exists at all past and future times.* Another form of this same statement is: *a motion once irrotational is always irrotational.* Without comment on Lagrange's result, Cauchy (1815)[2] gave a clear statement and correct proof of the following proposition: *In a continuous motion of such a fluid, a particle once in irrotational motion is at all times in irrotational motion.* The essential difference between the two enunciations is expressed by Lamb[3] as follows: "It is to be particularly noticed that this continued existence of a velocity-potential is predicated, not of regions of space, but of portions of matter. A portion of matter for which a velocity-potential exists moves about and carries this property with it, but the part of space which it originally occupied may, in the course of time, come to be occupied by matter which did not originally possess the property, and which therefore cannot have acquired it." Poisson (1829),[4] apparently unaware of Cauchy's result, gave a proof along

---

[1] [1917, 1, §II] [1921, 2, §815].
[2] [1947, 3, §2] [1947, 8, Ch. V, §5].

---

[1] [1783, 1, §§17–19] [1788, 1, Part II, Sect. 11, ¶¶16–17].
[2] [1827, 1, 1st part, Sect. II, §5].
[3] [1932, 1, §17].
[4] [1831, 1, ¶73] [1833, 1, §654].

the lines of that of Lagrange, but then stated that the theorem was not as general as usually believed. The flaw in the proof, he asserted, was in the assumption that if a function satisfy a differential equation it also must satisfy any equation deducible from it by differentiation, which is not necessarily the case. Poisson claimed that he had examples of convergent sine or exponential series which satisfied the equations of motion but contradicted the velocity-potential theorem. Power (1842)[5] attempted to re-establish the Lagrange form of the theorem by showing that the steps of Lagrange's original proof could be carried through equally well if the velocity at each point were expressed as a series of various unspecified not necessarily integral powers of t. Finally Stokes (1845)[6] stated that the velocity-potential theorem as applied to regions of space was incorrect, the Lagrange proof not being valid because it assumed a power series expansion for the velocity components as functions of time, while many functions which cannot be expanded in power series, such as $\exp(t^{-2})$ and $t \log t$, have the property of vanishing at $t = 0$ and not vanishing subsequently. He noticed that Power's method of proof applied equally to the equations of incompressible viscous fluids, for which the theorem itself was plainly false. He then called attention to the correct statement and proof of Cauchy and himself gave another proof.[7] Subsequent writers have adopted his opinion, but have usually proved the theorem as a special case of the Hankel-Kelvin circulation theorem (§49). The possible effect of waves, defined mathematically as propagating surfaces of discontinuity, in the generation of vorticity was discussed by Hadamard (1901). He demonstrated that in the case of a motion governed by the classical equations of gas dynamics, heat conduction being neglected, the presence of acceleration waves, sometimes called *sound waves* or *Mach waves,* in an initially barotropic motion does not destroy its barotropic character nor affect the circulation-preserving property,[8] while the passage of a shock wave generally renders an initially barotropic motion baroclinic and thus not only itself generates vorticity but also destroys the circulation-preserving property.[9]

---

[5] [1842, 1].

[6] [1845, 1, §§10–13] [1846, 1, §I] [1848, 2].

[7] Although both were apparently aware of the shortcoming of Lagrange's treatment, as well as its rectification, neither Helmholtz [1858, 1] nor Dirichlet [1860, 1] mentioned the French or English contributions to the subject, and in his notes on Dirichlet's incomplete MS Dedekind actually gave Cauchy's formula (94.3) without stating its authorship.

[8] [1901, 1, §4] [1903, 2, ¶¶254–255].

[9] [1903, 2, ¶256, and Note III]. In the engineering literature this fundamental result is usually attributed to various engineers.

The question raised by Stokes in regard to the possible extension of Lagrange's theorem to motions of incompressible viscous fluids initiated a second protracted controversy. St. Venant (1869)[10] gave a fuller indication of proof that in such fluids Lagrange's theorem follows from the assumption of an analytic solution; he concluded, however, that this fact simply indicated that *vorticity cannot be generated in the interior of a viscous incompressible fluid, subject to conservative extraneous force, but is necessarily diffused inward from the boundaries.* When, apparently ignorant of all prior results on this subject, Bresse (1880)[11] claimed an extension of Lagrange's theorem to incompressible viscous fluids, Boussinesq[12] hastened to point out St. Venant's priority and called attention to his statement of the predominant effect of the boundaries; Boussinesq contributed greatly to the clear understanding of the phenomenon by exhibiting a beautiful exact solution[13] representing a semi-infinite mass of fluid starting to move along a plane edge; in this

---

[10] [1869, 2].

[11] [1880, 1].

[12] [1880, 2–4]. *Cf.* [1932, **3**, §2.6].

[13] [1880, 2]. He had given other non-analytic solutions earlier [1868, **9**, §VII]. His result is discussed in [1925, **2**] [1929, **3**, Ch. X]. It is extended to a more general type of fluid by Viguier [1950, **4**].

To Professor Kuerti I owe another elegant example. The velocity field

$$\dot{r} = 0, \qquad \dot{x}_\theta = r\dot{\theta} = \frac{K}{2\pi r}\left(1 - e^{-\frac{r^2}{4\nu t}}\right), \qquad \dot{z} = 0,$$

an example of the simple vortices (48.4), is a solution of the Navier-Stokes equations for a fluid of kinematic viscosity ν. As $t \to 0+$, $\dot{\mathbf{x}}_\theta \to K/(2\pi r)$, and hence this flow is appropriate to the decay of an irrotational vortex of circulation K. Consider the circulation about the polar rectangle shown in Fig. 104.1. Since the flows along the straight sides cancel one another, we get

$$\oint d\mathbf{x}\cdot\dot{\mathbf{x}} = \frac{K}{2\pi}\,\Delta\theta\left(1 - e^{-\frac{r^2}{4\nu t}}\right)\Bigg|_{r_1}^{r_2}$$

$$= -\frac{K\Delta\theta}{2\pi}\,e^{-\frac{r^2}{4\nu t}}\Bigg|_{r_1}^{r_2}.$$

Also

$$\frac{\delta}{\delta t}\oint d\mathbf{x}\cdot\dot{\mathbf{x}} = \frac{\partial}{\partial t}\int d\mathbf{x}\cdot\dot{\mathbf{x}} = -\frac{K\Delta\theta}{8\pi\nu t^2}\left(r^2 e^{-\frac{r^2}{4\nu t}}\right)\Bigg|_{r_1}^{r_2},$$

and similar expressions can be obtained for the higher time derivatives of the circulation. As $t \to 0+$ the circulation and all its derivatives approach zero. But for any positive t, no matter how small, the circulation does not vanish, nor does any of its derivatives. Plainly these results carry over to any reducible circuit not including nor passing through the origin. As $t \to \infty$, the circulation and all its derivatives ap-

solution a non-analytic exponential of the type mentioned by Stokes occurs. Both Poincaré[14] and Hadamard[15] gave indications of proofs of the generalized Lagrange theorem; the former regarded this result as casting doubt upon the validity of the basic hypotheses of the theory, while the latter mentioned that *in incompressible viscous fluids vorticity is generated through a mechanism which cannot be represented by analytic functions.* Finally Duhem (1903)[16], after criticising all previous treat-

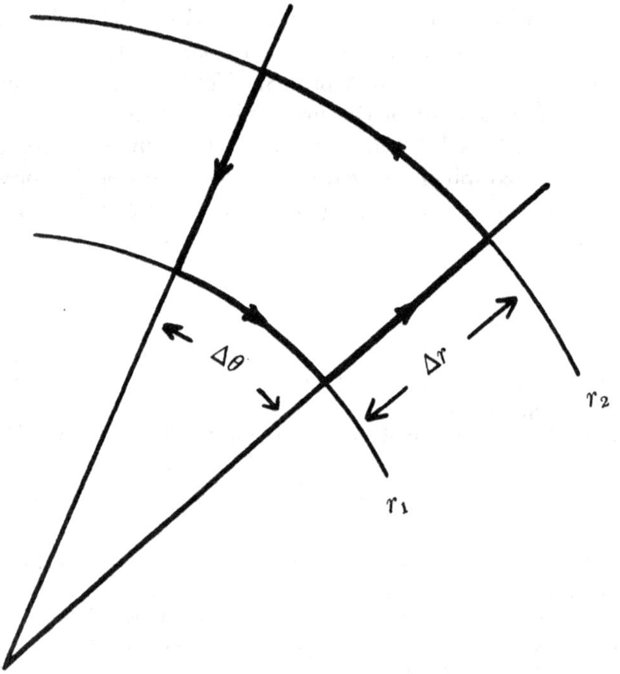

Fig. 104.1

proach zero analytically. The *creation* of vorticity by viscosity is a non-analytic process; its *decay* is analytic.

If we calculate $\mathfrak{W}_K$ we find by $(62.9)_1$ that

$$\mathfrak{W}_K = \frac{1}{\frac{4\nu t}{r^2}\left(e^{\frac{r^2}{4\nu t}} - 1\right) - 1}.$$

For a fixed place, $\mathfrak{W}_K$ *increases* steadily with time from $\mathfrak{W}_K = 0$ at $t = 0$ to $\mathfrak{W}_K = \infty$ at $t = \infty$! That is, while the rate of deformation and the vorticity separately decay in time, the former does so infinitely faster than the latter, so that the motion during the process of decay becomes steadily more rotational. Moreover, $\mathfrak{W}_K$ is independent of the strength of the vortex.

[14] [1893, 1, §152].

[15] [1901, 1, §4, footnote].

[16] [1903, **3**].

ments for lack of rigor, settled the controversy by a careful and exhaustive analysis which substantiated the remark of Hadamard. That his result is not well known, and, more surprising, that the generation of vorticity in viscous fluids is still not well understood, is evidenced by incomplete treatments given in text-books and treatises and by recent papers essentially repeating the errors of prior authors.

We shall now examine this subject from a purely kinematical point of view. We shall be able to present a complete analysis for the first problem, which concerns the case of circulation-preserving motions. The dynamical equations of viscous incompressible fluids, however, lie outside the scope of the present work, so for the second problem we shall be able to go no further than the kinematical foundation; fortunately this foundation suffices for a qualitative explanation, although it cannot render the resulting phenomenon determinate.

105. The velocity-potential theorem of Lagrange. Beginning by a recapitulation of Lagrange's argument, let us suppose that at each point of a stationary closed region of space the velocity be an analytic function of time:

$$\dot{\mathbf{x}} = \sum_{i=0}^{\infty} \dot{\mathbf{x}}_i(\mathbf{x})t^i, \qquad |t| < K. \tag{105.1}$$

Supposing further that term by term differentiation with respect to $\mathbf{x}$ be permitted, we have

$$\mathbf{w} = \sum_{i=0}^{\infty} \mathbf{w}_i(\mathbf{x})t^i, \qquad |t| < K, \tag{105.2}$$

where

$$\mathbf{w}_i(x) \equiv \operatorname{curl} \dot{\mathbf{x}}_i. \tag{105.3}$$

The d'Alembert-Euler condition (45.4) that the motion be circulation-preserving may be put into the form

$$\frac{\partial \mathbf{w}}{\partial t} + \dot{\mathbf{x}}\cdot\operatorname{grad}\mathbf{w} - \mathbf{w}\cdot\operatorname{grad}\dot{\mathbf{x}} + \mathbf{w}\operatorname{div}\dot{\mathbf{x}} = 0. \tag{105.4}$$

Let us call *regular* a velocity field such that for some value of K series expansions of the type (105.1) and (105.2) are valid and may be differentiated term by term with respect to the space variables. If (105.1) and (105.2) be substituted into (105.4), for a regular velocity field we obtain

$$\begin{aligned} 0 = {} & [\mathbf{w}_1 + \dot{\mathbf{x}}_0\cdot\operatorname{grad}\mathbf{w}_0 - \mathbf{w}_0\cdot\operatorname{grad}\dot{\mathbf{x}}_0 + \mathbf{w}_0\operatorname{div}\dot{\mathbf{x}}_0] \\ & + t[2\mathbf{w}_2 + \dot{\mathbf{x}}_1\cdot\operatorname{grad}\mathbf{w}_0 + \dot{\mathbf{x}}_0\cdot\operatorname{grad}\mathbf{w}_1 - \mathbf{w}_1\cdot\operatorname{grad}\dot{\mathbf{x}}_0 - \mathbf{w}_0\cdot\operatorname{grad}\dot{\mathbf{x}}_1] \\ & + t^2[3\mathbf{w}_3 + \ldots] + \ldots + t^n[n\mathbf{w}_n + \ldots] + \ldots; \end{aligned} \tag{105.5}$$

the coefficient of t^n is $n\mathbf{w}_n$ plus a linear homogeneous function of $\mathbf{w}_{n-1}$, $\mathbf{w}_{n-2}, \ldots \mathbf{w}_0$, and their gradients. Then the condition of circulation-preserving motion implies that throughout the region

$$\begin{aligned} &\mathbf{w}_1 + \dot{\mathbf{x}}_0 \cdot \operatorname{grad} \mathbf{w}_0 - \mathbf{w}_0 \cdot \operatorname{grad} \dot{\mathbf{x}}_0 + \mathbf{w}_0 \operatorname{div} \dot{\mathbf{x}}_0 = 0, \\ &2\mathbf{w}_2 + \dot{\mathbf{x}}_1 \cdot \operatorname{grad} \mathbf{w}_0 + \dot{\mathbf{x}}_0 \cdot \operatorname{grad} \mathbf{w}_1 - \mathbf{w}_1 \cdot \operatorname{grad} \dot{\mathbf{x}}_0 - \mathbf{w}_0 \cdot \operatorname{grad} \dot{\mathbf{x}}_1 = 0, \\ &\vdots \\ &n\mathbf{w}_n + \ldots = 0, \\ &\vdots, \end{aligned} \tag{105.6}$$

so long as $|t| < K$.

Suppose now that throughout the region $\mathbf{w} = 0$ when $t = 0$; that is, $\mathbf{w}_0 = 0$ and $\operatorname{grad} \mathbf{w}_0 = 0$. Then from the first of (105.6) follows

$$\mathbf{w}_1 = 0, \tag{105.7}$$

and since this condition holds at all points of the region, $\operatorname{grad} \mathbf{w}_1 = 0$. Proceeding similarly we find that $\mathbf{w}_n = 0$, and hence $\mathbf{w} = 0$ for $|t| < K$. Thus we have ***Lagrange's velocity-potential theorem***: *in a circulation preserving motion, a fixed region of space once in irrotational motion remains in irrotational motion so long as the velocity field be regular.* As a corollary, it follows that *in a circulation-preserving motion, when rotation enters a region originally in irrotational motion, it destroys the regularity of the velocity field.* The proof of Power replaces t^i in (105.1) by t^{λ_i}, where λ_i is not necessarily an integer, and indeed Lagrange's proof remains valid when t^i is replaced by various types of functions $f_i(t)$. But it is interesting to note in connection with the criticism of Poisson that the steps of the proof cannot be carried out if $f_i(t) = \sin it$.

Lagrange's proof is not erroneous; Lagrange did not state the restrictions to which his theorem was subject, perhaps expecting the reader to notice them in following the proof,[1] and the interpretation placed upon his statement by many readers, including Stokes, was certainly incorrect. The result, however, is not useful in itself, although in the light of other facts, to be discussed in the next section, it will prove suggestive of a very interesting line of thought. For the present we may remark that Poisson's and Stokes's objections to the use of power series for the velocity components are not really relevant in the case of circulation-preserving motions. Here it suffices to remark that a power series may have a finite radius of convergence. For example, suppose water be flowing in a uniform parallel stream with constant velocity. Then $\dot{\mathbf{x}} = \dot{\mathbf{x}}_0$ and all the requirements for Lagrange's theorem are

[1] *Note added in proof.* Further reading in the works of Lagrange has led me to regard the above statement as overly charitable.

satisfied. If now someone stir up a bucket of water and throw it in some miles up-stream from the observer, hours later this water will flow by, carrying its original rotation with it; by Lagrange's theorem, in so doing it destroys the regularity of the velocity field. This type of behavior is characteristic of solutions of partial differential equations admitting wave motions. A complete instantaneous knowledge of conditions in a region does not generally determine uniquely all the future or past motion there.

106. Five proofs of Cauchy's velocity-potential theorem. To Lagrange belongs the merit of having first called attention to the question of permanence of the velocity-potential and for discovering a minor result concerning it. The major theorem, however, which we may call ***Cauchy's velocity-potential theorem***, is quite different: *in a circulation-preserving motion, a particle once in irrotational motion is at all times in irrotational motion.* Cauchy's proof consists in the fact that if $\mathbf{W} = 0$ it follows from (94.3) that $\mathbf{w} = 0$. It is not only the first but also the best proof ever given, for (i) it does not require that $\mathbf{W} = 0$ in a region, but only at a single point, (ii) in case the hypothesis of continuity of motion $J \neq 0$ be violated, as for example when two particles originally separate come to occupy the same point, (94.3) becomes indeterminate, suggesting that in such a case vorticity may either be created or destroyed, and (iii) in case $\mathbf{W} \neq 0$ (94.3) gives us the value of $\mathbf{w}$ at all times, thus serving as much more than a proof of the velocity-potential theorem, being in fact a complete description of the mechanism of convection.

It is strange that although he possessed the d'Alembert-Euler equation for the vorticity, *viz.*

$$\dot{\mathbf{w}} = \mathbf{w} \cdot \text{grad } \dot{\mathbf{x}} - \mathbf{w} \text{ div } \dot{\mathbf{x}}, \tag{94.1}$$

Lagrange did not employ it in the deduction of the velocity-potential theorem. Instead of assuming $\dot{\mathbf{x}}(\mathbf{x}, t)$ to be an analytic function of t in a region, we now need only assume that $\mathbf{w}(\mathbf{X}, t)$ be an analytic function of t for a single particle $\mathbf{X}$, so that

$$\mathbf{w}(\mathbf{X}, t) = \sum_{i=0}^{\infty} \mathbf{w}_i(\mathbf{X}) t^i, \qquad |t| < l, \tag{106.1}$$

in place of (105.2). By successive material differentiations of (94.1) we obtain[1]

$$\begin{aligned} \ddot{\mathbf{w}} &= \dot{\mathbf{w}} \cdot \text{grad } \dot{\mathbf{x}} - \mathbf{w} \cdot \text{grad } \ddot{\mathbf{x}} - \dot{\mathbf{w}}\vartheta - \mathbf{w}\dot{\vartheta}, \\ \dddot{\mathbf{w}} &= \ddot{\mathbf{w}} \cdot \text{grad } \dot{\mathbf{x}} - \ddot{\mathbf{w}}\vartheta + \ldots, \quad \ldots . \end{aligned} \tag{106.2}$$

[1] A more elaborate treatment of this same analysis is given by Duhem [1903, **3**, Ch. I, §§1–2].

Hence if $\mathbf{W} = \mathbf{w}(\mathbf{X}, 0) = 0$, it follows that $\delta^n \mathbf{w}/\delta t^n = 0$ when $t = 0$. Since then

$$0 = \frac{1}{n!} \frac{\delta^n \mathbf{w}}{\delta t^n}\bigg|_{t=0} = \mathbf{w}_i(\mathbf{X}), \tag{106.3}$$

by (106.1) it follows that $\mathbf{w}(\mathbf{x}, t) = 0$ for $|t| < l$. The foregoing analysis is quite similar to that applied by Lagrange to the spatial expression $\mathbf{w}(\mathbf{x}, t)$, and it now leads to a weaker form of Cauchy's result; *viz.* that in a circulation-preserving motion a particle if once in irrotational motion remains in irrotational motion so long as the vorticity of that particle be an analytic function of the time. While this proof is far inferior to Cauchy's in that it requires the assumption of analyticity, it will be useful as a starting point for the discussion below of the possible extension of Lagrange's theorem to motions which are not circulation-preserving. Since it was indicated by Helmholtz,[2] we shall call it the *Lagrange-Helmholtz proof*. It is preferable to the original proof of Lagrange (§105) in that it requires only the single assumption of the analyticity of $\mathbf{w}$ as a function of time *for each particle*, rather than a combination of analyticity with respect to time and term by term differentiability with respect to space variables.

The equation (94.1) was rediscovered by Stokes, who based upon it a proof of Cauchy's velocity-potential theorem which in modern terminology is substantially the following. If $\mathbf{X}$ be held fixed and $\dot{\mathbf{x}}$ be regarded as given, then (94.1) is an ordinary differential equation of first order for $\mathbf{w}$ as a function of t. Clearly $\mathbf{w} = 0$ is a solution satisfying the condition $\mathbf{w} = 0$ when $t = 0$. Then by the uniqueness theorem for ordinary differential equations, $\mathbf{w} = 0$ is the only such solution. Hence from $\mathbf{w}(\mathbf{X}, 0) = 0$ it must follow that $\mathbf{w}(\mathbf{X}, t) = 0$ for a certain range of values $|t| < K$. An examination of the analytic conditions of the theorem yields sufficient restrictions upon the velocity gradient that $K = \infty$. This proof involves analytical difficulties which are quite foreign to the problem under consideration. It should be mentioned that Stokes gave a uniqueness proof for systems of linear ordinary differential equations of first order in this connection.

An elegant proof using the material form of the equations was noticed by Lamb (1879).[3] It consists in the remark that if $\dot{\mathbf{X}}(\mathbf{X}) = -\text{GRAD}\ \Phi(\mathbf{X})$, then by Weber's transformation in the form (103.4) we obtain

$$\dot{\mathbf{x}}(\mathbf{x}, t) = \text{grad}\ [\Psi - \Phi(\mathbf{X}(\mathbf{x}, t))]. \tag{106.4}$$

This proof has the advantage of expressing the velocity potential

---

[2] [1858, 1, §2].

[3] [1879, 1, §23]. *Cf.* [1921, 2, §731].

$\phi(x, t)$ in terms of the potential $\Phi(\mathbf{X})$ of the initial motion, for by (98.13) and (106.4) we have

$$\phi(\mathbf{x}, t) = \Phi(\mathbf{X}) + \int_0^t (\phi^* - \tfrac{1}{2}\dot{x}^2)\, dt + f(t), \tag{106.5}$$

ϕ^* being the acceleration potential, a relation which may also be established directly by combining $(36.4)_2$ and (48.3).

Kelvin's proof is the simplest of all, and at the same time not inferior to Cauchy's in directness and elegance. If the motion be irrotational in a region at $t = 0$, then the circulation about all circuits in this region vanishes at $t = 0$. But since the motion is circulation-preserving, the circulation about these same material circuits must always vanish Hence the same portion of the substance is in irrotational motion at all times. Like Cauchy's proof, Kelvin's requires only that the motion be continuous; analyticity is not necessary.

107. The non-analyticity of motions which are not circulation-preserving. So long as our attention be confined to circulation-preserving motions, Cauchy's and Kelvin's proofs of the velocity-potential theorem leave nothing to be desired. Stokes's objection that the Lagrange proof applies equally well to the equations of viscous incompressible fluids raises a new question, however: is there a velocity-potential theorem for certain types of motions which are not circulation-preserving? Cauchy's and Kelvin's proofs rely upon the condition of circulation-preserving motion in an essential way. The generalization (85.4) of Cauchy's vorticity formula shows that even if the diffusion mechanism do not operate in irrotational regions, *i.e.*, even if $\text{curl}\,(\text{grad}\,\mathbf{x}\cdot\ddot{\mathbf{x}}) = 0$ when $\mathbf{w} = 0$, it does not necessarily follow from $\mathbf{W} = 0$ that $\mathbf{w} = 0$ for all time. The Lagrange-Helmholtz proof, however, being based upon the assumption of analyticity, can be carried through in just the same way in any motion *where the spatial diffusion vector* curl $\ddot{\mathbf{x}}$ *vanishes when* $\mathbf{w}$ *vanishes. The real question, then, for these more general motions, is whether the assumption of analyticity be justified.* We shall now show that in an important special case, at least, it is not.

Thus far we have confined our attention to the differential equations governing vorticity, and have neglected the effect of the boundaries. In the special case of an isochoric motion the effect of the boundary conditions is made apparent by the first theorem of §37: *the only isochoric irrotational motion in which the material adheres to a finite stationary surface, however small, is a state of rest.* Now a motion starting from rest is irrotational in the initial instant. Therefore *for an isochoric motion, a velocity-potential theorem of Cauchy's type is incompatible with the adherence of the material to stationary boundaries.*[1]

By combining the results of the two preceding paragraphs we obtain the following ***non-analyticity theorem***: *in a medium in which the diffusion mechanism is such that the spatial diffusion vector* curl $\ddot{\mathbf{x}}$ *vanishes in a region of irrotational motion, if there be a finite stationary boundary, however small, to which the material adheres without slipping, then in an isochoric motion starting from rest the vorticity of some of the particles must fail to be an analytic function of time at the initial instant.*[2]

---

[1] Since Cauchy's velocity-potential theorem is valid for any continuously differentiable isochoric motion of a perfect fluid, subject to conservative extraneous force, it follows that *a motion of this class which represents a fluid starting from rest and adhering to a finite stationary surface cannot exist. Cf.* [1903, **3**, Ch. II, §2].

[2] For viscous incompressible fluids of uniform density and viscosity, subject to convervative extraneous force, by putting (49.5) into (78.3) we obtain ***St. Venant's diffusion equation*** [1869, **2**, §4, footnote]:

$$\dot{\mathbf{w}} = \mathbf{w}\cdot\text{grad}\,\dot{\mathbf{x}} - \nu\,\text{curl curl}\,\mathbf{w},$$

whence Bobylew [1873, **1**, §2] concluded that Lagrange's theorem does not hold for these fluids. In case a *whole region* be in irrotational motion, however, it follows that $\dot{\mathbf{w}} = 0$ throughout the region. By successive material differentiations we obtain equations similar to (106.2), but slightly more complicated, and hence the theorem: *in a motion of a homogeneous viscous incompressible fluid subject to conservative extraneous force, a material volume which is in irrotational motion at one instant remains in irrotational motion so long as the vorticity be an analytic function of time for each of its particles.* The same result is easily deduced also from the Poincaré-Bjerknes-Jaffé circulation theorem for viscous fluids [1893, **1**, §150] [1902, **1**, footnote] [1920, **1**]. It was this theorem, in one form or another, to which Stokes, St. Venant, Bresse, Boussinesq, Poincaré, and Hadamard referred (§104) and which both St. Venant and Boussinesq correctly stated to be of little or no application to flow problems because the customarily imposed condition of adherence to boundaries makes irrotational motion altogether impossible when the boundaries are stationary. Thus Boussinesq conjectured and Duhem gave a partial proof of the following ***non-analyticity theorem for motions of viscous liquids***: *in a motion of a homogeneous viscous fluid subject to conservative extraneous force and starting from rest, if there be a finite stationary boundary to which the liquid adheres without slipping, there must be some particles whose vorticity is not an analytic function of time at the initial instant.* This result is a special case of the kinematical theorem given in the text above. Carstoiu [1947, **4**, §2] points out that from the special case of (85.3) appropriate to viscous incompressible fluids it is very plain that an initially irrotational flow of such a fluid will generally not remain irrotational.

Index of Symbols

Only symbols used in more than one section are listed here. Within a single section sometimes these same symbols are defined and used in a different sense. Numbers refer to equations or sections where the symbol is defined or first used. General conventions:

Latin minuscule indices generally indicate tensor character with respect to transformations of the spatial co-ordinates (§14). *E.g.*, $\dot{x}^i$, Δ_{ij}. Indices set in the middle of the line denote physical components (§1).

Latin majuscule indices indicate tensor character with respect to transformations of the material co-ordinates (§14). *E.g.*, W^I.

Both types of indices may occur in a single symbol, *e.g.*, $x^i{}_{,I}$, $X^I{}_{,i}$.

The usual comma notation for covariant and contravariant differentiation is employed.

Bold face Latin letters, *e.g.* $\dot{\mathbf{x}}$, $\mathbf{W}$, stand for matrices of tensor components according to a scheme explained in §1. The Gibbs vector symbols $\cdot$, $\times$, $\Sigma.$, $\mathbf{\Sigma}_\times$ are defined in §1.

The vector notations grad, div, curl, ∇^2 refer to differentiation with respect to spatial co-ordinates. The notations GRAD, DIV, CURL refer to differentiation with respect to material co-ordinates (See §§3, 14).

The time differentiations $\partial/\partial t$ and D/Dt are defined in §14. The material derivative, denoted either by a superposed dot or by $\delta/\delta t$, is defined in §18.

German minuscules: curves, surfaces, and volumes in space, $\mathfrak{c}$, $\mathfrak{s}$, $\mathfrak{v}$, etc.

German majuscules: curves, surfaces, and volumes in the material: $\mathfrak{C}$, $\mathfrak{S}$, $\mathfrak{V}$, etc.

Script capitals: averages or quantities defined by integrals, $\mathcal{W}$, $\mathcal{W}_n$, etc.

Special symbols

ϵ_{ijk}, ϵ_{IJK} permutation symbols . . . 1
ζ . . . . (74.1)
η specific entropy 17
θ temperature 17
ϑ expansion . . . . (24.4)
Θ . . . . (9.2)
λ, μ viscosities 17
$\boldsymbol{\pi}$ vector potential of velocity (16.11)
$\boldsymbol{\pi}^*$ vector potential of acceleration . . . . (16.13)
ϕ { velocity-potential . . . (36.2)
ϕ { scalar potential of velocity (16.11)
ϕ^* { acceleration-potential . (48.1)
ϕ^* { scalar potential of acceleration . . . . (16.13)
ρ density 17
σ flexion-potential . . . . . (44.6)
υ extraneous force potential (17.3)
χ . . . . (51.8)
$\boldsymbol{\omega}$ angular velocity . . . . (16.8)
$\tilde{\boldsymbol{\omega}}$ local angular velocity . . (32.2)
Ω abnormality . . . . . . (9.2)
$\emptyset$ general tensor 4
$\emptyset_{\mathfrak{n}}$, $\emptyset_{\mathfrak{t}}$ normal and tangential projections of $\emptyset$ . . . . (4.5)
$\emptyset_{(n)}$ n^{th} moment of $\emptyset$ . . . (6.6)
$\{\mathfrak{b}^{(n)}\emptyset\}$ . . . . (4.12)
$\mathcal{W}$ total vorticity . . . . . (63.11)
$\mathcal{W}_n$ n^{th} moment of vorticity (65.1)
$\mathfrak{W}_{\mathrm{K}}$ kinematical vorticity number . . . . (56.3)
$\mathfrak{W}_{\mathrm{D}}$ dynamical vorticity number . . . . (62.3)

List of Works Quoted

1687 1. I. Newton, *Philosophiae Naturalis Principia Mathematica*, London; 3rd ed., ed. H. Pemberton, London (1726) = repr. Glasgow (1871); trans. A. Motte, *Sir Isaac Newton's Mathematical Principles of Natural Philosophy and his System of the World*, London (1729); revised ed., Berkeley (1934).

1736 1. L. Euler, *Mechanica sive Motus Scientia Analytice Exposita* **1**, Petropoli = *Opera* (2) **1**.

1738 1. D. Bernoulli, *Hydrodynamica sive de Viribus et Motibus Fluidorum Commentarii*, Argentorati.

1743 1. J. Bernoulli, *Hydraulica nunc primum detecta ac demonstrata directe ex fundamentis pure mechanicis*, *Opera* **4**, 387–493. Virtually the same work appears under different titles in Comm. Acad. Sci. Petrop. **9** (1737), 3–49 (1744); **10** (1738), 207–260 (1747).

1744 1. J. L. d'Alembert, *Traité de l'Équilibre et du Mouvement des Fluides pour servir de Suite au Traité de Dynamique*, Paris; 2nd ed., 1770.

1745 1. L. Euler, *Neue Grundsätze der Artillerie, aus dem Englischen des Herrn Benjamin Robins übersetzt und mit vielen Anmerkungen versehen*, Berlin = *Opera* (2) **14**, 1–409.

1752 1. J. L. d'Alembert, *Essai d'une Nouvelle Théorie de la Resistance des Fluides*, Paris.

1757 1. L. Euler, *Principes généraux de l'état d'équilibre des fluides*, Hist. Acad. Berlin **1755**, 217–273.

2. L. Euler, *Principes généraux du mouvement des fluides*, Hist. Acad. Berlin **1755**, 274–315.

3. L. Euler, *Continuation des recherches sur la théorie du mouvement des fluides*, Hist. Acad. Berlin **1755**, 316–361.

1761 1. L. Euler, *Principia motus fluidorum* (1752–1755), Novi. Comm. Acad. Sci. Petrop. **6** (1756–1757), 271–311.

2. J. L. d'Alembert, *Remarques sur les lois du mouvement des fluides*, Opusc. **1**, 137–168.

1762 1. Lettre de M. Euler à M. de La Grange, *Recherches sur la propagation des ébranlemens dans une milieu élastique*, Misc. Taur. 2^2 (1760–1761), 1–10 = *Opera* (2) **10**, 255–263. This letter is dated 1 January 1760. The small part omitted in the original publication may be found in Euler's *Opera Postuma* **1**, 561; the entire letter in *Oeuvres de Lagrange* **14**, 178–188.

2. J. L. Lagrange, *Nouvelles recherches sur la nature et la propagation du son*, Misc. Taur. 2^2 (1760–1761), 11–172 = *Oeuvres* **1**, 151–316.

3. J. L. Lagrange, *Application de la méthode exposée dans le mémoire précédent à la solution de differents problèmes de dynamique*, Misc. Taur. 2^2 (1760–1761), 196–298 = *Oeuvres* **1**, 365–468.

1766 1. L. EULER, *Supplément aux recherches sur la propagation du son*, Mém. Acad. Sci. Berlin **15** (1759), 210–240 = *Opera* (3) **1**, 452–483.

1767 1. L. EULER, *Recherches sur le mouvement des rivières* (1751), Mém. Acad. Sci. Berlin **16** (1760), 101–118.

1769 1. L. EULER, *Sectio prima de statu aequilibrii fluidorum*, Novi Comm. Petrop. **13** (1768), 305–416.

1770 1. L. EULER, *Sectio secunda de principiis motus fluidorum*, Novi Comm. Acad. Sci. Petrop. **14** (1769), 270–386. By a printer's error the volume is dated 1759.

2. L. EULER, *Institutionum Calculi Integralis Volumen Tertium*, Petropolis = *Opera* (1) **13**.

1771 1. L. EULER, *Sectio tertia de motu fluidorum lineari potissimum aquae*, Novi Comm. Petrop. **15** (1770), 219–360.

1772 1. L. EULER, *Sectio quarta de motu aeris in tubis*, Novi Comm. Petrop. **16** (1770), 281–425.

1783 1. J. . LAGRANGE, *Mémoire sur la théorie du mouvement des fluides*, Nouv. Mem. Acad. Berlin **1781**, 151–198 = *Oeuvres* **4**, 695–748.

1787 1. G. MONGE, *Supplément, où l'on fait voir que les équations aux différences ordinaires, pour lesquelles les conditions d'intégrabilité ne sont pas satisfaites, sont susceptibles d'une véritable intégration, et que c'est de cette intégration que depend celle des équations aux différences partielles élevées*, Mém. Acad. Sci. Paris **1784**, 502–576.

1788 1. J. L. LAGRANGE, *Méchanique Analitique*, Paris. (See [1815, **1**].)

1806 1. L. EULER, *Die Gesetze des Gleichgewichts und der Bewegung flüssiger Körper*, Leipzig, being a trans., with some changes and additions, by H. W. BRANDES, of [1769, **1**], [1770, **1**], [1771, **1**], and [1772, **1**].

1813 1. C. F. GAUSS, *Theoria attractionis corporum sphaerodicorum ellipticorum homogeneorum methodo nova tractata*, Comm. Soc. Sci. Gotting. **2** = *Werke* **5**, 3–22.

1815 1. 2nd ed. of vol. 2 of [1788, **1**]. *Oeuvres* **12** is the 5th ed.

1818 1. J. F. PFAFF, *Methodus generalis, aequationes differentiarum partialium, nec non aequationes differentiales vulgares, utrasque primi ordinis, inter quotcunque variabiles, complete integrandi* (1815), Abh. Akad. Wiss. Berlin **1814–1815**, 76–136; trans. G. KOWALEWSKI, *Allgemeine Methode, partielle Differentialgleichungen zu integrieren*, Ostwald's Klass. **129**, Leipzig, 1902.

1823 1. A.-L. CAUCHY, *Recherches sur l'équilibre et le mouvement intérieur des corps solides ou fluides, élastiques ou non-élastiques*, Bull. Soc. Philomath., 9–13.

2. A.-L. CAUCHY, *Mémoire sur une espèce particulière de mouvement des fluides*, J. École Poly. **12**, cahier 19, 204–214 = *Oeuvres* (2) **1**, 264–274.

1826 1. A. M. AMPÈRE, *Mémoire sur la Théorie Mathématique des Phénomènes Electrodynamiques Uniquement Déduite de l'Expérience*, Paris.

1827 1. A.-L. CAUCHY, *Théorie de la propagation des ondes à la surface d'un fluide pésant d'une profondeur indéfinie* (1815), Mém. Divers Savants (2) **1** (1816), 3–312 = *Oeuvres* (1) **1**, 5–318.

2. A.-L. CAUCHY, *Sur la condensation et la dilatation des corps solides*, Ex. Math. **2** = *Oeuvres* (2) **7**, 60–78.

3. C.-L.-M.-H. NAVIER, *Mémoire sur les lois du mouvement des fluides* (1822), Mém. Acad. Sci. Inst. France (2) **6**, 389–440.

1828 1. A.-L. CAUCHY, *Sur quelques théorèmes relatifs à la condensation ou à la dilatation des corps*, Ex. Math. **3** = *Oeuvres* (2) **8**, 9–35.

2. G. GREEN, *An Essay on the Application of Mathematical Analysis to the Theories of Electricity and Magnetism*, Nottingham = J. Reine Angew. Math. **39** (1850), 73–89; **44** (1852), 356–374; **47** (1854), 161–221 = *Papers*, 3–115.

1831 1. S.-D. POISSON, *Mémoire sur les équations générales de l'équilibre et du mouvement des corps solides élastiques et des fluides* (1829), J. École Poly. **13**, cahier 20, 1–174.

2. OSTROGRADSKY, *Note sur une intégrale qui se rencontre dans le calcul de l'attraction des sphéroïdes*, Mém. Acad. Sci. St. Pétersb. Ser. Math. (6) **1**, 39–53.

1833 1. S.-D. POISSON, *Traité de Mécanique*, 2nd ed., Paris.

2. J. FOURIER, *Sur le mouvement de la chaleur dans les fluides*, Mém. Acad. Sci. Inst. France (2) **12**, 507–530 = *Oeuvres* **2**, 595–614.

1835 1. G. CORIOLIS, *Mémoire sur la manière d'établir les différents principes de la mécanique pour des systèmes de corps, en les considérant comme des assemblages de molécules*, J. École Poly. **15**, cahier 24, 93–132.

2. G. CORIOLIS, *Mémoire sur les équations du mouvement relatif des systèmes de corps*, J. École. Poly. **15**, cahier 24, 142–154.

1837 1. S. EARNSHAW, *On fluid motion, so far as it is expressed by the equation of continuity*, Trans. Cambr. Phil. Soc. **6**, (1836–1838), 203–233.

1841 1. A.-L. CAUCHY, *Mémoire sur les dilatations, les condensations, et les rotations produites par un changement de forme dans un système de points matériels*, Ex. d'Anal. Phys. Math. **2** = *Oeuvres* (2) **12**, 343–377.

2. A. F. SVANBERG, *On fluides rörelse*, K. Vetenskaps-Acad. Handlingar (1839), 139–154; trans. *Sur le mouvement des fluides*, J. Reine. Angew. Math. **24** (1842), 153–163.

1842 1. J. POWER, *On the truth of the hydrodynamical theorem, that if* $u\,dx + v\,dy + w\,dz$ *be a complete differential with respect to* x, y, z *at any one instant, it is always so*, Trans. Cambr. Phil. Soc. **7** (1839–1842), 455–464.

2. G. G. STOKES, *On the steady motion of incompressible fluids*, Trans. Cambr. Phil. Soc. **7** (1839–1842), 439–453 = *Papers* **1**, 1–16.

3. J. CHALLIS, *A general investigation of the differential equations applicable to the motion of fluids*, Trans. Cambr. Phil. Soc. **7** (1839–1842), 371–396.

1845 1. G. G. STOKES, *On the theories of the internal friction of fluids in motion, and the equilibrium and motion of elastic solids*, Trans. Cambr. Phil. Soc. **8** (1844–1849), 287–319 = *Papers* **1**, 75–129.

1846 1. G. G. STOKES, *Report on recent researches in hydrodynamics*, Rep. Brit. Assn. Pt. I, 1*ff.* = *Papers* **1**, 151–187.

1848 1. W. THOMSON (LORD KELVIN), *Notes on hydrodynamics* (2), *On the equation of the bounding surface*, Cambr. Dubl. Math. J. **3**, 89–93 = *Papers* **1**, 83–87.

2. G. G. STOKES, *Notes on hydrodynamics* (4), *Demonstration of a fundamental theorem*, Cambr. Dubl. Math. J. **3**, 209–219 = (with added notes) *Papers* **2**, 36–50.

3. J. MACCULLAGH, *An essay towards a dynamical theory of crystalline reflexion and refraction* (1839), Trans. R. Irish Acad. Sci. **21**, 17–50 = *Works*, 145–184.

1849 1. W. THOMSON (LORD KELVIN), *Notes on hydrodynamics* (5), *On the vis-viva of a liquid in motion*, Cambr. Dubl. Math. J. **4**, 90–94 = *Papers* **1**, 107–112.

1850 1. W. THOMSON (LORD KELVIN), *A mathematical theory of magnetism*, Abstr. Papers R. Soc. London **5** (1843–1850), 975–978. (Abstract of [1851, **3**].)

2. P. TARDY, *Some observations on a new equation in hydrodynamics*, Phil. Mag. **36**, 171–178.

1851 1. G. G. STOKES, *On the dynamical theory of diffraction* (1849), Trans. Cambr. Phil. Soc. **9**^{1}, 1–62 = *Papers* **2**, 243–328.

2. G. G. STOKES, *On the effect of the internal friction of fluids on the motion of pendulums* (1850), Trans. Cambr. Phil. Soc. **9**^{2}, 8–106 = *Papers* **3**, 1–141.

3. W. THOMSON (LORD KELVIN), *A mathematical theory of magnetism*, Phil. Trans. R. Soc. London **141**, 243–285 = *Papers Electr. Magn.*, §§432–523.

1854 1. G. G. STOKES, *Smith's Prize Examination Papers* (Feb. 1854) = *Papers* **5**, 320–322.

1857 1. A. CLEBSCH, *Über eine allgemeine Transformation der hydrodynamischen Gleichungen*, J. Reine Angew. Math. **54**, 293–312.

2. B. RIEMANN, *Lehrsätze aus der Analysis Situs für die Theorie der Integrale von zweigliedrigen vollständigen Differentialien*, J. Reine Angew. Math. **54**, 105–109 = *Werke*, 2nd ed., 91–96.

1858 1. H. HELMHOLTZ, *Über Integrale der hydrodynamischen Gleichungen, welche den Wirbelbewegungen entsprechen*, J. Reine Angew. Math. **55**, 25–55 = *Wiss. Abh.* **1**, 101–134; trans. P. G. TAIT, *On integrals of the hydrodynamical equations, which express vortex-motion*, Phil. Mag. (4) **33** (1867), 485–512.

1859 1. A. CLEBSCH, *Über die Integration der hydrodynamischen Gleichungen*, J. Reine Angew. Math. **57**, 1–10.

1860 1. G. LEJEUNE DIRICHLET, *Untersuchungen über ein Problem der Hydrodynamik*, Gött. Abh. Math. Cl. **8** (1858–1859), 3–42 = J. Reine Angew. Math. **58**, 181–216 (1861) = *Werke* **2**, 263–301.

1861 1. H. HANKEL, *Zur allgemeinen Theorie der Bewegung der Flüssigkeiten*, Göttingen.

1862 1. L. EULER, *Anleitung zur Natur-Lehre, worin die Gründe zu Erklärung aller in der Natur sich ereignenden Begebenheiten und Veränderungen festgesetzt werden* (probably written between 1755 and 1759), *Opera Postuma* **2**, 449–560 = *Opera* (3) **1**, 16–178.

2. Letter, dated 27 October 1759, from EULER to LAGRANGE, *Opera Postuma* **2**, 559–561 = *Oeuvres de Lagrange* **14**, 164–170.

1863 1. G. ROCH, *Anwendung der Potentialausdrücke auf die Theorie der molekular-physikalischen Fernewirkungen und der Bewegung der Elektricität in Leitern*, J. Reine Angew. Math. **61**, 283–308.

1866 1. G. DARBOUX, *Sur les surfaces orthogonales*, Ann. École Norm. (1) **3**, 97–141.

1867 1. W. THOMSON (LORD KELVIN) & P. G. TAIT, *Treatise on Natural Philosophy*, Part I, Cambridge; new ed., in 2 parts, Part I (1879), Part II (1883).

2. J. BERTRAND, *Théorème relatif au mouvement le plus général d'un fluide*, C.R. Acad. Sci. Paris **66**, 1227–1230.

1868 1. H. WEBER, *Ueber eine Transformation der hydrodynamischen Gleichungen*, J. Reine Angew. Math. **68**, 286–292.

2. H. HELMHOLTZ, *Sur le mouvement le plus général d'un fluide. Réponse à une communication précédente de M. J. Bertrand*, C.R. Acad. Sci. Paris **67**, 221–225 = *Wiss. Abh.* **1**, 135–139.

3. J. BERTRAND, *Note relative à la théorie des fluides. Réponse à la communication de M. Helmholtz*, C.R. Acad. Sci. Paris **67**, 267–269.

4. J. BERTRAND, *Observations nouvelles sur un mémoire de M. Helmholtz*, C.R. Acad. Sci. Paris **67**, 469–472.

5. H. HELMHOLTZ, *Sur le mouvement des fluides. Deuxième reponse à M. J. Bertrand*, C.R. Acad. Sci. Paris **67**, 754–757 = *Wiss. Abh.* **1**, 140–144.

6. J. BERTRAND, *Réponse à la note de M. Helmholtz*, C.R. Acad. Sci. Paris **67**, 773–775.

7. H. HELMHOLTZ, *Réponse à la note de M. J. Bertrand, du 19 Octobre*, C.R. Acad. Sci. Paris **67**, 1034–1035 = *Wiss. Abh.* **1**, 145.

8. Anonymous, Les Mondes **17**, 620–623.

9. J. BOUSSINESQ, *Sur l'influence des frottements dans les mouvements regulières des fluides*, J. Math. Pures Appl. (2) **13**, 377–438.

1869 1. W. THOMSON (LORD KELVIN), *On vortex motion*, Trans. R. Soc. Edinb. **25**, 217–260 = *Papers* **4**, 13–66.

2. A.-J.-C.-B. DE ST. VENANT, *Problème des mouvements que peuvent prendre les divers points d'une masse liquide, ou solide ductile, contenue dans un vase à parois verticales, pendant son écoulement par un orifice horizontal intérieur*, C.R. Acad. Sci. Paris **68**, 221–237.

1870 1. J. WARREN, *Note on a fundamental theorem in hydrodynamics*, Q. J. Math. **10**, 128–129.

2. W. VELTMANN, *Die Helmholtz'sche Theorie der Flüssigkeitswirbel*, Z. Math. Phys. **15**, 450–463.

1871 1. E. BELTRAMI, *Sui principi fondamentali della idrodinamica*, Mem. Acc. Sci. Bologna (3) **1**, 431–476; **2** (1872), 381–437; **3** (1873), 349–407; **5** (1874), 443–484 = *Ricerche sulla cinematica dei fluidi*, *Opere* **2**, 202–379.

2. H. Durrande, *Extrait d'une théorie du déplacement d'une figure qui se deforme*, C.R. Acad. Sci. Paris **73**, 736–738.

1873 1. D. Bobylew, *Einige Betrachtungen über die Gleichungen der Hydrodynamik*, Math. Ann. **6**, 72–84.

2. J. C. Maxwell, *A Treatise on Electricity and Magnetism*, 2 vols., Oxford; 3rd ed., 1892.

1874 1. E. J. Nanson, *Note on hydrodynamics*, Mess. Math. **3**, 120–121.

1876 1. G. Kirchhoff, *Vorlesungen über mathematische Physik: Mechanik*, Leipzig; 2nd ed., 1877; 3rd ed., 1883.

2. Cotterill, *On the distribution of energy in a mass of liquid in a state of steady motion*, Phil. Mag. (5) **1**, 108–111.

3. Н. Е. Жуковскій, *Кинематика жидкаго тѣла*, Мат. Сборн. **8**, 1–79, 163–238 = repr. separ., Moscow (1876) = *Собр. Соч.* **3**.

1877 1. H. Lamb, *Note on a theorem in hydrodynamics*, Mess. Math. **7** (1877–1878), 41–42.

2. J. Trowbridge, *On liquid vortex-rings*, Phil. Mag. (5) **3**, 290–295.

1878 1. E. J. Nanson, *Note on hydrodynamics*, Mess. Math. **7** (1877–1878), 182–185.

2. H. Lamb, *On the conditions for steady motion of a fluid*, Proc. London Math. Soc. **9** (1877–1878), 91–92.

3. W. K. Clifford, *Elements of Dynamic, an Introduction to the Study of Motion and Rest in Solid and Fluid Bodies*, I. *Kinematic*, London.

4. J. Müller, *Einleitung in die Hydrodynamik*, Vierteljahrsschrift Naturf. Ges. Zürich **23**, 129–159, 242–265.

1879 1. H. Lamb, *A Treatise on the Mathematical Theory of the Motion of Fluids*, Cambridge. (See [1895, **1**] [1916, **2**] [1924, **1**] [1932, **1**].)

2. E. Betti, *Teorica delle Forze Newtoniane e sue Applicazioni all' Elettrostatica e al Magnetismo*, Pisa.

3. A. R. Forsyth, *On the motion of a viscous incompressible fluid*, Mess. Math. **9** (1879–1880), 134–139.

1880 1. Bresse, *Fonction des vitesses; extension des théorèmes de Lagrange au cas d'un fluide imparfait*, C.R. Acad. Sci. Paris **90**, 501–504.

2. J. Boussinesq, *Sur la manière dont les frottements entrent en jeu dans un fluide qui sort de l'état de repos, et sur leur effet pour empêcher l'existence d'une fonction des vitesses*, C.R. Acad. Sci. Paris **90**, 736–739.

3. Bresse, *Réponse à une note de M. J. Boussinesq*, C.R. Acad. Sci. Paris **90**, 857–858.

4. J. Boussinesq, *Quelques considérations à l'appui d'une note du 29 mars, sur l'impossibilité d'admettre, en général, une fonction des vitesses dans toute question d'hydraulique où les frottements ont un rôle notable*, C.R. Acad. Sci. Paris **90**, 967–969.

5. T. Craig, *Motion of viscous fluids*, J. Franklin Inst. **110**, 217–227.

6. T. Craig, *On steady motion in an incompressible viscous fluid*, Phil. Mag. (5) **10**, 342–357.

7. T. Craig, *On certain possible cases of steady motion in a viscous fluid*, Amer. J. Math. **3**, 269–293.

8. H. A. ROWLAND, *On the motion of a perfect incompressible fluid when no solid bodies are present*, Amer. J. Math. **3**, 226–268.

1881 **1.** M. J. M. HILL, *Some properties of the equations of hydrodynamics*, Q. J. Math. **17**, 1–20, 168–174.

2. T. CRAIG, *Methods and Results, General Properties of the Equations of Steady Motion*, U.S. Treasury Dept. (Coast and Geod. Survey) Document **71**, Washington.

3. J. W. GIBBS, *Elements of Vector Analysis*, Part I, New Haven = *Works* **2**, 17–36.

4. И. ГРОМЕКА *Нѣкоторые случаи движенія несжимаемой жидкости*, Kazan = *Собр. Соч.*, 76–148.

1883 **1.** J. J. THOMSON, *A Treatise on the Motion of Vortex Rings*, London.

1884 **1.** J. W. GIBBS, *Elements of Vector Analysis*, Part 2, New Haven = *Works* **2**, 50–90.

1885 **1.** M. J. M. HILL, *On some general equations which include the equations of hydrodynamics* (1883), Trans. Cambr. Phil. Soc. **14**, 1–29.

2. H. LAMB, *Proof of a hydrodynamical theorem*, Mess. Math. **14**, 87–92.

1887 **1.** E. BOGGIO-LERA, *Sulla cinematica dei mezzi continui*, Il Nuovo Cimento (3) **22**, 63–69, 143–149, 231–240; **23** (1888), 158–162; **24** (1888), 41–55 = Ann. R. Scuola Norm. Pisa **4**, 53–99.

1888 **1.** A. B. BASSET, *A Treatise on Hydrodynamics*, 2 vols., Cambridge.

1889 **1.** E. BELTRAMI, *Considerazione idrodinamiche*, Rend. Ist. Lombardo (2) **22**, 121–130 = *Opere* **4**, 300–309.

2. G. MORERA, *Sui moti elicoidali dei fluidi*, Rend. Lincei (4) **5**, 611–617.

1890 **1.** M. LEVY, *L'hydrodynamique moderne et l'hypothèse des actions à distance*, Rev. Gen. Sci. Pures Appl. **1**, 721–728.

1892 **1.** W. KILLING, *Über die Grundlagen der Geometrie*, J. Reine Angew. Math. **109**, 121–186.

1893 **1.** H. POINCARÉ, *Leçons sur la Théorie des Tourbillons*, Paris.

1895 **1.** H. LAMB, *Hydrodynamics*, Cambridge (2nd ed. of [1879, **1**], see [1906, **1**] [1916, **2**] [1924, **1**] [1932, **1**].)

2. J. R. SCHÜTZ, *Über die Herstellung von Wirbelbewegungen in idealen Flüssigkeiten durch conservative Kräfte*, Ann. der Physik (n.F.) **56**, 144–147.

1896 **1.** L. SILBERSTEIN, *Ueber die Entstehung von Wirbelbewegung in einer reibungsloser Flüssigkeit*, C.R. Acad. Sci. Cracovie, 280–290.

2. E. & F. COSSERAT, *Sur la théorie de l'élasticité*, Ann. Toulouse **10**, I 1–I 116.

1897 **1.** P. APPELL, *Sur les équations de l'hydrodynamique et la théorie des tourbillons*, J. Math. Pures Appl. (5) **3**, 5–16.

2. A. FÖPPL, *Die Geometrie der Wirbelfelder*, Leipzig.

1898 **1.** V. BJERKNES, *Ueber die Bildung von Circulationsbewegungen und Wirbeln in reibungslosen Flüssigkeiten*, Vidensk. Skrift., No. 5.

2. V. BJERKNES, *Ueber einen hydrodynamischen Fundamentalsatz und seine Anwendung besonders auf die Dynamik der Atmosphäre und des Weltmeeres*, K. Svenska Vet. Hand. (2) **31**, No. 4.

1899 **1.** P. APPELL, *Lignes correspondentes dans la déformation d'un milieu;*

extension des théorèmes sur les tourbillons, J. Math. Pures Appl. (5) **5**, 137–153.

1900 1. K. Zorawski, *Ueber die Erhaltung der Wirbelbewegung*, C.R. Acad. Sci. Cracovie, 335–341.

2. V. Bjerknes, *Das dynamische Princip der Cirkulationsbewegungen in der Atmosphäre*, Meteor. Z. **17**, 97–106, 145–146.

1901 1. J. Hadamard, *Sur la propagation des ondes*, Bull. Soc. Math. France **29**, 50–60.

2. P. Duhem, *Sur les équations de l'hydrodynamique*, Ann. Toulouse (2), **3**, 253–279.

3. J. Weingarten, *Über die geometrische Bedingungen, denen die Unstetigkeiten der Derivierten eines Systems dreier stetigen Funktionen des ortes unterworfen sind, und ihre Bedeutung in der Theorie der Wirbelbewegung*, Archiv Math. Phys. (3) **1**, 27–33.

1902 1. V. Bjerknes, *Cirkulation relativ zu der Erde*, Meteor. Z. **37**, 97–108.

1903 1. P. Appell, *Sur quelques fonctions de point dans le mouvement d'un fluide*, J. Math. Pures Appl. (5) **9**, 5–19. (An abstract is given in C.R. Acad. Sci. Paris **136**, 186–187 (1903).)

2. J. Hadamard, *Leçons sur la Propagation des Ondes et les Équations de l'Hydrodynamique*, Paris.

3. P. Duhem, *Recherches sur l'hydrodynamique, Cinquième partie, le théorème de Lagrange et les conditions aux limites*, Ann. Toulouse (2) **5**, 353–376 = *Recherches sur l'Hydrodynamique*, **2**, Paris, 99–122 (1904).

4. P. Duhem, *Sur quelques formules de cinématique utiles dans la théorie générale de l'élasticité*, C.R. Acad. Sci. Paris **136**, 139–141.

5. O. Reynolds, *The Sub-Mechanics of the Universe*, *Papers* **3**.

6. P. Appell, *Sur quelques fonctions et vecteurs de point contenant uniquement les dérivées premières des composantes de la vitesse*, Bull. Soc. Math. France **31**, 68–73.

1905 1. G. Jaumann, *Die Grundlagen der Bewegungslehre von einem modernen Standpunkte aus*, Leipzig.

1906 1. 3rd ed. of [1879, **1**]. (See [1895, **1**] [1916, **2**] [1924, **1**] [1932, **1**].)

1908 1. W. Stekloff, *Sur la théorie des tourbillons*, Ann. Fac. Sci. Toulouse (2) **10**, 271–334.

1909 1. J. W. Gibbs & E. B. Wilson, *Vector Analysis*, New Haven.

1911 1. E. Vessiot, *Sur les transformations infinitésimales et la cinématique des milieux continus*, Bull. Sci. Math. (2) $\mathbf{35}^1$, 233–244.

2. G. Hamel, *Zum Turbulenzproblem*, Nachr. Ges. Wiss. Gött. Math.-Phys. Kl., 261–270.

1913 1. J. W. Strutt (Lord Rayleigh), *On the motion of a viscous fluid*, Phil. Mag. **26**, 776–786 = *Papers* **6**, 187–196.

1916 1. J. Spielrein, *Lehrbuch der Vektorrechnung nach den Bedürfnissen in der Technischen Mechanik und Elektrizitätslehre*, Stuttgart.

2. 4th ed. of [1879, **1**]. (See [1895, **1**] [1906, **1**] [1924, **1**] [1932, **1**].)

1917 1. P. Appell, *Sur une extension des équations de la théorie des tourbillons et des équations de Weber*, C.R. Acad. Sci. Paris **164**, 71–74.

1918 1. U. Crudeli, *Le formule del Cauchy e i fluidi viscosi*, Rend. Lincei (5) **27**², 49–52.

1919 1. L. Lecornu, *Sur les tourbillons d'une veine fluide*, C.R. Acad. Sci. Paris **168**, 923–926.

1920 1. G. Jaffé, *Bemerkung über die Entstehung von Wirbeln in Flüssigkeiten*, Phys. Z. **21**, 541–543.

1921 1. G. Jaffé, *Über den Transport von Vektorgrössen, mit Anwendung auf Wirbelbewegung in reibenden Flüssigkeiten*, Phys. Z. **22**, 180–183.

2. P. Appell, *Traité de Mécanique Rationnelle*, **3**, *Équilibre et Mouvement des Milieux Continus*, 3rd ed., Paris.

1922 1. A. Guglielmi, *Sul moto vorticoso dei liquidi*, Atti. R. Ist. Veneto **81**², 289–314.

1923 1. U. Cisotti, *Considerazione sulla nota formula idrodinamica di Daniele Bernoulli*, Boll. Un. Mat. Ital. **2**, 125–128.

2. U. Cisotti, *Sul carattere necessariamente vorticoso dei moti regolari, permanenti di un fluido qualsiasi in ambienti limitati, eppure in quiete all'infinito*, Boll. Un. Mat. Ital. **2**, 170–172.

1924 1. 5th ed. of [1879, **1**]. (See [1895, **1**] [1906, **1**] [1916, **2**] [1932, **1**].)

2. B. Caldonazzo, *Sulla geometria differenziale di superficie aventi interesse idrodinamico*, Rend. Lincei (5^{A}) **33**², 396–400.

1925 1. L. Lichtenstein, *Über einige Existenzprobleme der Hydrodynamik homogener, unzusammendruckbarer, reibungsloser Flüssigkeiten und die Helmholtzschen Wirbelsätze*, Math. Z. **23**, 89–154.

2. M. Roy, *Sur les singularités d'un passage à la limite effectué dans la solution d'une équation aux dérivées partielles*, Nouv. Annales de Math. (5) **3**, 321–327.

3. B. Caldonazzo, *Un' osservazione a proposito del teorema di Bernoulli*, Boll. Un. Mat. Ital. **4**, 1–3.

1926 1. E. Hoppe, *Geschichte der Physik*, Hdbuch der Phys. **1**, 1–179.

1927 1. E. L. Ince, *Ordinary Differential Equations*, London.

2. L. Lichtenstein, *Über einige Existenzprobleme der Hydrodynamik* II: *Nichthomogene, unzusammenruckbare, reibungslose Flüssigkeiten*, Math. Z. **26**, 196–323.

3. A. Masotti, *Osservazioni sui moti di un fluido nei quali è stazionaria la distribuzione del vortice*, Rend. Lincei (6) **6**, 224–228.

4. M. Pastori, *Sulle superficie ortogonali a una congreunza normale di curve*, Rend. Ist. Lombardo (2) **60**, 111–119.

5. O. Veblen, *Invariants of Quadratic Differential Forms*, Cambr. Tracts **24**, Cambridge.

6. G. Bouligand, *Sur le signe de la pression dans un liquide pesant, en mouvement irrotationnel*, Verh. 2. Int. Congr. Techn. Mech. (1926), 460–461.

7. G. Bouligand, *Un théorème relatif à la pression au sein d'un liquide parfait en mouvement irrotationnel*, J. Math. Pures Appl. (9) **6**, 427–433.

1928 1. H. Jeffreys, *The equations of viscous motion and the circulation theorem*, Proc. Cambr. Phil. Soc. **24**, 477–479.

2. M. LAGALLY, *Vorlesungen über Vektor-Rechnung*, Leipzig.

1929 1. L. LICHTENSTEIN, *Grundlagen der Hydromechanik*, Berlin.

2. O. D. KELLOGG, *Foundations of Potential Theory*, Berlin.

3. H. VILLAT, *Leçons sur l'Hydrodynamique*, Paris.

4. D. POMPEIU, *Sur la condition des vitesses dans un fluide incompressible*, Bull. Math. Phys. École Poly. Bucarest **1**, 42–43.

1930 1. H. VILLAT, *Leçons sur la Théorie des Tourbillons*, Paris.

2. E. KAMKE, *Differentialgleichungen reeller Funktionen*, Leipzig.

1931 1. F. SBRANA, *Sulla validità del teorema di Bernoulli per un fluido reale*, Boll. Un. Mat. Ital. **10**, 77–78.

2. P. BURGATTI, *Intorno a una formula generale di trasformazione di un integrale di spazio in uno di superficie e alle sue varie deduzioni*, Boll. Un. Mat. Ital. **10**, 1–5.

3. A. J. MCCONNELL, *Applications of the Absolute Differential Calculus*, London & Glasgow.

1932 1. 6th ed. of [1879, **1**]. (See [1895, **1**] [1906, **1**] [1916, **2**] [1924, **1**].)

2. M. LELLI, *Sulla estensione dei teoremi di Helmholtz e di Lagrange al moto dei fluidi viscosi*, Atti Congr. Int. Mat. 1928, Bologna, **6**, 247–249.

3. H. BATEMAN, *Part II, Motion of an incompressible viscous fluid*, Bull. Nat. Res. Council **84**.

1933 1. H. B. PHILLIPS, *Vector Analysis*, New York.

1936 1. L. CROCCO, *Una nuova funzione di corrente per lo studio del moto rotazionale dei gas*, Rend. Lincei (6^A) **23**, 115–124; trans. *Eine neue Stromfunktion für die Erforschung der Bewegung der Gase mit Rotation*, Z. Angew. Math. Mech. **17**, (1937), 1–7.

2. G. HAMEL, *Ein allgemeiner Satz über den Druck bei der Bewegung volumbeständiger Flüssigkeiten*, Monatshefte Math. Phys. **43**, 345–363.

1937 1. G. HAMEL, *Potentialströmungen mit konstanter Geschwindigkeit*, Sitzungsber. Preuss. Akad. Wiss. Phys.-Math. Kl., 5–20.

2. M. LAGALLY, *Die zweite Invariante des Verzerrungstensors*, Z. Angew. Math. Mech. **17**, 80–84.

3. G. LAMPARIELLO, *Varietà sostanziali nel moto di un sistema continuo*, Rend. Lincei (6a) **15**, 383–387.

4. T. v. KÁRMÁN, *On the statistical theory of turbulence*, Proc. Nat. Acad. Sci. **23**, 98–105.

1940 1. T. LEVI-CIVITA, *Nozione adimensionale di vortice e sua applicazione alle onde trocoidale di Gerstner*, Acta Pontif. Acad. Sci. **4**, 23–30.

2. R. BALLABH, *Superposable fluid motions*, Proc. Benares Math. Soc. (2) **2**, 69–79.

3. R. BALLABH, *Self superposable fluid motions of the type $\xi = \lambda u$, etc.*, Proc. Benares Math. Soc. (2) **2**, 85–89.

4. H. WEYL, *The method of orthogonal projection in potential theory*, Duke Math. J. **7**, 411–444.

1941 1. M. MUNK, *On some vortex theorems of hydrodynamics*, J. Aero. Aci. **5**, 90–96.

2. G. HAMEL, *Über die Potentialströmungen zäher Flüssigkeiten*, Z. Angew. Math. Mech. **21**, 129–139.

1942 1. H. ERTEL, *Ein neuer hydrodynamischer Erhaltungssatz*, Die Naturwissenschaften **30**, 543–544.

2. H. ERTEL, *Ein neuer hydrodynamischer Wirbelsatz*, Meteor. Z. **59**, 277–281.

3. H. ERTEL, *Über das Verhältnis des neuen hydrodynamischen Wirbelsatzes zum Zirkulationssatz von V. Bjerknes*, Meteor. Z. **59**, 385–387.

4. H. ERTEL, *Über hydrodynamische Wirbelsätze*, Phys. Z. **43**, 526–529.

5. H. GÖRTLER & K. WIEGHART, *Über eine gewisse Klasse von Strömungen zäher Flüssigkeiten und eine Kennzeichnung der Poiseuille-Strömung*, Math. Z. **48**, 247–250.

1944 1. C. JACOB, *Sur une interprétation de l'équation de continuité hydrodynamique*, Bull. Math. Soc. Roumaine Sci. **46**, 81–90.

1946 1. I. CARSTOIU, *Généralization des formules de Helmholtz et de Cauchy pour un fluide visqueux incompressible*, C.R. Acad. Sci. Paris **223**, 1095–1096.

2. I. CARSTOIU, *Sur le vecteur tourbillon de l'accelération et les fonctions qui s'y rattachent*, Bull. Cl. Sci. Acad. Roumaine **29**, 207–214.

3. V. VÂLCOVICI, *Sur une interprétation cinématique du tourbillon et sur la rotation des directions principales de la déformation*, Mathematica (Timişoara) **22**, 57–65.

4. I. CARSTOIU, *Sur un mouvement fluide de Beltrami*, Bull. Cl. Sci. Acad. Roumaine **28**, 270–272.

5. I. CARSTOIU, *Sur la possibilité des mouvements tourbillonaires à Ω = const. d'un fluide parfait incompressible*, Bull. Cl. Sci. Acad. Roumaine **28**, 503–504.

6. I. CARSTOIU, *Sur le mouvement tourbillonaire à Ω = const. d'un fluide parfait incompressible*, Bull. Cl. Sci. Acad. Roumaine **28**, 589–592.

1947 1. C. TRUESDELL & R. PRIM, *Zorawski's kinematic theorems*, U.S. Naval Ordnance Lab. Mem. **9354**.

2. L. BRAND, *Vector and Tensor Analysis*, New York.

3. I. CARSTOIU, *De la circulation dans un fluide visqueux incompressible*, C.R. Acad. Sci. Paris **224**, 534–535.

4. I. CARSTOIU, *Sur la possibilité des mouvements irrotationnels d'un fluide visqueux incompressible*, C.R. Acad. Sci. Paris **225**, 664–666.

5. L. CASTOLDI, *Sopra una proprietà dei moti permanenti di fluidi incomprimibili in cui le linee di corrente formano una congruenza normale di linee isotache*, Rend. Accad. Lincei (8) **3**, 333–337.

6. B. HICKS, P. GUENTHER, & R. WASSERMAN, *New formulation of the equations for compressible flow*, Q. Appl. Math. **5**, 357–361.

7. Anon., *N. E. Joukovsky*, Prikl. Mat. Mekh. (2) **11**, 9–26.

8. I. CARSTOIU, *Recherches sur la théorie des tourbillons*, MS thesis, Univ. Paris.

1948 1. C. TRUESDELL, *On the total vorticity of motion of a continuous medium*, Phys. Rev. (2) **73**, 510–512.

2. C. TRUESDELL, *The kinematics of vorticity*, U.S. Naval Ordnance Lab. Mem. **9591**.

3. C. TRUESDELL, *Généralisation de la formule de Cauchy et des théorèmes de Helmholtz au mouvement d'un milieu continu quelconque*, C.R. Acad. Sci. Paris **227**, 757–759.

4. C. TRUESDELL, *Une formule pour le vecteur tourbillon d'un fluide visqueux élastique*, C.R. Acad. Sci. Paris **227**, 821–823.

5. L. CASTOLDI, *Superficie e linee di Bernoulli nel moto stazionario di un fluido reale*, Atti Accad. Ligure **4**, (1947), 21–25.

6. R. C. PRIM, *On doubly-laminar flow fields having a constant velocity magnitude along each stream-line*, U.S. Naval Ordnance Lab. Mem. **9762**.

7. A. BILIMOVITCH, *Aires et volumes vélocidiques et hodographiques dans un mouvement du fluide*, Acad. Serbe Sci. Publ. Inst. Math. **2**, 37–52.

8. R. PRIM, *Extension of Crocco's theorems to flows having a non-uniform stagnation enthalpy*, Phys. Rev. (2) **73**, 186.

9. P. NEMÉNYI & R. PRIM, *Some properties of rotational flow of a perfect gas*, Proc. Nat. Acad. Sci. U.S.A. **34**, 119–124. *Erratum*, ibid. **35** (1949), 116.

10. C. TRUESDELL, *On the transfer of energy in continuous media*, Phys. Rev. (2) **73**, 513–515.

11. С. Г. ПОПОВ, *О винтовых движениях идеальной жидкости*, Вестник Москов. Универс. No. 8, 35–47.

12. R. BALLABH, *On coincidence of vortex and stream lines in ideal liquids*, Ganita **1**, 1–4.

13. J. A. STRANG, *Superposable fluid motions*, Comm. Fac. Sci. Ankara **1**, 1–32.

14. J.-J. MOREAU, *Sur deux théorèmes généraux de la dynamique d'un milieu incompressible illimité*, C.R. Acad. Sci. Paris **226**, 1420–1422.

1949 1. C. TRUESDELL, *The effect of viscosity on circulation*, J. Meteor. **6**, 61–62.

2. C. TRUESDELL, *Trois conférences faites à la Sorbonne*, U.S. Naval Research Lab. Theoret. Mech. Sect. Mem. **3836–1**.

3. C. TRUESDELL, *Deux formes de la transformation de Green*, C.R. Acad. Sci. Paris **229**, 1199–1200.

4. P. NEMÉNYI & R. PRIM, *On the steady Beltrami flow of a perfect gas*, Proc. 7th Int. Congr. Appl. Mech. (1948), **2**, 300–314.

5. G. SUPINO, *Sul moto irrotazionale dei liquidi viscosi*, I., Rend. Lincei (8) **6**, 615–620.

6. H. ERTEL & H. KÖHLER, *Ein Theorem über die stationäre Wirbelbewegung kompressibler Flüssigkeiten*, Z. Angew. Math. Mech. **29**, 109–113.

7. R. BERKER, *Sur certaines propriétés du rotationnel d'un champ vectoriel qui est nul sur la frontière de son domaine de définition*, Bull. Sci. Math. (2) **73**, 163–176. Abstract, C.R. Acad. Sci. Paris **228**, 1630–1632.

8. B. L. HICKS, *On the characterization of fields of diabatic flow*, Q. Appl. Math. **6**, 405–416.

9. R. Prim, *A note on the substitution principle for steady gas flow*, J. Appl. Phys. **20**, 448–450.
10. H. Ertel & C.-G. Rossby, *Ein neuer Erhaltungssatz der Hydrodynamik,* Sitzungsber. Akad. Wiss. Berlin Math.-nat. Kl. **1949**, 1–11.
11. H. Ertel & C.-G. Rossby, *A new conservation theorem of hydrodynamics*, Geofisica Pura e Appl. **14**, 189–193.
12. A. N. Ergun, *Some cases of superposable fluid motions*, Comm. Fac. Sci. Ankara **2**, 48–88.
13. J.-J. Moreau, *Sur l'interprétation tourbillonaire des surfaces de glissement*, C.R. Acad. Sci. Paris **228**, 1923–1925.
14. J.-J. Moreau, *Sur la dynamique d'un écoulement rotationnel*, C.R. Acad. Sci. Paris **229**, 100–102.

1950 1. R. Prim & C. Truesdell, *A derivation of Zorawski's criterion for permanent vector-lines*, Proc. Amer. Math. Soc. **1**, 32–34.
2. C. Truesdell, *Bernoulli's theorem for viscous compressible fluids*, Phys. Rev. **77**, 535–536.
3. C. Truesdell, *On the balance between deformation and rotation in the motion of a continuous medium*, J. Washington Acad. Sci. **40**, 313–317.
4. G. Viguier, *Quelques aperçus sur un problème de M. J. Boussinesq*, Bull. Cl. Sci. Acad. R. Belg. (5) **36**, 71–76.
5. R. Dugas, *Histoire de la Mécanique*, Neuchâtel.
6. J. L. Synge, *Note on the kinematics of plane viscous motion*, Q. Appl. Math. **8**, 107–108.
7. J. Delval, *Le principe de la moindre contrainte appliqué à la dynamique des fluides incompressibles*, Acad. R. Belg. Bull. Cl. Sci. (5) **36**, 639–648 (1950).
8. L. Castoldi, *Linee sostanziali nel moto di un continuo deformabile e moti con linee di flusso (e di corrente) "sostanzialmente permanente,"* Rend. Ist. Lombardo (3) **14**, 259–264.
9. J.-J. Moreau, *Relations générales directes entre les actions aérodynamiques et les éléments tourbillonaires*, Actes Congr. Internat. Méc. Poitiers **4**, 6pp.
10. H. Ertel, *Ein Theorem über asynchron-periodische Wirbelbewegungen kompressibler Flüssigkeiten*, Misc. Acad. Berol. **1**, 62–68.

1951 1. C. Truesdell, *Vorticity averages*, Canad. J. Math. **3**, 69–86.
2. C. Truesdell, *A form of Green's transformation*, Am. J. Math. **73**, 43–47.
3. C. Truesdell, *Caractérisation des champs vectoriels qui s'annulent sur une frontière fermée*, C.R. Acad. Sci. Paris **232**, 1277–1279.
4. C. Truesdell, *Analogue tri-dimensionnel au théorème de M. Synge sur les champs vectoriels plans qui s'annulent sur une frontière fermée*, C.R. Acad. Sci. Paris **232**, 1396–1397.
5. C. Truesdell, *Vereinheitlichung und Verallgemeinerung der Wirbelsätze ebener und rotationssymmetrischer Gasbewegungen*, Z. Angew. Math. Mech. **31**, 65–71. An abstract under the title *A new vorticity*

theorem appears in Proc. Int. Congr. Math. 1950, **1**, 639–640 (1952).

6. C. Truesdell, *On Ertel's vorticity theorem*, Z. Angew Math. Phys. **2**, 109–114.
7. C. Truesdell, *Proof that Ertel's vorticity theorem holds in average for any medium suffering no tangential acceleration on the boundary*, Geofis. Pura Appl. **19**, No. 3–4, 1–3.
8. C. Truesdell, *On the equation of the bounding surface*, Bull. Tech. Univ. Istanbul **3**, 71–77.
9. J. L. Synge, *On permanent vector-lines in n dimensions*, Proc. Am. Math. Soc. **2**, 370–372.
10. J. L. Synge, *Conditions satisfied by the expansion and vorticity of a viscous fluid in a fixed container*, Q. Appl. Math. **9**, 319–322.
11. F. van den Dungen, *Note on the Hamel-Synge theorem*, Q. Appl. Math. **9**, 203–204.
12. O. Bjørgum, *On Beltrami vector fields and flows, Part I*, Univ. Bergen Årbok **1951**, Naturv. rekke No. 1.
13. L. A. Santaló, *On permanent vector-varieties in n dimensions*, Portug. Math. **10**, 125–127.
14. С. Г. Попов, *Замечание об интегралах бернулли и лагранжа (коши)*, Москов. Гос. Унив. Зап. **152**, Мех. **3**, 43–46.

1952 1. C. Truesdell, *The mechanical foundations of elasticity and fluid dynamics*, J. Rational Mech. Anal. **1**, 125–300; **2**, 593–616 (1953).
2. C. Truesdell, *Vorticity and the Thermodynamic State in a Gas Flow*, Mém. Sci. Math. No. 119, Paris.
3. J.-J. Moreau, *Bilan dynamique d'un écoulement rotationnel*, J. Math. Pures Appl. (9) **31**, 355–375; **32**, 1–78 (1953).
4. R. C. Prim, *Steady rotational flow of ideal gases* (1949), J. Rational Mech. Anal. **1**, 425–497.
5. H. Ertel, *Über die physikalische Bedeutung von Funktionen, welche in der Clebsch-Transformation der hydrodynamischen Gleichungen auftreten*, Sitzber. Akad. Wiss. Berlin Kl. Math. allg. Naturw. **1952**, No. 3.
6. N. Coburn, *Intrinsic relations satisfied by the vorticity and velocity vectors in fluid flow theory*, Michigan Math. J. **1**, 113–130.

1953 1. C. Truesdell, *Two measures of vorticity*, J. Rational Mech. Anal. **2**, 173–217. (Partial abstract in Proc. Int. Congr. Theor. Appl. Mech. Istanbul. 1952.)
2. C. Truesdell, *The physical components of vectors and tensors*, Z. angew. Math. Mech. **33**, 345–356.
3. O. Bjørgum & T. Godal, *On Beltrami vector fields and flows*, Part II, Univ. Bergen Årbok **1952**, Naturv. rekke No. 13.
4. C. Truesdell, *Generalization of a geometrical theorem of Euler*, Comm. Mat. Helv., **27**, 233–234.
5. C. Truesdell, *La velocità massima nel moto di Gromeka-Beltrami*, Rend. Lincei (8) **13** (1952), 378–379.

6. L. N. Howard, *Constant speed flows*, Thesis, MS in Princeton University Library.

7. A. Bilimovitch, *Sur l'homogénéisation des équations de nature vélocidique*, Acad. Serbe Sci. Publ. Inst. Math. **5**, 29–34.

8. H. Masuda, *A new proof of Lagrange's theorem in hydrodynamics*, J. Phys. Soc. Japan **8**, 390–393.

1954 **1.** C. Truesdell, *Rational fluid mechanics, 1687–1765*, Introduction to L. Euleri Opera Omnia (2) **12**, forthcoming.

Index of Authors Quoted

Numbers refer to sections, footnotes, and bibliography.

Index of Matters Treated

Numbers refer to chapters, sections, and footnotes

A CATALOG OF SELECTED

DOVER BOOKS

IN SCIENCE AND MATHEMATICS

Physics

THEORETICAL NUCLEAR PHYSICS, John M. Blatt and Victor F. Weisskopf. An uncommonly clear and cogent investigation and correlation of key aspects of theoretical nuclear physics by leading experts: the nucleus, nuclear forces, nuclear spectroscopy, two-, three- and four-body problems, nuclear reactions, beta-decay and nuclear shell structure. 896pp. 5 3/8 x 8 1/2. 0-486-66827-4

QUANTUM THEORY, David Bohm. This advanced undergraduate-level text presents the quantum theory in terms of qualitative and imaginative concepts, followed by specific applications worked out in mathematical detail. 655pp. 5 3/8 x 8 1/2. 0-486-65969-0

ATOMIC PHYSICS AND HUMAN KNOWLEDGE, Niels Bohr. Articles and speeches by the Nobel Prize–winning physicist, dating from 1934 to 1958, offer philosophical explorations of the relevance of atomic physics to many areas of human endeavor. 1961 edition. 112pp. 5 3/8 x 8 1/2. 0-486-47928-5

COSMOLOGY, Hermann Bondi. A co-developer of the steady-state theory explores his conception of the expanding universe. This historic book was among the first to present cosmology as a separate branch of physics. 1961 edition. 192pp. 5 3/8 x 8 1/2. 0-486-47483-6

LECTURES ON QUANTUM MECHANICS, Paul A. M. Dirac. Four concise, brilliant lectures on mathematical methods in quantum mechanics from Nobel Prize-winning quantum pioneer build on idea of visualizing quantum theory through the use of classical mechanics. 96pp. 5 3/8 x 8 1/2. 0-486-41713-1

THE PRINCIPLE OF RELATIVITY, Albert Einstein and Frances A. Davis. Eleven papers that forged the general and special theories of relativity include seven papers by Einstein, two by Lorentz, and one each by Minkowski and Weyl. 1923 edition. 240pp. 5 3/8 x 8 1/2. 0-486-60081-5

PHYSICS OF WAVES, William C. Elmore and Mark A. Heald. Ideal as a classroom text or for individual study, this unique one-volume overview of classical wave theory covers wave phenomena of acoustics, optics, electromagnetic radiations, and more. 477pp. 5 3/8 x 8 1/2. 0-486-64926-1

THERMODYNAMICS, Enrico Fermi. In this classic of modern science, the Nobel Laureate presents a clear treatment of systems, the First and Second Laws of Thermodynamics, entropy, thermodynamic potentials, and much more. Calculus required. 160pp. 5 3/8 x 8 1/2. 0-486-60361-X

QUANTUM THEORY OF MANY-PARTICLE SYSTEMS, Alexander L. Fetter and John Dirk Walecka. Self-contained treatment of nonrelativistic many-particle systems discusses both formalism and applications in terms of ground-state (zero-temperature) formalism, finite-temperature formalism, canonical transformations, and applications to physical systems. 1971 edition. 640pp. 5 3/8 x 8 1/2. 0-486-42827-3

QUANTUM MECHANICS AND PATH INTEGRALS: Emended Edition, Richard P. Feynman and Albert R. Hibbs. Emended by Daniel F. Styer. The Nobel Prize–winning physicist presents unique insights into his theory and its applications. Feynman starts with fundamentals and advances to the perturbation method, quantum electrodynamics, and statistical mechanics. 1965 edition, emended in 2005. 384pp. 6 1/8 x 9 1/4. 0-486-47722-3

Physics

INTRODUCTION TO MODERN OPTICS, Grant R. Fowles. A complete basic undergraduate course in modern optics for students in physics, technology, and engineering. The first half deals with classical physical optics; the second, quantum nature of light. Solutions. 336pp. 5 3/8 x 8 1/2. 0-486-65957-7

THE QUANTUM THEORY OF RADIATION: Third Edition, W. Heitler. The first comprehensive treatment of quantum physics in any language, this classic introduction to basic theory remains highly recommended and widely used, both as a text and as a reference. 1954 edition. 464pp. 5 3/8 x 8 1/2. 0-486-64558-4

QUANTUM FIELD THEORY, Claude Itzykson and Jean-Bernard Zuber. This comprehensive text begins with the standard quantization of electrodynamics and perturbative renormalization, advancing to functional methods, relativistic bound states, broken symmetries, nonabelian gauge fields, and asymptotic behavior. 1980 edition. 752pp. 6 1/2 x 9 1/4. 0-486-44568-2

FOUNDATIONS OF POTENTIAL THERY, Oliver D. Kellogg. Introduction to fundamentals of potential functions covers the force of gravity, fields of force, potentials, harmonic functions, electric images and Green's function, sequences of harmonic functions, fundamental existence theorems, and much more. 400pp. 5 3/8 x 8 1/2. 0-486-60144-7

FUNDAMENTALS OF MATHEMATICAL PHYSICS, Edgar A. Kraut. Indispensable for students of modern physics, this text provides the necessary background in mathematics to study the concepts of electromagnetic theory and quantum mechanics. 1967 edition. 480pp. 6 1/2 x 9 1/4. 0-486-45809-1

GEOMETRY AND LIGHT: The Science of Invisibility, Ulf Leonhardt and Thomas Philbin. Suitable for advanced undergraduate and graduate students of engineering, physics, and mathematics and scientific researchers of all types, this is the first authoritative text on invisibility and the science behind it. More than 100 full-color illustrations, plus exercises with solutions. 2010 edition. 288pp. 7 x 9 1/4. 0-486-47693-6

QUANTUM MECHANICS: New Approaches to Selected Topics, Harry J. Lipkin. Acclaimed as "excellent" (*Nature*) and "very original and refreshing" (*Physics Today*), these studies examine the Mössbauer effect, many-body quantum mechanics, scattering theory, Feynman diagrams, and relativistic quantum mechanics. 1973 edition. 480pp. 5 3/8 x 8 1/2. 0-486-45893-8

THEORY OF HEAT, James Clerk Maxwell. This classic sets forth the fundamentals of thermodynamics and kinetic theory simply enough to be understood by beginners, yet with enough subtlety to appeal to more advanced readers, too. 352pp. 5 3/8 x 8 1/2. 0-486-41735-2

QUANTUM MECHANICS, Albert Messiah. Subjects include formalism and its interpretation, analysis of simple systems, symmetries and invariance, methods of approximation, elements of relativistic quantum mechanics, much more. "Strongly recommended." – *American Journal of Physics.* 1152pp. 5 3/8 x 8 1/2. 0-486-40924-4

RELATIVISTIC QUANTUM FIELDS, Charles Nash. This graduate-level text contains techniques for performing calculations in quantum field theory. It focuses chiefly on the dimensional method and the renormalization group methods. Additional topics include functional integration and differentiation. 1978 edition. 240pp. 5 3/8 x 8 1/2. 0-486-47752-5

Physics

MATHEMATICAL TOOLS FOR PHYSICS, James Nearing. Encouraging students' development of intuition, this original work begins with a review of basic mathematics and advances to infinite series, complex algebra, differential equations, Fourier series, and more. 2010 edition. 496pp. 6 1/8 x 9 1/4. 0-486-48212-X

TREATISE ON THERMODYNAMICS, Max Planck. Great classic, still one of the best introductions to thermodynamics. Fundamentals, first and second principles of thermodynamics, applications to special states of equilibrium, more. Numerous worked examples. 1917 edition. 297pp. 5 3/8 x 8. 0-486-66371-X

AN INTRODUCTION TO RELATIVISTIC QUANTUM FIELD THEORY, Silvan S. Schweber. Complete, systematic, and self-contained, this text introduces modern quantum field theory. "Combines thorough knowledge with a high degree of didactic ability and a delightful style." – *Mathematical Reviews.* 1961 edition. 928pp. 5 3/8 x 8 1/2. 0-486-44228-4

THE ELECTROMAGNETIC FIELD, Albert Shadowitz. Comprehensive undergraduate text covers basics of electric and magnetic fields, building up to electromagnetic theory. Related topics include relativity theory. Over 900 problems, some with solutions. 1975 edition. 768pp. 5 5/8 x 8 1/4. 0-486-65660-8

THE PRINCIPLES OF STATISTICAL MECHANICS, Richard C. Tolman. Definitive treatise offers a concise exposition of classical statistical mechanics and a thorough elucidation of quantum statistical mechanics, plus applications of statistical mechanics to thermodynamic behavior. 1930 edition. 704pp. 5 5/8 x 8 1/4. 0-486-63896-0

INTRODUCTION TO THE PHYSICS OF FLUIDS AND SOLIDS, James S. Trefil. This interesting, informative survey by a well-known science author ranges from classical physics and geophysical topics, from the rings of Saturn and the rotation of the galaxy to underground nuclear tests. 1975 edition. 320pp. 5 3/8 x 8 1/2. 0-486-47437-2

STATISTICAL PHYSICS, Gregory H. Wannier. Classic text combines thermodynamics, statistical mechanics, and kinetic theory in one unified presentation. Topics include equilibrium statistics of special systems, kinetic theory, transport coefficients, and fluctuations. Problems with solutions. 1966 edition. 532pp. 5 3/8 x 8 1/2. 0-486-65401-X

SPACE, TIME, MATTER, Hermann Weyl. Excellent introduction probes deeply into Euclidean space, Riemann's space, Einstein's general relativity, gravitational waves and energy, and laws of conservation. "A classic of physics." – *British Journal for Philosophy and Science.* 330pp. 5 3/8 x 8 1/2. 0-486-60267-2

RANDOM VIBRATIONS: Theory and Practice, Paul H. Wirsching, Thomas L. Paez and Keith Ortiz. Comprehensive text and reference covers topics in probability, statistics, and random processes, plus methods for analyzing and controlling random vibrations. Suitable for graduate students and mechanical, structural, and aerospace engineers. 1995 edition. 464pp. 5 3/8 x 8 1/2. 0-486-45015-5

PHYSICS OF SHOCK WAVES AND HIGH-TEMPERATURE HYDRO DYNAMIC PHENOMENA, Ya B. Zel'dovich and Yu P. Raizer. Physical, chemical processes in gases at high temperatures are focus of outstanding text, which combines material from gas dynamics, shock-wave theory, thermodynamics and statistical physics, other fields. 284 illustrations. 1966–1967 edition. 944pp. 6 1/8 x 9 1/4. 0-486-42002-7